The Role of Chemical Markers and Chemometrics in the Identification of Grasses Used as Food in Pre-Agrarian South West Asia

Michelle Cave

BAR International Series 1277
2004

Published in 2016 by
BAR Publishing, Oxford

BAR International Series 1277

The Role of Chemical Markers and Chemometrics in the Identification of Grasses Used as Food in Pre-Agrarian South West Asia

ISBN 978 1 84171 630 5

BAR Publishing is the trading name of British Archaeological Reports (Oxford) Ltd. British Archaeological Reports was first incorporated in 1974 to publish the BAR Series, International and British. In 1992 Hadrian Books Ltd became part of the BAR group. This volume was originally published by Archaeopress in conjunction with British Archaeological Reports (Oxford) Ltd / Hadrian Books Ltd, the Series principal publisher, in 2004. This present volume is published by BAR Publishing, 2016.

Printed in England

BAR
PUBLISHING

BAR titles are available from:

BAR Publishing
122 Banbury Rd, Oxford, OX2 7BP, UK
EMAIL info@barpublishing.com
PHONE +44 (0)1865 310431
FAX +44 (0)1865 316916
www.barpublishing.com

ABSTRACT

This thesis explores the way in which novel chemical criteria can be used to identify charred remains of grains of small-grained grasses used as food by pre-agrarian hunter-gatherers in south-western Asia but which have hitherto rarely been identified with any precision.

The grass family Gramineae or Poaceae, is the most diverse, abundant and widespread family of higher plants on the planet. Grasses correspondingly have enormous ecological and economic importance worldwide. Their importance is reflected in the prominent role of grain from wild grasses in hunter-gatherer subsistence. In order to reconstruct past subsistence practices and diet, especially of arid-zone hunter-gatherers, it is important to identify the remains of grasses recovered from archaeological sites. However, the recovered grass remains are most often charred, therefore the interpretive potential can be realized only if these charred remains are accurately identified at the level of genus and, in some cases, species. There are enormous problems in identifying charred remains, particularly when relying totally on gross morphological criteria. There is therefore a need for alternative criteria, such as that utilized by chemical analytical techniques.

The core rationale in applying the different chemical techniques is the same throughout: grains are taken from modern grasses of known identity and spanning a spectrum of taxa likely to include all the charred ancient specimens to be identified (the unknowns). These modern grains are then analysed to generate spectra. Equivalent spectra from unknowns are then compared with those from the modern grains to effect an identification. Standard practice has hitherto involved comparing the two sets of spectra (know and unknowns) by visual inspection; i.e. "by eye". However, identifications based on such comparisons are inevitably to some degree untestable and unrepeatable, and this represents a long-standing problem in chemistry generally. In the present project I have therefore explored the use of chemometrics: i.e. the use of statistical systems to compare spectra in a manner that is rigorously testable and repeatable. This is an entirely new development, and has never previously been applied in the analysis of archaeological data.

TABLE OF CONTENTS **Page No.**

LIST OF FIGURES

Chapter 8

LIST OF TABLES

Page No.

Chapter 4

Chapter 7

Chapter 8

ACKNOWLEDGMENTS

Gwynne Vaughn Botanical Sciences Studentship.
Central Research Fund for a dedicated computer system.
NERC; five day European Chemometrics Workshop and Conference in Bristol.
NERC. SCAN Chemometrics Computer Software. Graduate School for funding to attend an eight-day conference in Mexico to meet chemists working with New World Plants.
British School at Ankara and the Graduate School for funding to travel to Turkey to study the school's plant reference collection.
The UCL student union for providing both free and subsidized childcare for the summer school breaks.
Many thanks to my main supervisor Gordon Hillman, who gave me both the support and the inspiration to complete this task.
Also thanks to John Evans whose guidance with chemistry kept me on the right track.
Thank you also to Mike Czwarno who taught me to how work with a computer and not be afraid of offending it.
Also thank you to Kevin Reeves who helped me in the lab with both technical support and seventies rock music.
Many thanks and hugs to Vanessa, my sister who has always been there.
Big thank you to Karen in Oz who just came when I needed her help, as she has always done.
A special thank you to my friend Ruth Prior whose many hours of listening and litres of cappuccino helped make the task less daunting.
Also to my many friends who were there when things were at their toughest; thank you to Yemi, fellow mum who inspired me to keep going, Robert, Graham, Giles, Alan and Tess.
Most of all thank you and big hugs to my daughter Gabriela who has put up with endless poverty without complaint so I can finish.
And, of course, my fiancé Chris who has endured many hours of my self doubt and come through sane, all my love.
Thank you to you all.

PREFACE

At the point of inception of the project to be discussed in the following pages, it was evident that several distinct modes of analysis would be required if each of the key aspects of the design were to be covered. As a consequence, a multidisciplinary approach was deemed necessary, within which all areas of research would be treated with equal importance, and each would make its specific contribution to the overall project.

Although the multidisciplinary approach adopted for the project as a whole seemed to be the most appropriate, it did mean that certain constraints had to be placed on the written material. The most prominent of these was the inclusion of definitions for both words and concepts in a greater number and a more detailed form than would normally be required or expected. This requirement would, inevitably, result in the sacrifice of a certain degree of textual elegance. Nonetheless, while many definitions are useful for the understanding of the more complex parts of the project, there has to be a point where the inclusion of various simpler ones becomes absurd. It is not unreasonable to assume a certain level of competence in the major areas covered by the research, as it was clearly designed for those who work in these fields. The multidisciplinary character can only extend so far; it is simply not possible to render material contained in the project accessible to *everyone* who may come into contact with it.

Consequently, I have included as many definitions as was practicable, in addition to paragraphs, page numbers and instructions for the most effective way to read both volumes. If the reader still requires further background information, then it will be necessary to use the bibliography, and to refer to those original papers and texts in which the chemistry, mathematics, computing and archaeobotany are discussed in detail.

The initial design phase entailed the making of decisions as to precisely which disciplines and techniques would be utilized. This phase included the learning of what were, for me, entirely new skills, which would be required in order to utilize data and information from previous, known research. Fortunately I had John Evans from the University of East London as one of my supervisors; his laboratory had covered a lot of ground that would have been time-consuming for me to replicate. The most important of area involved the charring of the modern seeds under numerous conditions in order to test the reliability of the comparison between charred and uncharred seeds. Due to the thoroughness with which the work of Evans and his team was carried out, Hillman and myself decided that it would be unnecessary for me to repeat this work in the context of the present research.

As it was, the extensive laboratory work required three years, including, as it did, the sample preparation, the soxhlet extraction for both solvents and the Fourier transform infrared analysis. A further period was then required for the chemometric analysis of the acquired spectral data in its entirety, and for the final writing up of the completed project. During this time I was, furthermore, seeking out information from other workers, reading as much background information as time allowed and attending conferences to meet fellow chemists working in similar and related fields. As a consequence, there was simply not enough time or money to include every possible contingency.

One further requirement imposed by the selection of a multidisciplinary approach is the necessity to provide clear and concise explanations of a range of new or unfamiliar concepts, both in introductory paragraphs and in captions accompanying graphic printouts. Such a requirement was particularly acute in the case of the appendices, as these contained all the detailed explanations for the chemometric statistical procedures. Each statistical method carried out via SCAN generated results in the form of tables and graphs, and, as these were vital to the understanding of the results, it was necessary to provide a detailed account of each methodological step. In the interests of thoroughness, each technique was supplied with an introduction and a step-by-step outline. The need for detail was rendered still more acute by the fact that the SCAN programme does not allow additions or deletions to the completed graphs and tables it generates. This stipulation is written into the programme in order to preclude any creative modifications that may serve to enhance the results. (Each table and graph within the chapters, 'Final Results' and 'Discussion of Results' has been given a caption beneath the relevant figure or table, where there was space on the page. For reasons discussed above, they could not be added to the full-scale, colour SCAN printouts in the appendix.)

As a consequence of these considerations, it is strongly suggested that the text and appendices be read in tandem, in order that detailed explanations and descriptions be available at the appropriate moment. Utilizing the text in such a fashion will assist in avoiding confusion regarding the meanings of complex and unfamiliar terms and procedures.

One of the fundamental problems encountered in the design of any research consists in deciding precisely what work is to be carried out within the constraints of time and budget imposed upon one by prevailing circumstances. Furthermore, a timescale conceived under ideal conditions may be subject to changes forced upon it by unforeseen contingencies. In the case of this project, there were two significant factors that would lead me to modify the schedule I had originally envisaged, and create, in addition, financial burdens over and above those I had anticipated.

Both of these contingencies involved people whose role was to supervise the research. The first came about owing to the fact that, at the end of my first year, my second supervisor left the U.K. to pursue fieldwork in North America. Since his specific input had entailed the teaching of the core discipline of chemometrics, and also of the advanced computational skills required by the project's statistical imperatives, his departure came as something of a blow. The second eventuality was still more unfortunate, and occurred at the close of my second year when my primary supervisor, who had, during the course of our work together, become a personal friend, developed a serious and debilitating illness. This latter circumstance was to some degree mitigated by the fact that, his condition being a progressive one, I was still able to benefit from his judgment, experience and expertise, albeit with a reduced regularity. These events were to impose a degree of hardship and austerity upon a project which had commenced in an atmosphere of some conviviality, but it is perhaps a credit to my teachers that I was able to surmount the ensuing difficulties and to produce a completed work which promises to be of some scientific utility and value. I would like to acknowledge my debt to them here.

By way of explanation, I have tried to deal with any questions which may arise in respect to the structure of the thesis, particularly as it entails such a diverse selection of disciplines. In order to aid the understanding of the plant species employed, both the modern and ancient have been listed twice: once, at the end of the text in numerical order, and also at the beginning of the appendix, here organized by four groupings: order, family, genus and species. The appendices comprise several programmes which had been linked up for purposes of data translation. In certain instances, it was necessary to 'force' the programmes to function in combination, in order to optimise the output. The appendices involve, therefore, the graphic and tabular culmination of these processes of linkage, and must be read in tandem with the main text. This is essential as it facilitates the understanding of both the numerical research and the procedural definitions and issues.

CHAPTER ONE

INTRODUCTION

This thesis explores the way in which novel chemical criteria can be used to identify charred remains of grains of small-grained grasses which were used as food by pre-agrarian hunter-gatherers in South Western Asia, but which have hitherto rarely been identified with any precision. It utilizes a form of statistical analysis which has never previously been applied to chemical extractions of archaeological material, with the aim of mathematically matching species of unknowns within a model created from a series of modern, identified specimens.

Although chemometrics is a relatively new discipline, it is slowly being used in many fields of investigation where instrumental analysis, such as FTIR, HPLC, GCMS, UV and so on are utilized for chemical analysis. Archaeological science is just one of the disciplines finding this type of analysis extremely useful.

1.0 Why target wild grasses?

The grass family Graminae or Poacae, is the most diverse, abundant and widespread family of higher plants on the planet. Grasses correspondingly have enormous ecological and economic importance worldwide. Their importance is reflected in the prominent role of grain from wild grasses in hunter-gatherer subsistence in the arid and semi-arid zones of both Old and New Worlds, as evidenced by the ethnographic record from recent hunter-gatherers and the abundance of grass grains among the food plants recovered from pre-agrarian sites. Examples of recent hunter-gatherers include the Owen's Valley Paiute in North America, who made heavy use of wild grass grains from both porcupine grass and a type of lyme grass (Stewart 1934); the Bagundji of the Darling Basin in Australia who used wild millets as a caloric staple (Allen 1974:in Hillman 1989); and the Alyawara tribe from the Great Western Desert in Australia who utilized a range of wild grass grains to make a type of unleavened bread or 'damper'(Cable & French 1942). Damper is the name for a type of unleavened bread, which is made by grinding the grass grains to flour, mixing it with water and, where possible, a small measure of salt to form a sticky dough. This is then placed amongst the red-hot wood of the campfire to cook. Similarly, among prehistoric hunter-gatherers there are examples worldwide of the use of grass-grain as food, this ranges from Mesolithic North Germany where there is evidence for the use of flote-grass, Glyceria, fluitans, to Southwestern. Asia (e.g. Edwards, College *et al.* 1988; Hillman *et al.* 1989; Nesbitt 1997), and Archaic sites in North America such as Hidden Cave in Nevada where there is evidence for the use of a wide range of wild grass-grains as food.

The heavy use of grain from wild grasses by many pastoralists and even some cultivators in recent times further underlines the fact that wild grass grains are (or were) widely available and favoured as food (e.g. Cable and French 1942 for the Mongols. Harlan 1989 for Saharan groups).

1.1 Why target hunter-gatherer sites in South Western Asia?

South Western Asia is an area where pre-agrarian hunter-gatherers appear to have made particularly heavy use of wild grass-grain as food. The grasses include several species from genera such as *Stipa*, the feather grasses, and *Hordeum*, the wall-barley grasses (Hillman *et al.* 1989, Hillman in Moore *et al.* in press. Nesbitt 1997). Several of these grasses were to become the major crops of the local Neolithic and later periods, (e.g. Colledge, 1994; Zohary and Hopf 1993), eventually leading to the cereal crops of the present day. By examining the past use of these wild grasses, it becomes possible to gain an understanding of not only a major component of pre-agrarian subsistence, but also the circumstances leading to their eventual cultivation and domestication. The site of Abu Hureyra has produced rich assemblages of charred remains of wild food grasses, as have other pre-agrarian hunter-gatherer sites including Tell Mureybit which like Abu Hureyra, is located on the Middle Euphrates), Quermez Dere in Northern Iraq (Nesbitt 1997) and Ohalo 11 in Israel on the shores of Galilee. Although this project has used mainly the grass-grain remains from Abu Hureyra, it should be stressed from the outset that these other sites could well provide some ancient specimens of species not present at Abu Hureyra, as well as offering further opportunities to explore the potential of the techniques developed in this project.

The other advantage of sites of South Western Asia is that the grass remains have been preserved by charring. (The effects of charring on extracts are discussed in chapter 6 with reference to McLaren's work). The process of charring has effectively preserved the internal chemical markers present in the ancient grains. By contrast, preservation by water-logging or desiccation, which has been the burial environment for many archaeological sites in other regions, often allows the degeneration of these compounds. (The effects of other types of environmental conditions on biomolecules are discussed in detail in chapter 4, with reference to the work of Eglinton and Logan). However, these markers preserved by charring remain viable for analysis only where the processes of recovery and post-excavation storage have been sensitive to the requirements of those carrying out such chemical analysis, as they can easily be contaminated or even destroyed. Generally there has been a high standard of recovery, as most excavators have used flotation with water rather than alternative liquids, and have avoided froth flotation with organic frothing agents and systems such as paraffin flotation. As flushing systems of water-flotation have been used since the late 1960's, there has been quite an accumulation of charred seed remains from many locations, and these are generally well suited to chemical analysis. In addition, it should be noted that, while for the purposes of morphological identification the possession of intact seed samples is desirable, the chemical method is able to utilize fragmented specimens with equal facility.

Because grass-grain appears to have been such an important component of hunter- gatherers diet in South Western Asia, any attempt to reconstruct past subsistence practices and diet, on sites in this area clearly relies heavily on accurate identification of the remains of grasses. On such sites, grass remains can also provide information on other aspects of past human ecology. For example, many grass genera and species grow only where there are particular patterns of topography, water-availability, and soil type, and in particular types of plant community i.e. specific grasses can be regarded as diagnostic of specific conditions. Properly identified grass remains from sites in South Western Asia can therefore provide evidence of not only the classes of resource environment that were available within the economic catchment of the site, but also clues to where the site occupants went to gather grain. On this basis it is theoretically possible to reconstruct past patterns of seasonal movement between the areas concerned, and/ of possible patterns of exchange between groups.

1.2 Problems associated with identification

Archaeobotanists face a problem. On the one hand, the interpretive potential of these charred remains can be realized only if they are accurately identified at the level of genus or species. On the other hand they are extremely difficult to identify using conventional systems, because the grass group is taxonomically complex (see, e.g., Clayton and Renvoize 1989), and, when preserved by charring, the grains round off and can thereafter loose any remaining features due to the effects of abrasion post charring (Nesbitt 1997).

True, on some European sites where remains of grass-grains are preserved by water-logging, it has often proved possible to identify grains to species level from the patterns of surface cells, i.e. from histological or micro-morphological criteria (e.g. Korber-Grohne 1978, 1982; Korber-Grohne & Piening 1980), but when the remains are preserved by charring, such cell patterns are seldom visible (e.g. Colledge 1988). Most archaeobotanists have therefore relied on gross morphological criteria that have seldom been precisely defined, and even more rarely subjected to systematic study. While (at the point I began this project) an attempt to systematise the use of gross morphology had been started by Kislev 1997, this targetted primarily the use of size and gross proportions, and these seem to be of uncertain diagnostic potential. (At this point in time Nesbitt had not started his research re-evaluating the role of morphological criteria for identifying grass grains.)

There was therefore a need for alternative criteria to help overcome problems of identifying charred remains of these key food plants, especially in the all-to-common cases where all traces of useable morphological features are lost. Ideally, we needed criteria which a) were independent of morphology and histological criteria, whether traditional or novel, and b) could be used to identify remains that were so badly preserved that morphological criteria are, in any case, unusable (Hillman *et al.* 1994; Jones *et al.* 1996). Chemical criteria met both these requirements.

Evans *et al.* (1993) initially explored the diagnostic potential of various forms of chromatographic and spectroscopic analyses (principally IR, TLC and HPLC) to identify invisible residues of organic materials of ancient pots. At about the same time, Hillman, Jones & Robins (1982) sought to use chemical criteria to identify charred remains of grains of grasses, particularly wild and domestic cereals. More specifically, they sought to determine the taxonomic level (working down from sub-family, to tribe, genus, species and sub-species) at which it ceased to be possible to distinguish between adjacent taxa using pyrolysis mass spectrometry, and they then attempted to apply the resulting model to identifying critical remains of wild cereals from Abu Hureyra. Thereafter infrared spectroscopy was used to identify charred remains of wild and domestic cereals from sites in South Western Asia (McLaren *et al.* 1993. Hillman *et al.* 1994). However, these explorations of the potential of chemical criteria in identifying charred remains were not extended to the much broader range of small-grained food grasses that characterize so many pre-agrarian hunter-gatherer sites in this same region. This was clearly a much bigger undertaking.

It was against this background that I conducted a preliminary feasibility study targeting the small -grained barley grasses and using chemical criteria generated by standard infrared spectroscopy (Cave 1992). For this feasibility study, the small-grained barley grasses were selected because they are the most abundant class of grass-grain on epipalaeolithic sites such as Abu Hureyra (Hillman *et al.* 1989). The work formed the springboard for the present project which now a) targets examples from all the more important taxa likely to have been gathered as food and preserved in charred form on pre-agrarian archaeological sites in South Western Asia, and b) uses a more sophisticated form of infrared analysis: FTIR (Fourier Transform Infrared Spectroscopy). It should be noted that, in the meantime, Mark Nesbitt has undertaken a comprehensive exploration of ***morph-ological*** criteria in the identification of the same broad spectrum of SW Asian grass grains (Nesbitt 1997).

1.2.1 The core rationale of the analysis

The core rationale in applying the different chemical analytical techniques is the same throughout: grains are taken from modern grasses of known identity and spanning a spectrum of taxa likely to include as many as possible of the charred ancient specimens to be identified (the unknowns). These modern grains are then analysed by FTIR to generate spectra. Equivalent FTIR spectra from the unknowns are then compared with those from the modern grains to effect an identification. Standard practice has hitherto involved comparing the two sets of spectra (known and unknown) by visual inspection; i.e. "by eye". However, identifications based on such comparisons are inevitably largely untestable and scientifically unrepeatable. This represents a long-standing problem in chemistry generally, and it becomes wholly unmanageable when very large numbers of taxa are involved, as in this project. In the present project 1 have therefore explored the use of chemometrics: the application of statistical systems to compare spectra in a manner that is both rigorously testable and repeatable. This is an entirely new development, and has never before been applied in the analysis of archaeological data.

More specifically, the chemometrical systems have allowed the hierarchical clustering of the spectra generated by analyses of the grains from modern grasses of known identity. If, then, in analysing the spectra from ancient specimens of ***unknown*** identity, the chemometrical system recognizes a sufficient number of significant similarities, it allows them to be allocated within the pattern of clusters representing spectra from the grasses of ***known*** identity, and the degree of affinity between ancient specimens and named modern specimens can then be measured. Every step of the procedure is therefore testable and repeatable.

1.3 Summary of the core objectives

1.3.1 To select ancient specimens of grass grains a) from one or more pre-agrarian, Epipalaeolithic hunter-gatherer sites where grass grains (gathered from wild stands) appear to have served as food; b) one or more early agrarian, Neolithic sites where there is some possibility that the same grasses continued to be gathered as food, or where closely related grasses became established as weeds of domestic crops.

A site is recognized as agrarian when evidence of grain planting is revealed, such as some types of tools, signs of seed storage, or perhaps even a store of seeds which could be identified as varying from the many types of small seeded wild grasses normally found close to the habitation site.

1.3.2 To select modern grains of grasses of known identity and representing a) all the major sub-families, tribes, and genera of grasses of South Western Asia likely to have served as hunter-gatherer foods, and to be present in the plant remains from the selected sites, b) a range of species within certain key genera which are likely to have been particularly important in hunter-gatherer subsistence at these same sites.

1.3.3 To extract from both ancient and modern specimens a range of compounds including chemical markers that are potentially diagnostic taxonomically, using (separately) two solvents (hexane and chloroform) of increasing polarity likely to extract a different range of markers in either case.

1.3.4 To examine the extracts spectroscopically using FTIR.

1.3.5 To build up a reference library of spectra from the modern specimens, for the purpose of a) visual comparison, and b) more importantly, to use in the construction of a matrix for statistical modelling, for the numerical identification of large numbers of unknowns.

1.3.6 If possible, to interpret the spectral differences in terms of molecular structure, particularly for the esters of fatty acids and other chemical species that are potentially diagnostic taxonomically.

1.3.7 To use chemometrical systems to cluster the spectra from the modern-grain extracts, in order to explore the data to assess the potential for regression analysis. In this context, this demonstrates the possibility of the modelling of the modern learning set, which acts as the internal calibration for the test set, which is numerically compared to this model. A model is constructed from data of known origin, it is refined and validated, so that the lesser known data can be compared to it. (See later sections in chapter 5 for explanations of model, regression analysis and modern learning set).

1.3.8 To compare this pattern of clusters with conventional groupings defined by grass taxonomists (based on large numbers of morphological and histological criteria of the grass flowering and vegetative structures), and on this basis, to assess the potential of these chemical markers to identify grasses (at least modern ones) on a basis directly relateable to recognized taxonomic hierarchies.

1.3.9 To use chemometrical systems to compare the spectra from ancient specimens of **unknown** identity with those of **known** identity, and to locate them on the pattern of clusters (generated in 1. 1. 7, above), and, on this basis, measure their closeness of fit to modern taxa. This is carried out via the creation of a model using the spectra of the modern specimens converted to numbers and put on a spreadsheet, which are then transformed into matrix format. By performing a number of analyses including factor analysis, principal component analysis, principal component regression and finally the nonlinear alternating conditional expectation regression analysis on the matrices, it is possible to compare spectra from ancient specimens with those of modern ones. However, to ensure that the model is actually capable of carrying out the identifications prior to the matrix comparisons, the calculations for the goodness-of-prediction and the goodness-of-fit are checked, if they are at the correct levels of confidence then it is possible to proceed with the final part of the identification. That is the comparison of the matrix of the unknowns with that of the knowns, this should literally show where the unknowns lie within the model and hence which genera or species they ultimately belong to.

1.3.10 On this basis, to offer identifications as many as possible of the ancient specimens, and, in cases where their morphology had already allowed suggestion of a possible identity (generally at the level of only tribe or genus), to compare this identification with that indicated by the chemical analytical procedures.

1.3.11 Assess the future potential of the chemical system used in this project to provide independent identification of charred remains of ancient foodstuffs in general. The archaeological samples used are charred; these are the best preserved samples from the site in question, and they are subjected to comparison with the modern uncharred, as this is how the samples are gathered.

As the work carried out by McLaren *et al.* on the charring of modern seeds clearly demonstrated, charring has no effect on the extracts; therefore, the comparison of the two in this form was methodologically sound. (See Chapter 4 for a full explanation.)

Overall the core objectives of the project have been achieved chemically, numerically and taxonomically, utilizing all the techniques originally envisaged. The basic design, was modified as the project progressed as, occasionally, results appeared at a level of significance not entirely anticipated. This is to be expected where the analysis is using a goodness-of-fit and goodness-of-prediction for the basic modelling capability, and allows for the direction to be altered, in order to optimise the overall result. Direction in this context refers to the overall type of statistical analysis that can be used, and the selection of an alternative type if there is another which best fits the data and utilizes it more efficiently.

1.4 What level of identification

As previously mentioned, the grass family is huge and complex. Given the relatively non- descript morphology of caryopses preserved by charring, therefore it was considered that identification at even coarse levels such as sub-family or tribe would be helpful in resolving some of the key archaeological questions. Nevertheless, the more precise the identification, the broader the range of questions that can be addressed with each successive level of identification, allowing a higher level of refinement. The work therefore aimed at higher levels of precision wherever possible. In the end, the system developed in this project proved to be capable of identifying charred remains of grass caryopses down to the level of genus, and often down to species. Such precision clearly allows us to address even greater specificity.

CHAPTER 2

STUDY MATERIALS

2.0 The modern (reference) specimens

The modern specimens came from those species likely to have been a) present in the Abu Hureyra area, particularly during the early (hunter-gatherer) phase of its occupancy, and b) used as food, or, thereafter, invading early crops as weeds. The local vegetation has changed substantially since this time as a result of climate change and intensive farming, so many of the species likely to have been used at Abu Hureyra no longer grow in the area. As a result, earlier work on the identification of plant remains from these earlier sites has involved Hillman and others in collecting modern specimens of seeds from not only Syria, but also from Jordan, Israel, Egypt, Turkey and Central Asia.

Since growing conditions can effect the biochemical content in terms of triacylglycerol type and quantity, it was also necessary that the modern samples be collected from habitats with edaphic conditions as similar as possible to those likely to been available around Abu Hureyra.

The modern reference specimens were taken from Hillman's comprehensive seed collection at the Institute of Archaeology, which he has long made available for general use. This collection covered an enormous range of species collected from Syria, Turkey, Greece, Jordan, Egypt, Turkmenistan, and thus represented the vast majority of the species likely to be present in the archaeological samples. Other specimens came from collections made by Mark Nesbitt in Turkey and Israel. These specimens became available in the course of the project when Nesbitt established a separate grass seed collection in the Archaeobotany laboratory of the Institute of Archaeology. This collection eventually incorporated all the grass specimens from Hillman's own S.W. Asia seed collection.

These samples were identified, from their morphological characteristics, by the collectors themselves (Hillman & Nesbitt) and, in some cases, by botanists specializing in particular genera. This ensured that the model developed from chemical criteria was based, from the outset, on specimens that had been accurately identified from those morphological criteria on which their taxonomy has been based. This secure taxonomic foundation is clearly vital to the subsequent comparison of the unknowns, both visually and via the chemometrics, for hierarchical clustering, and thence the model building via regression analysis.

2.1 The archaeological specimens

The archaeological materials used in this project were those recovered from the Epipalaeolithic period at the Tell site of Abu Hureyra on the Middle Euphrates in central Syria. This is a large mound, around half a kilometer in diameter, which was originally excavated in 1972 and 1973 by A.M.T. Moore, but now lies beneath Lake Assad. It was located at the edge of the Euphrates Valley and was occupied during the Epipalaeolithic through to the Neolithic, producing a large assemblage of plant material, the remains of which have been largely preserved by charring. These remains included charred grains of a wide range of wild grasses from the Epipalaeolithic. It is these specimens which provided most of the archaeological samples for this project.

2.1.1 Abu Hureyra and food plants

A vast range of steppe grasses produce edible seeds, not just those which were the precursors of the modern cultigens, but many other varieties such as the perennial feather grasses *Stipa* and *Stipa agrostis* spp, characteristic of steppe, woodland steppe and the glades of arid-zone park-woodland. Even during the early part of the Epipalacolithic occupation when plant resources were plentiful, it was likely that the wild, small seeded grasses provided a food source rich in carbohydrate, oils, proteins, vitamins and minerals; and, even though they incur higher processing costs, their employment in this role was probably still greater during the dry and cold conditions which seem to have characterized the ensuing Younger Dryas period.

Charred seed remains from the site of Abu Hureyra indicate the utilization of a whole range of such wild grasses, from the outset of occupation at c.9500bc. The remains are dated with unusual accuracy due to the large number of C-14 samples, 22 radiocarbon dates, 20 of which are AMS dates based on samples of bone and grain. Abu Hureyra also lends itself to study, as few sites have been studied so closely with respect to: a) the role of wild plant foods in the Epipalaeolithic economy; b) the emergence of cultivation during this same period; and c) the further development of agriculture during the ensuing Neolithic (Hillman, in Moore *et al*, in press).

The chemical analysis here aims to look at a small portion of the grass seed remains, identified by Hillman, and deriving mainly from Abu Hureyra. It aims to specify the identity of the plants used, and to compare this identification to that based on morphological examination. This required a wet chemistry procedure for extracting the samples, and an instrumental technique, such as infrared spectroscopy, for examining them. Spectroscopy was required in order to view structural information, which could be related to species identification.

Studying the extracts of these grasses enables a clearer identification of the plants used, possibly as far as genus/species. Once identities have been established, it is then possible to look at where these plants were located with respect to the home sites of the hunter gatherer people, and for what purpose they may have been used. Plants appeared to have played an important role in the diet especially when meat was not readily available.

2.2 Consequences

As indicated above, the study of this material can suggest possible uses of those plant resources available to the hunter-gatherers, and contribute to an enhanced understanding of the manner in which, with the passage of time, they became cultivators. It can also direct the attentions of archaeobotanists toward a number of species now extinct in the region of study. The techniques deployed by chemical analysis constitute an additional tool to be brought to bear upon the field, and can add to the vast array of information already gathered by workers in other disciplines. It is able to utilize certain specific aspects of the sample material not readily available to the non-chemist.

CHAPTER 3

THE ANALYTICAL TECHNIQUES SELECTED

3.0 Introduction

The selection of the instrumental analysis is fundamental to the experimental design. It dictates the form the results are to take, whether quantitative or qualitative in nature. Therefore, it is essential that the chosen system not only suits all possible facets of the project, but also conforms to certain criteria:

1. It must be time efficient, and allow analyses of very large numbers of specimens within the allotted time, while still providing adequate information for identification.

2. It must be cost effective, and allow these large numbers of analyses within the limited budget.

3. The allocated system must be able to generate data which are themselves amenable to analyses ranging from hierarchical clustering to regression analysis, And thus allow the creation of a calibration system, without excessive manipulation and the loss of the core content of the results.

4. The system must be sufficiently flexible to allow it to be specifically adapted to the project as a whole. For example, calibration specific to this project requires sample and instrument standardization, i.e. the instrumental parameters or machine settings are to be identical for each set of analyses and each group of samples are to be handled in an identical manner under identical conditions. These set parameters that are consequently built into the experimental design, should ideally remain unaltered for the duration of the project. The instrument standardization is the setting of all experimental parameters within the analyses, whereby all the measurements made by the FTIR and the chemometrics programme are identical for each period of analysis.

While any equivalent set of instrumentation would provide comparable results, the standardization protocols adopted in the project were important for purposes of experimental accuracy and consistency. If another researcher had used the FTIR for analysis of some other compound, the system would have required recalibration on each occasion.

The use of the FTIR system, as opposed to a more complex alternative such as a GC Mass Spectrometer, meant that the system could be devoted entirely to the imperatives of my own project, thereby allowing for an economy and concentration of time and labour, and the experimental optimisation of the instruments. (For a full discussion of instrument standardization, see chapter 5.5.1.)

5. It would be desirable to select a system, complete with a statistical capability, that could be packaged and made widely available for routine use in archaeology generally, a system suitable for use by archaeological scientists not necessarily highly trained in either chemistry or mathematics.

Although there are a number of analytical techniques, they are not always appropriate for the project being designed, as they often provide extremely detailed information which may be more than is actually required for the material being analysed, or the questions being asked.

Taking this into account, Fourier Transform Infrared was considered the most appropriate analytical technique, especially the new generation equipment employed here. Primarily due to its availability, but also because it was able to provide the necessary experimental requirements. GC/MS was a technique considered unnecessary as the dedicated system principally due to its cost, but also because it can often provide more detail than required for the type of material being analysed. Especially where archaeological specimens are concerned, as it is often difficult to compare these results with those of the modern organics libraries, frequently incorporated within the software of these systems.

Although the providing of immense quantities of information does not, in itself, render a technique inappropriate, it may lead to the spending of very long periods engaged in the work of interpretation, employing time that could be better utilized on data which is more relevant to the samples.

3.1 Fourier Transform Infrared compared with dual-beam grating system

The type of infrared analysis carried out by previous workers would frequently utilize the older, dual beam grating systems, which had limited efficiency, and generally lacked the dedicated computer compatibility.
Although the lack of computer compatibility is not the cause of inefficiency in the older systems, it does mean that they lack the ability to store large quantities of spectral data. These systems, possessing no solid-state memory, are unable to store large data sets, and are thus precluded from utilizing programmes- such as chemometrics packages- which manipulate such sets. The order of numerical analysis undertaken in the course of this research is simply beyond the capabilities of such instruments.

Therefore, for highly efficient data handling and clearer, more precise graphics, it was vital that a more sophisticated system be used. The Fourier Transform Infrared Perkin Elmer 2000 Series was selected for this research, as it possessed the capacity to carry out all of the essential data storage and manipulations required by the chemometric techniques employed in the project. In addition, the 2000 series is equipped with an improved method of sample presentation, providing both the required software and allowing for the use of disposable cards. The system also possesses a sealed chamber which can be purged with nitrogen, this feature enabling the standardization of the samples. This standardization was necessary for the construction of an efficient calibration system to which the unknown samples would be compared, both chemically and numerically.

3.2 Problems with the older systems

In the older type dual-beam system, the second chamber held a reference disk, from which a ratio was taken between the sample and the blank. Although spectrum manipulations were possible, they were much more mechanical, and consequently less responsive to standardization. Put crudely, such a characteristic rendered each operation slightly different, making standardization effectively impossible. In the dual-beam situation, the chamber is exposed to the ambience of the room, and thus allows for the introduction of variables unique to each experimental occasion.

The dual beam system overcomes the variations in the day to day environment, as the results taken are not used for any form of quantitative analysis. Consequently, as long as the sample analysis is carried out correctly and the results are of a good quality, the constant experimental standardization is not necessary.

The classical system is, nevertheless, sufficient if the final results from 'knowns' and 'unknowns' were to be compared by visual means. This method involves the overlaying of the archaeological spectra by the spectra of the known species.

In this project, however, the spectra were not analysed by the overlay technique, even though the results from both systems were given as chemical structural information i.e. the peaks of each spectrum represents bond frequencies, which are absorption patterns given by the structure of the different bonds, according to type and location within the compound.

Some of the most fundamental variations between the two systems, that reveal the advantages of the FTIR over the grating or dispersive form, occur in respect of the sample presentation. This is particularly in evidence where sensitivity is concerned, simply because there are more chemical problems being encountered, problems for which the grating spectrometer is inadequate. These include the measurement of weak bands, which were a feature of many of the archaeological samples. The measurement of weak bands requires an extremely high resolution, which the conventional system is simply unable to achieve. In order to acquire a high resolution in the representation of spectral data, more energy is required to be directed to the detector; this enables smaller and less dramatic peaks to be defined.

One of the major components of the grating spectrometer consists of the monochromator, and it is here that the principal disadvantage of the instrument lies. The monochromator is incapable of measuring IR spectra over a broad frequency range at acceptable standards of resolution. This fact is due largely to the mechanics of the monochromator, which possesses narrow slits at both its entrance and exit, and these limit the frequency range of radiation reaching the detector. The entrance slit defines the beam of light entering the monochromator, so that only a sufficiently collimated beam is incident on the grating (which is essentially a prism). Following the dispersal of this beam, the second slit permits only the one small frequency of interest to reach the detector, while the remainder of the beam is either reflected or absorbed by the jaws of the slit. The narrower these apertures are, the higher the frequency resolution of the spectrometer. However, as the resolution increases, so too does the amount of radiation wasted through absorption on the slit, with a corresponding decrease in energy reaching the detector.

If the sample analysis requires high resolution, grating spectroscopy is not going to be adequate because not enough energy is reaching the detector to allow for this. Even low resolution systems operating between 4000 and $400cm^{-1}$ with a resolution of $8cm^{-1}$, the amount of energy reaching the detector is around 8:3600, or about 0.2%, and should this resolution be increased to $1cm^{-1}$, then the amount of energy loss is even greater, about 99.97%. Accordingly, another form of energy detection is required if high resolution with low energy loss is to be attained, conditions necessary to ensure that weaker bands are measured satisfactorily.

The more efficient system comes in the form of the interferometer, one of the principal components of the Fourier Transform Infrared instrument. This splits the beam of radiation into two, and, after introducing a path difference, recombines them. The division of the beam is achieved with a beam splitter transmitting about 50% and reflecting about 50% of the total radiation. One part of this beam goes to a fixed mirror and the other to a movable mirror that creates the varying path difference. Once these two beams are recombined, an interference pattern (the sum of the cosine waves for all frequencies present) is obtained as the path difference is varied via the mirror. The interference created by the recombination causes the energy reaching the detector to vary as the path differences are modified; this interference variation is called the interferogram. The recombined beam then passes through the sample to the detector, and the ensuing spectral information is gained from the visibility curve of the interference fringes, generated by the interferometer.

The resultant interferogram is then transformed into the customary spectrum, showing energy as a function of frequency, by means of the mathematical process known as Fourier Transformation.

There are two other principal advantages of the FTIR

over dispersive spectroscopy: one derives from what is termed *Fellgett's advantage*, which simply means that data from all the spectral frequencies (the radiation) are measured simultaneously during the complete measurement. (See Perkin-Elmer lab Manual.) This results in a reduced measurment time, and, as a consequence, enhances the sensitivity of the signal to noise ratio.

The other basic advantage is derived from what is called *Jacquinot's advantage,* a term which refers to the increased spectral signal to noise ratio resulting from the increased signal at the detector. (See Perkin-Elmer Lab Manual.) Combined, the two give greater resolution, while extraneous noise is simultaneously reduced over the whole desired spectral range. In quantitative terms, this increase of sensitivity for the FT is equal to the square root of the number of resolution elements in the spectral range been covered.

Grating spectrometers have a typical mid-infrared range covering the region between 400-4000cm^{-1} , normally utilizing three gratings which change throughout the scan run, but do not lose frequency accuracy. For the same range, meanwhile, the FTIR spectrometer employs one beam splitter, with a frequency range for greater than 80% efficiency centered at 2000cm^{-1}, between the regions of 600-3400cm^{-1} , and for 40% efficiency within the range of 3600-400cm^{-1} . At the extremes, the efficiency becomes progressively worse, so it is often a disadvantage to use FTIR for measurements at spectrum extremities, where they are unlikely to produce the superior results. However, the other advantages of this system far out weigh these fringe discrepancies. The analyst being familiar with the material being analysed in terms of frequency range, which is set prior to scanning, there is no necessity to utilize the end extremes.

3.3 Comparative systems

Although there are a number of analytical techniques, they are not always appropriate for the project being designed, as they often provide extremely detailed information which may be more than is actually required for the material being analysed, or the questions being asked. These systems include High Pressure Liquid Chromatography (HPLC), Pyrolysis Gas Chromatography (PGC), Gas Chromatography/Mass Spectroscopy (GC/MS) and classical Infrared (IR), although the use of these systems may be extremely advantageous in some circumstances, a brief explanation of the techniques and why they were not selected is provided, in order to demonstrate the preference for FTIR as the dedicated system for the project.

1. The first of these techniques is **High Pressure Liquid Chromatography (HPLC)**, an extremely useful and accurate method for organic analysis. It is primarily a method of separation, which relies upon the differential distribution of the components of a mixture between a stationary microparticulate silica column-packing phase, and a mobile aqueous-organic mixture (buffer or organic solvent) phase.

The HPLC mobile phase can be an aqueous-organic mixture, a buffer solution or a mixture of organic solvents, depending on the chromatographic method and on the detector that is used. The stationary phases, conversely, comprise microparticulate column packings; they are commonly uniform, porous silicate particles, with spherical or irregular shape and nominal diameter of 10, 5, 3*Um*. Different separation mechanisms can be realized by bonding different chemical groups to the surface of the silicate particles to produce bonded phases. (Lindsay: chapt.1. 1987)

The sample is introduced into the chromatographic column via an injection unit; however, it is important that the sample and the mobile phase are at the right temperature before being introduced to the column, and in order to do this efficiently, the mobile phase is passed through a preheating coil at high pressure and controlled rate, before it reaches the injection point. After this process, the separated solutes then begin to pass slowly through the column; the time taken for this to occur is called the retention time which is recorded by an in-line detector.

Although this may be an excellent method for separating mixtures, it may not provide all the detail necessary for a clear identification of the separated compounds because retention data on its own is not specific, as many different solutes can have identical retention times for a particular set of conditions. However, the new generation systems developed by Perkin Elmer have tackled many of the fundamental problems of the older systems, including more sophisticated computer software. Consequently, HPLC could become the leading technique used for this type of analysis. The second type of chromatography to be discussed, Gas Chromatography (GC), also relies upon retention times as a result of the separation process. (Lindsay 1987)

2. **Gas chromatography** is very similar to HPLC: there is a stationary phase, a solid adsorbent and a mobile gaseous phase, usually nitrogen or helium. The sample is also treated in much the same way: it is injected, vaporized by the heating system, diluted with carrier gas, and then a small fraction is delivered to the column while the remainder is vented to the atmosphere.

The sample mixture utilizes elution for separation into its constituents in the same as those in HPLC, by relying upon the component's differential distribution between the stationary phase and the mobile one moving past it. This operation is controlled by the component's vapour-pressures and their sorption by the stationary phase, which differs slightly between the two techniques. In GC there is only one phase, the stationary phase, which is available for interaction with the sample molecules, whereas in HPLC both phases can interact selectively with the sample.

In the GC the mixture is examined in a vapour phase which must be stable, if this does not occur the substances have to be converted to derivatives that are thermally stable. As a consequence only a small number of compounds are suitable for GC without some form of modification, such as methylation and silylation which renders compounds, for example, carboxylic acids adequately volatile for chromatography. Where, by contrast, HPLC is better able to handle less volatile substances, as there is no need for modification. However, in some circumstances the third technique for discussion, pyrolysis gas chromatography, is better suited. (Willet 1991)

3. Pyrolysis Gas Chromatography (PGC) is ideal for involatile compounds that have relative molecular masses which create very large dispersion forces between the molecules in the liquid phase. This is a significant problem, as it cannot be solved simply by changing the temperature or making chemical derivatives. The only way to overcome this complication is to break up the large molecules into smaller, more volatile fragments that can be analysed chromatographically.

Although PGC is used for analysing non-volatile compounds, it is not applicable to all as it works most efficiently with compounds that, on pyrolysis, produce as few fragments as possible. Pyrolysis is essentially a heating process that occurs at the very beginning of analysis. A 'pyrolyser' is fitted on top of the column in place of the injection head; a sample is inserted into the heated enclosure where pyrolysis is carried out, after which the products of this reaction are swept by the carrier gas straight onto the column for chromatographic analysis.

Once the system reaches the chromatography stage it is susceptible to the same problems as the standard GC, in terms of specific compound identification, and would benefit enormously from a coupled spectrometer. The linking together of these two techniques is to be the final system of analysis to be discussed.(Jones 1982)

4. Gas chromatography/mass spectroscopy. The direct linking of chromatography and spectrometry has proven to be extremely useful in the identification of complex mixtures. The two systems are linked via an interface which ensures: a) temperature-compatibility within the chambers, and b) the efficient transfer of the trapped samples through the system as they elute from the GC analysis. Once these molecules are in the chamber they are a) converted to ions, then b) examined by a mass analyser, which distinguishes the newly formed ions according to their mass to charge ratios (*m/z*): the mass of an ion is related to the relative mass of the initial molecule.

GC/MS analysis considers two aspects of a sample, its ionization and fragmentation patterns. Ionization is generally achieved by directing a beam of electrons at a GC vapourized sample, as this often involves more energy than is actually required the excess energy is usually transferred to the ions as they are being formed. As a result of this excess rotational and/or vibrational energy, these newly formed ions exist in an electronically excited state, and are, consequently, often unstable and prone to dissociation. If this excess energy is sufficiently large, it will lead to vibrations of amplitude greater than the elastic limits of some of the bonds, leading to further dissociation or fragmentation.

Basically the mass spectrometer is required to measure the *m/z* values of the ions; an ion with its specific value will traverse the magnetic field and reach the detector, which is effectively the ion's final destination. Once this result has been achieved, the outcome is recorded and stored by the computer, and then visually displayed in the form of a spectrum.

The spectrum of the results from a GC/MS analysis is often very complex and requires not only a great deal of time to interpret, but also requires a reasonable level of proficiency on the part of the analyst, specifically when complex samples have been run through the system.

The complexity of the spectra relates to the mass of information which can be derived from the analysis of a single sample. Although a library could be built in a similar fashion to that of an FTIR spectral library, it would take much more time, expertise and sample analysis. If the project had much more time available, in addition to much more funding, and a computer with the appropriate software (such as 'CD-write'), it would be possible. The ideal situation would be a technique somewhere "in between" in terms of complexity and time consumption, such as HPLC.

A spectrum is read according to its fragmentation pattern of the ions in terms of *m/z* values; therefore, in order to demonstrate a simple spectrum, methanol has been selected as a basic example with its four major ions:

m/z value	ion
32	CH_3OH^+
31	CH_2OH^+
29	CHO^+
15	CH_3^+

The first of these is *m/z* 32 which is a molecular ion; the others are fragment ions which came about by: **a)** *m/z* 31 losing an hydrogen atom from the molecular ion, where one of the C-H bonds was broken which caused its formulation as CH_2OH^+; **b)** 5 *m/z* 29 loosing of a molecule of hydrogen from CH_2OH^+which gave the CHO^+ formulation; **c)** and finally *m/z* 15 the CH_3^+ ion, formed by the molecular ions with the correct energy for the fragmentation of the C-O bond rather than C-H bond cleavage.

As mentioned above this is just a minor example of a mass spectrum using a simple compound, it produces masses of data, especially from complex compounds, much of which is superfluous particularly when dealing with ancient materials, as it is often difficult to compare these results with those of the modern organics libraries, frequently incorporated within the software of these systems.

All told GC/MS is an excellent system given the correct type of analysis, however because it is extremely time consuming and expensive, this may present major problems when an experiment needs a dedicated system if the project is going to run over a substantial length of time.

A dedicated system for the instrumental analysis is useful and practical, as it means the required parameters can be set and maintained. It also means that the analysis can be carried out every time a sample batch is ready, and allows for a greater number of analyses to be carried out in the given time as the experimenter is not waiting for allocated time on a shared system.

Taking all these requirements into account, Fourier Transform Infrared was considered the most appropriate analytical technique, for its availability, cost effectiveness and above all its ability to provide the right answers.

However, given the opportunity I would have chosen HPLC as the main system for analysis if there had been one which could have been totally dedicated to this project for its duration. It is a superior system of analysis, especially the new generation systems now available from Perkin Elmer, which are computer controlled with dedicated software for particular tasks, such as compound comparison and chemometric analysis. The problem being of course there was not one available for the required time; due to the number of samples that needed analysing, share time on a borrowed system was not a viable option, therefore the FTIR system was utilized. In its most optimised form the FTIR system chosen was more than adequate for the project, which was demonstrated by the accurate data provided by the final analytical output.

3.4 Derivation of Spectra and Analyte Identification

In essence, an infrared spectrum is derived due to the absorption of electromagnetic energy in the infrared region of the spectrum. This type of radiation does not have sufficient energy to cause electron excitation; however, it does induce atoms and groups of atoms of the organic compound being analysed to vibrate faster about the covalent bonds that connect them. As a consequence of this, peaks are produced related to the functional groups present; these groups are effectively the structural information, or the basic chemical make up of the compound. The location of these infrared absorption bands or peaks derived on the spectrum can be specified in frequency units by its wave number $\overline{\nu}$, measured in reciprocal centimeters (cm^{-1}), or by wavelength λ, measured in micrometers.

With regard to the structural information of the analyte, it is not always possible to determine the composition wholly from consideration of the IR spectrum, even though assignments of absorption bands for these are sufficiently well developed to provide clues to some structural features of the molecule. When dealing with complex mixtures of compounds, especially where the sample is of unknown content, caution must be taken when assigning an identity. A broader approach may be the first step, as it is possible to see, with a high degree of confidence, whether a substance is aromatic, aliphatic and whether it contains OH, C=O or COOH groups. Specific features of unsaturation may also be identified, where recognition of the C=C bond is obvious. Since IR spectra arise from atomic and molecular vibrations, they are characteristic of most materials; therefore, precautions must be taken to limit the recorded spectrum to that of the substance being examined: that is, by eliminating, where possible, outside noise such as CO_2 and water vapour from the sample chamber.

3.5 Interpretation of IR Spectra: Earlier Work

Traditionally, infrared analysis was used as a qualitative technique for the visual comparison of peaks, using some form of atlas. This entailed the comparison of test samples with those of known GC assayed samples drawn from a commercially produced organics spectral library. Although this is a well-utilized technique in chemistry, it has been primarily utilized in the field of quality control, as the requirement here is for structural information as the primary data source, without the complications of calibration.

The use of IR as a system of quality control is quite well known and has proven to be very efficient in many fields. This is particularly the case where the compounds being analysed have a known structure and spectral "map". The pharmaceutical industry is one major area where quality control is regularly used; when a drug has been synthesized and tested, it is important that it be manufactured at a constant level of purity. This purity is often maintained by FTIR analysis of either the major components of the drug and/or the end product, these spectra are then kept as controls for subsequent batches. Once the purity has been established, each new batch is compared spectrally to ensure identical quality is produced at all times.

3.5.1 Pattern Recognition

This spectral comparison simply involves the overlaying of a sample spectrum with one from an organics library, in order to view both the similarities and the differences in peak presence and location. It works satisfactorily where comparisons are being made between materials of similar origins, and where the specimens are providing aliquots large enough to produce all the diagnostic peaks for the targeted chemical species. However, the technique rests purely upon pattern recognition of the graphical output, and does not utilize in any form the numerical value of either the spectrum as a whole, or the individual peaks. Therefore, accuracy in finer detail may be in doubt, as it is unlikely that two analysts will see the smaller, less defined regions with the same discerning criterion.

3.5.2 Interpretative Opinion

Pattern recognition is functional, but with limited application, as there comes a point when it may become difficult to distinguish between a true result and one which is purely 'experimenter's opinion'. In cases where the sheer quantity of data renders a purely visual means of comparison impossible, it is all too easy to begin to find the desired result wherever one looks. At its most extreme, this situation can involve a forcing of the data, in which circumstances the result of observation clearly become unreliable.

Consequently, if accuracy of identification is to be achieved, another form of analysis must be considered, and what is required is one that is systematic, quantitative (to the optimum degree practicable), and wholly representative of the data being analysed. The method must also be objective, or, as far as is humanly possible, free of that subjective individual interpretation which leads to 'experimenter bias'. Finally, it must be a procedure which can interpret, extrapolate and transform the data, without loss of the original IR content.

3.5.3 Limitations of Infrared Spectroscopy

There are several positive aspects to the classical form of Infrared analysis, which include its economical character in terms of cost and duration, and its user-friendliness. Often, however, analysts will require something further, especially in cases where samples are ambiguous in nature.

FTIR spectroscopy has produced advances that enable the system to be semi-quantitative, allowing for the introduction of some form of calibration, even though it is unable to split samples into the individual compound fractions.

The splitting of the sample into its individual components refers to a chromatographic technique, the extracts would still be the same material they would just be in a more refined form. Different species would only yield different components if their basic biochemistry was different, other than that the extracts should be basically the same.

The technique permits an acceptable level of standardization of the sample analysis, even if the resultant data is still only structural, and not specific enough to be totally diagnostic where complex unknowns are concerned.

The structural information refers to the spectrum of a sample which clearly indicates the specific functional groups and important bonds, which when combined make up a particular type of substance, in this case the triacylglycerols. It is not as refined as information from a chromatographic source such as GC or HPLC, but it is nonetheless still extremely important and very useful information.

Nevertheless, it is able to detect differences between specimens; often quite small distinctions can be evident. This capability allows the introduction of a complimentary analysis, which utilizes these spectral differences in a numerical form.

3.6 Numerical/Statistical Approach To Interpretation of Data

The introduction of a mathematical analysis was to enable the large amount of spectral data to be analysed systematically, via a procedure which was repeatable for each data set. Use of the subtle differences observed in the spectral results allowed a realistic expectation of some indication of species separation. By translating the spectral information into a numerical format, it was thought that a greater accuracy would be achieved to this end.

3.6.1 Advantages/Disadvantages In Numerical Translation

Due to the obvious inherent problems caused by the logistics of employing visual analysis to specify differences and similarities between spectra, a numerical approach to interpretation was considered the best adjunct. The numerical method offers a semi-quantitative approach which may be applied uniformly across the entirety of the assembled data, while at the same time eliminating errors induced by fatigue and by experimenter bias.

Problems potentially deriving from the use of statistical techniques would tend to stem from poor or inappropriate selection, in which the chosen techniques are unsuited to the type of data involved. In such a case, numerical and graphic output would be without real significance since they would not bear comparison to the original spectral data. Great care was taken, therefore, in selecting appropriate techniques.

3.6.2 Archaeological Data and Statistical Analysis

Previous research has utilized methods of interpretation which require a great deal of speculation, purely because most archaeological data is so indistinct. The application of a more rigid system of interpretation was intended to try to give a more definitive approach, whereby specific criteria could be devised to which the data could be assigned. This would then reduce ambiguity by allowing a higher level of standardization in the final groupings and identification of results.

It was therefore decided that a statistical approach, normally utilized in analytical chemistry, would best fit the required analysis. This approach consists of the set of techniques entailed in the discipline of *chemometrics*, and will be fully discussed in chapter 5.

CHAPTER 4

BIOCHEMISTRY OF PLANT LIPIDS

4.0 Introduction

This chapter presents a basic background to the biochemistry required to understand the analyses central to the thesis.

The triacylglycerol components of the plant lipids were targeted as likely to represent a potentially diagnostic species-specific or genus-specific group of compounds. Unpublished research has suggested that the removal of diagnostic lengths of DNA from charred archaeological seed remains is problematic owing to the deteriorated state of the material. Polysaccharides are also non-diagnostic, owing to their complexity and the length of the carbon chains. Therefore triacylglycerols were considered viable alternatives because they are:

1. Taxonomically diagnostic (at least at generic level)
2. Easily extracted from the intact seed tissues
3. Not subject to denaturing by the heating process involved in preservation by charring.

The latter consideration meant that a number of phenolic compounds were precluded at the outset of the investigation, (Butler, pers com.)

4.1 Seed Lipids

Quantitatively, triacylglycerols form the major amount of lipid in mature seeds, constituting between 10 and 70% of dry weight tissue, whereas phospholipids and glycolipids normally represent less than 2%. However, many members of the *Graminae* may have higher ratios of polar lipid (Glycolipids and phospholipids) to triacylglycerol (neutral lipids), than most other classes of plant. For example, Fisher shows the triacylglycerol content of wheat to be 1-2% in the endosperm, 8- 1 5% in the germ, about 6% in the bran and an average of 6% for the whole kernel. He concluded that total seed extracts would contain a substantial proportion of the polar lipids. (Fisher 1962 in Galliard & Mercer 1975) This situation changes somewhat when analysing immature seeds, which contain far less lipid than mature tissue. Of the small lipid fraction present in an immature seed, a minor proportion is triacylglycerol, the remainder being phospholipid and glycolipid.

4.2 The Lipids

The lipids are natural fatty substances that are soluble in non-polar organic solvents and insoluble in water. These are distinguished from various other cellular components such as the proteins, carbohydrates and nucleic acids. The acyl lipids employed in this project are all derivatives of long-chain fatty acids of the form R-COOH. These fatty acids occur in nature mostly as glycerides; that is, esters of glycerol, or as esters of other 5 long-chain alcohols. Simple lipids are the most abundant fatty acid derivatives, occurring in both animal and plant tissues. The most predominant of this class (simple lipids) are the esters of fatty acids with glycerol and cholesterol, most often occurring as esters of the trihydric alcohol, glycerol (*1,2,3-trihydroxy propane* or structural formula CH_2OH CH(OH) CH_2OH).

The esterification of the fatty acids with glycerol may produce five different derivatives. All of these, with the exception of the triacylglycerols, are partial glycerides, as not all the -OH- positions are esteffied: **Fig. 4.1**

1.
CH_2O-CO-R^1
R^2 C-O-O-CH
CH_2O-CO-R^3

Triacylglycerol

2.
CH_2OCOR^1
$CHOCOR^2$
CH_2OH

1,2-diglyceride(1,2- diacylglycerol)

3.
CH_2OCOR^1
CHO
CH_2OCOR^2

1,3-diglyceride(1,3-diacylglycerol)

4.
CH_2OH
$CHOCOR^1$
CH_2OH

2-monoglyceride(2-monoglyceride)

5.
CH_2OCOR^1
CHOH
CH_2OH

1-monoglyceride (1-monoacylglycerol)

Fig. 4.1 The R^1, R^2 and R^3 positions are long-chain alkyl groups of the fatty acids, which may differ in both length and degree of unsaturation. Saturation refers to those structures that contain no double bonds [other than the carbonyl (C=O)] of the ester functional group, for example palmitic acid [16] CH_3 $[CH_2]_{14}COOH$. **(Galliard & Mercer 1975)**

The esterified triacylglcerols of plant seed oils have properties that depend on the original fatty acids present, their relative amounts **and** their positions in the molecule. Over 200 different fatty acids have been isolated from higher and lower plants, many of which are unusual and only occur in a few plant species. The widely distributed acids that are classified as **major** fatty acids are few in number, but are normally found in lipids from all parts of all plants. The most prevalent and important of these major lipids are palmitate oleate and stearate. The group classified as **minor** fatty acids are relatively common in

Table 4.1 (Christie, 1973) **Structures of the Major Fatty Acids**		
Lauric	12:0	$CH_3[CH_2]_{10}$ COOH
myristic	14:0	$CH_3[CH_2]_{12}$ COOH
Palmitic	16:0	$CH_3[CH_2]_{14}$ COOH
Stearic	18:0	$CH_3[CH_2]_{16}$ COOH
Oleic	18:1(9*c*)	$CH_3[CH_2]_7$ CH:CH$[CH_2]_7$COOH
Linoleic	18:2(9*c* 12*c*)	$CH_3[CH_2]_3[CH_2CH:CH]_2[CH_2]_7$COOH
Linolenic	18:3(9*c* 12*c* 15*c*)	$CH_3[CH_2CH:CH]_3[CH_2]_7$COOH

terms of distribution, but they generally only occur in minute proportions, as a consequence they are difficult to both isolate and detect with certainty. They have, therefore, been excluded from the project at this point. However, preliminary work suggests that they are most likely to be species-specific, and may thus represent valuable diagnostic indicators when the appropriate technical means become more reliable.

4.3 Major Fatty Acids

The major fatty acids are all saturated or unsaturated monocarboxylic acids, generally containing an aliphatic, even number carbon chain, with the carboxyl group at one extremity and double bonds in the unsaturated of the *cis* configuration in specific positions relative to the carboxyl group. The saturated homologues lauric, myristic, palmitic and stearic acids all occur in plants; lauric and myristic acid are most extensively represented in the seed oils of the small families of *Lauraceae* and *Myristicaceae*. Palmitic acid is the most widely occurring of all the saturated acids, and is found in practically all fats analysed. Whereas stearic acid is less common than palmitic acid; although it occurs in most vegetable oils, there are only a few where it actually forms the major component.

These four (lauric, myristic, palmitic and stearic), together with the abundant unsaturated analogues, oleic, linoleic and *a* - linolenic, constitute the seven most prolific acids found in lipids from all parts of all plants, often with palmitic and oleic predominating. Of these the monoenoic fatty acid oleic is the most abundant, occurring in almost all fats.

Fatty acids rarely, if ever, occur in the free form naturally; they are, consequently, usually analysed in the ester form. Esterification occurs at the carboxylic end (COOH) of the molecule, with the COOH losing a Hydrogen atom and the alcohol losing an hydroxyl group (OH radical). Thus, the process of esterification gives off water as a by-product.

4.3.1 Fatty Acid distribution

Triacylglycerol formation has been shown not to be a random process: the fatty acids exhibit tendencies to associate in specific positions. That is; the 1 and 3 positions, although not biochemically equivalent, tend to accumulate the saturated fatty acids during biosynthesis. The secondary hydroxyl group correspondingly shows a preference for the unsaturated acids. As a consequence of these formation patterns, many of the vegetable oils have a 95-100% chance of containing oleic, linoleic or linolenic in the 2- position. (Galliard and Mercer 1975).

Esterification forming the triacylglycerol

CH_2OH	$CH_3[CH_2]_{16}COOH$	stearic
CHOH	$CH_3[CH_2]_7CH:CH(CH_2)_7COOH$	oleic
CH_2OH	$CH_3[CH_2]_{16}COOH$	stearic
glycerol	fatty acid	

$CH_2[OCO]\,(CH_2)_{16}\,CH_3$

$CH[OCO]\,(CH_2)_7CH:CH(CH_2)_7CH_3$

$CH_2[OCO]\,(CH_2)_{16}\,CH_3$

Triacylglycerol

Fig. 4.2 (Galliard & Mercer 1975). Fatty acids are distinguished from one another by the character of their R group, and by whether they are saturated or unsaturated. In the major saturated fatty acids, the R group is an aliphatic hydrocarbon chain, while the functional component falls into a series represented by $CH_3(CH_2)_nCOOH$ (see Nomenclature, below).

Unsaturated fatty acids present a more complicated situation, as they are characterized by the presence of one or more ethylenic double bonds (-CH:CH-), with varied positions of both singular and multiple double bonds in the aliphatic chain. There are even differences between isomers of the same fatty acid, which is represented steriochemically by the shifting of the H bonds around the alkene (C=C) group, from *cis* to *trans*.

The formal steriochemical definition of *cis* and *trans* bonding is given below, and represents two isomeric forms of the ethylenic bond:

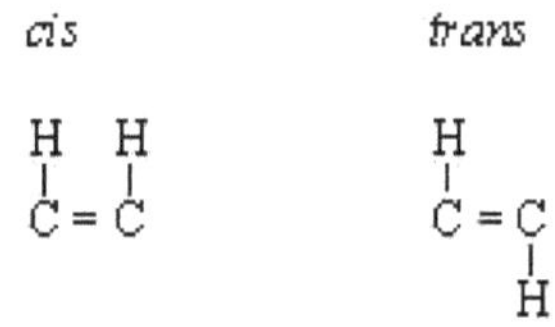

Fig 4.3 (Christie 1973)

An example of this isomerization can be demonstrated with oleic acid, which is the most abundant of all fatty acids and accounts for some 40% of all naturally occurring acids. In its normal ester form, oleic is a *cis* 9-monosaturated C_{18}acid with a melting point 16.3°C, while in the *trans* form it becomes elaidic acid with a melting point of 43.7°C. The *trans* form allows the acid to be packed more economically within storage tissue, and appears to be only found in animals.

4.3.1.1 Nomenclature

As with saturated fatty acids, a general formula can be devised to represent unsaturation which incorporates the double bonds (including mono-saturation).

$CH_3(CH_2)_x(CH{:}CHCH_2)_y(CH_2)_zCOOH$
where x=1,4,7 or 10

Fig. 4.4 (Christie 1973)

Y =2 to 6 is shorthand to indicate the chain length where double bonds occur ($CH{:}CHCH_2$)
z = is variable, and represents the number of plain (CH_2) bonds in the chain.

For example, this nomenclature satifies most of the frequent orders of unsaturation;

Linoleic x=4	y=2	z=6 (*cis*)
Linolenic x=1	y=3	z=6 (*cis*)
λ-linolenic x=4	y=3	z=3 (*cis*)

Fig. 4.5 (Christie 1973)

The above three examples are found only in plant fats. λ -linolenic is an essential precursor in animal diet for the synthesis of the physiologically important Arachidonic acid

(C_{20}; x=4, y=4, z=2). **Fig.4.6 (Christie 1973)**

Although most fatty acids conform to this system of characterization, there are a number of unsystematic characteristics which can be classified in their own groups:

1. Branching
2. Hydroxylation
3. Acetylenic (-C:C-) bonding

Branching is quite rare, and occurs generally as a simple methyl branch on either the penultimate carbon (forming an iso-acid) or on the next adjacent carbon (forming an antieso acid) (see above). **Fig. 4.7 (Christie 1973)**

When **Hydroxylation** occurs, the carbon atom to which the hydroxyl group is attached is asymmetric. That is, the four attached valence groups differ from each other, thereby allowing the molecule to take up one or two opposed, optically active forms. An example of an hydroxy acid, ricinoleic acid (a 12-hydroxy compound), may be compared to linoleic acid.

Other acids may also be found in trace quantities with the hydroxy group in the 4 or 5 position, which, when isolated from the acylglycerol tend to undergo ring closure, losing water and forming lactones. For example: **Fig. 4.9 (Galliard & Mercer 1975)**

4.4 Fatty Acid Composition Of Glycerides

Naturally occurring fats contain mixtures of acyl glycerols. While difficulties can arise in determining precisely which acyl glycerols are present in any particular fat, a small group of predominant fatty acids are normally to be found. Within this group, variations occur in the relative proportions present between different fats; and these differences in proportion provide the most straightforward method of distinguishing between fats, and of identifying and specifying a single, given fat. Such determination of percentage composition requires an appropriate technique of separation, such as gas chromatography, which may permit the resolution of closely related isomers. For this specific research project, however, the identification of materials by the aforesaid technique was not an option, as explained in the previous chapter. The ratio of major fatty acids present in a single plant species depends on the identity and growth history of the plant, the part of the plant sampled, and the identity and function of the acyl lipids analysed. For example, a plant grown at low temperature has a higher degree of

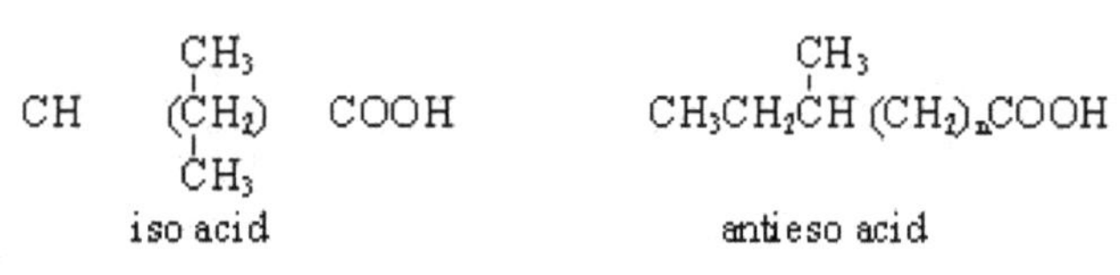

Fig.4.7

ricinoleic acid	$CH_3(CH_2)_4CH_2\,CHOHCH_2\,CH_2CH{:}CH(CH_2)_7COOH$	(9*c*)
linoleic acid	$CH_3(CH_2)_4CH{:}CHCH_2\,CH{:}CH(CH_2)_7COOH$	(9*c*, 12*c*)

Fig.4.8 (Galliard & Mercer 1975)

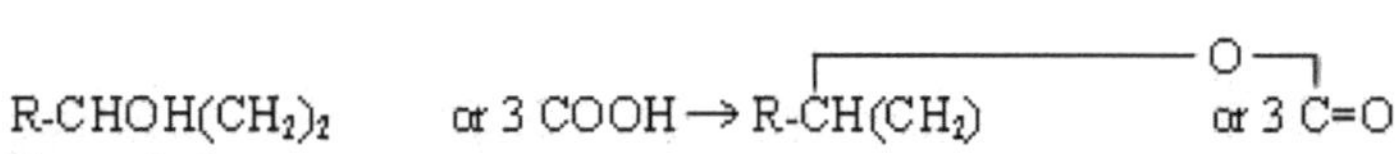

Fig. 4.9

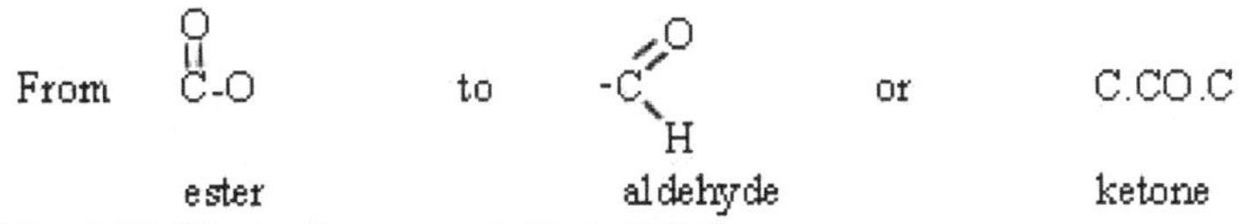

Fig. 4.10 (Pavia, Lampan & Kriz 1979)

saturation of the fatty acid mixtures than one grown at higher temperatures (Galliard and Mercer 1975). Additionally, different species may show an accumulation of unusual acids also found as triacylglycerols in a few specific seed oils. Generally, when this accumulation occurs in seeds they could be seen as species markers. However, these unusual fatty acids are of limited utility as species indicators. This is because, within a phenotype, seeds accumulate them differentially from plant to plant.

This differential accumulation greatly reduces the viability of unusual fatty acids as an indicator of species, as in some cases the detectable presence can be affected by environmental conditions or maturity of the seeds being analysed. Consequently, the level may be either too low to detect or not present at all, a situation that can occur not only from species to species, but plant to plant.

4.5 Factors Which May Effect Lipid Proportion

The total amount of lipid extracted from fresh seeds varies in quantity; however, when dealing with small sample groups such as those utilized in this project, the amount extracted is likely to be quite small. Moreover, systematic experimental errors inherent in wet chemistry may further reduce this amount, in some cases to almost non-diagnostic levels. As that is the situation when dealing with the modern samples, it is likely that the charred archaeological remains, being dry and fragile, may suffer an even further reduction in lipid content.

4.5.1 Extract Deterioration and Contamination

Fatty acid esters are susceptible to deterioration after extraction. Specifically, light- sensitive denaturing and crystallization is problematic. Samples extracted from fresh material and spotted on potassium bromide (KBr) discs displayed deterioration effects after three months. It is likely that the KBr-held samples deteriorated more rapidly than samples held on disposable cards because the KBr-held samples suffered greater from exposure to the atmosphere. This hypothesis was confirmed by a second series of FTIR scans on a single set of KBr-mounted samples, which indicated a shift in the major functional groups after several months. The carbonyl (C=O) peak shifted from1750-1730cm^{-1} to 1725-1710cm^{-1}, and no complementary peaks were visible in the 1300-1000cm^{-1} (C-O) region in the aged samples. These peak shifts indicated a deterioration of the triacylglycerols to aldehydes or ketones, with the aldehydes displaying characteristic set of two low frequency C-H stretching peaks around 2700-2800cm^{-1}.

Stoichemically, this deterioration takes the form displayed in the above **figure 4.10**.

After several months, KBr-mounted samples were discoloured and opaque; they also appeared to have crystallized. When analysed again on the FTIR, they clearly showed increased noise, which was probably due to scattering caused by the greater opacity of the discs.[1]

The deterioration of the extracts on KBr. discs was a problem requiring investigation. To select an unreliable method could lead, in the most serious instance, to the loss of irreplaceable samples, and result in time-consuming and repetitious laboratory work undertaken in order to replace them. Fortunately, the work of Dr. Raymond White from the National Gallery on the deterioration of oil paints and the component lipids is well known. As White uses various methods of instrumental analysis including FTIR, it was considered prudent to seek his advice regarding the problems of crystallization and discolouration of the extracts on KBr. discs. The aim was to confirm the results of the deterioration in order to select the most reliable method of sample holding. White's work with oils has demonstrated the effects of light, air and heat on the bond structures, and confirmed the prevalence those effects observed during the present project. As a consequence, it was possible to select a more reliable method of both sample holding and storage. Such a capability was essential, in view of the need to work on the samples at a later stage.

Ideally, fatty acid esters removed from the plant matrix should be stored under argon, in a cool, dark environment. As this was not practical, the samples were spotted onto disposable mid-range IR cards, and stored in acid-free tissue in a cool, dark cupboard. Such a method of storage appeared to be quite adequate, and was shown to be so by the condition of the samples when they were re-analysed. Furthermore, as the cards are not hygro-scopic in nature, opacity (and hence increased noise) was also eradicated. A second series of scans carried out on these samples one year after initial mounting showed little or no changes in the major functional group positions within the spectra. These final analyses demonstrated that the cards, in this particular case, were more effective for sample preservation than the discs.

[1] Personal communication with Raymond White of the National Gallery regarding his work on the deterioration of lipids in light and air, specifically the discolouration and crystallization of samples on KBr. Discs over a period of several months.

While it is essential for the research to take account of those factors which may influence the proportion of lipids present in the samples, it is equally important to consider the mechanisms and environments which may render the lipids and other compounds structurally able to survive or deteriorate. Knowledge of conditions obtaining when the constituent organisms of these compounds were interred will assist in both the efficient recovery of their remains, and a more thoroughgoing understanding of the recovered materials. These materials may, effectively, be chemical residues or the by-products of bacterial metabolism.

4.6 Molecular preservation

Molecular Preservation explores the relations between the molecular composition of living organisms and the nature of fossil organic matter through consideration of the mechanisms of decay and preservation'.

All living organisms contain essential biopolymers, nucleic acids, proteins and carbohydrates. They may also contain a variety of additional organic compounds made up of carbon, hydrogen, nitrogen, oxygen, sulphur, phosphorous and other elements, amongst which the fat-soluble lipids are important components. Biomolecular palaeontology involves the systematic study of the molecular record of ancient life as it is inscribed in fossils and sediments. Studies have demonstrated that the fossil record contains large quantities of molecular debris derived from biomolecular input. The preservation potential of biopolymers and other individual biomolecules varies considerably with molecular structure, depositional environment and diagenetic history. Biomolecular palaeontology has much to offer in revealing the processes which result in the preservation or decay of original biomolecules. It presents us with the possibility of employing fossil molecular structures for the assessment of palaeo-environments and the study of evolution.

4.7 Molecular composition of biomass

The overall preservation, potentials of differing compound classes or groups of biomolecules, in terms of their susceptibility to attack on the links that bind the atoms of the molecules together, are highly dependent upon the physical state of the organic materials in question. These considerations must also include the siting of the materials within mineral matrices or tissue membranes at the molecular level itself.

While some of these biomolecules, such as nucleic acids, proteins, carbohydrates and lignin are not of direct relevance to the research under discussion, it is necessary nonetheless to consider the presence of their particular biomass, in addition to the basic nature of their processes of deterioration. In simple terms, that is, to consider upon what constituents or combinations thereof might their manner of deterioration be dependent? Such considerations do, in fact, bear upon the specific biomolecules studied in the course of this research, i.e. what directly causes this deterioration, as these factors may be important with respect to the breakdown of the biomolecules relevant to this research.

4.8 Nucleic Acids - DNA and RNA.

Viewed as a simple structure, DNA is a single chain macro-molecule comprising deoxyribose sugar units, linked by phosphate ester groupings at the 3 and 5 positions, with each unit carrying either a purine or pyrimidine base. RNA is similar in structure, carrying the ribose sugar in place of the deoxygenated form. In order to consider the preservation and destruction of this macromolecule, it is necessary to focus on that part of the structure upon which the chain is dependent i.e. the phosphate ester link. Eglinton and Logan suggest that, although a great deal of further work is required in order to gain a proper understanding of the underlying processes of preservation/destruction of these molecules, processes of oxidation appears to engender greater levels of damage than do these of reduction. Oxidation has the effect of cleaving the phosphate linkages, which creates shorter strands. In the early stages of fossilization, however, during which oxidation is minimal there appears to be a less extensive assault on these links, suggesting that the binding of DNA to mineral surfaces may lead to a higher level of preservation. (Eglinton and Logan, 1991.)

4.9 Proteins - Amino acids.

Proteins are quite lengthy and often complex biomolecules. They form linear polymers alpha-amino acid residues which are linked by the peptide bond. Altogether, there are some 20 amino acids, be possessing the L-configuration during life, and which go to make up the proteins of all known organisms. Once subjected to a process of degradation such as the hydrolysis which may occur during fossilization, the peptide links are liable to be broken, resulting in progressively shorter amino acid chains or units. However, this process does not necessarily mean the total loss of information. Particularly in those circumstances where the hydrolysis is non- random, the short chains may still be useful for protein recognition. There are other processes of degradation which may result in the total or partial loss of protein identity, one of which has been utilized as a technique of identification; namely, that of racemization or the deterioration of the naturally occurring L-form of amino acids into the D-form. This involves the chiral alpha-carbons of the amino acid units being subjected to epimerization, resulting in the gradual racemization of the protein, in which these 'habitually' occurring L-form amino acids deteriorate to the D-form. Macko and Engel have employed this method to determine the origins of amino acids in fossil samples. (The reference to habitual L-forming amino acids is the regular or constant way this occurs). In addition, they have demonstrated that fossil systems are not closed systems. The process of racemization may be retarded in organic materials in the fossil matrix due to the presence of non-racemic amino acids which are present because of the actual preservation of a percentage of the original protein structure. (Macko and Engel, 1991.)

The process of epimerization and the rate at which it can cause racemization to proceed has been found to be closely dependent upon both environmental conditions and the structural nature of the amino acids. In the case, of amino acid structure, if the resulting proteins are tightly folded onto themselves- as with myoglobin- the action of polar denaturing agents or attacking enzymes may not penetrate to the interior of the amino acid residues. This creates an effective resistance to modification during fossilization and diagenesis.

As regards the environment, dehydration may lead to the greater preservation of protein structures, while atmospheric air and moisture can create high rates of degradation.

4.10 Carbohydrates - mono-polysaccharides

The carbohydrates are found in two forms: the first consists of the simple water-soluble mono-saccharides, usually of a C5 or C6 structure; they are readily susceptible to deterioration, having a low survival rate. The second form consists of the complex, long chained polysaccharides, which are insoluble in water. They are produced by the condensation of the hydroxyl of one molecule with the carbonal group of another as ketal or acetyl branched structures. The hydroxyl groups and the readily hydrolyzed acetyl links on the exterior surfaces of these closely packed molecular assemblages are susceptible to attack by extra-cellular enzymes. However, evidence has been produced to demonstrate that some polysaccharides can be preserved in certain examples of fossilized tissue.

4.11 Lignin

This biopolymer is found in vascular plants, which may vary in structure according to species. The monomer structure comprises hydroxy-propyl-benzene (C9) units, oxygenated in the alkyl group and a benzene ring, they form structures through a number of ether links and carbon- carbon bonds. This structure renders them resistant to alteration during the early phase of diagenesis, but the process demethylation of the methoxy groups substituent on the benzene ring may engender some degradation. In spite of this fact, however, lignin is thought to be one of the most resistant biomolecules found in fossil plant debris.

4.12 Lipids

It is not my intention here to enter into a detailed discussion of the nature of lipids, full descriptions having been made elsewhere. Nonetheless, it is important to take account of those elements within the structure of the lipids which may be resistant and remain well-preserved in the context of the fossil environment. The structure of fossil lipids has, in fact, enabled them to count amongst the best preserved of biomolecules: their aliphatic nature renders them insoluble in water, a fact which enhances their preservation potential. It is true that the hydrogenation of double bonds, the aromatisation of rings and/or loss of functional groupings during diagenesis may result in some degree of modification. However, the utilization of lipids as 'bio-markers' has derived from the possibility of recognizing the original precursor bio-lipid within the structure of the modified form of geo-lipid. It is this capacity which allows for their use, as bio-markers in the assessment of palaeo-environmental conditions and the process of diagenesis. (Brassell & Eglinton,1986.)

When situated in the context of the living organism, many of the compounds are linked into larger molecules; these include the ethers or esters, which go to make up storage lipids. These storage lipids are the triacylglycerol esters, and are the specific biomolecules which constitute the foundation upon which this research is erected. Information regarding the preservation and deterioration of these compounds is, accordingly, of capital importance; the crucial factor being dependent upon their ability to maintain structural integrity over the passage of time, notwithstanding the exigencies of the external environment. There are, nonetheless, various factors which may have adverse effects on certain of the lipid components: for example, the action of hydrolysis the triacylglycerol esters, which can increase amounts of water-soluble free fatty acids. These latter compounds are much more susceptible to the action of microbial enzymes and water-enhanced chemical reactions.

It is the formation of free fatty acids which has generated high levels of interest in the context of the human body; specifically, the formation of adipocine, found in corpses which have been interred for long periods. Interest has centred upon the origin of this material whose major components consist of calcium salts, which are also present in the free fatty acids of the triacylglycerols. The essential appearance of adipocine consists of waxy, coagulated lumps of fat resembling early forms of soap. The author was shown similar material while in South America; it had been excavated by local 'sifters' from deep within the interior of long standing refuse heaps. While the precise origin of the material was unknown, large numbers of the local population had encountered it when digging deep beneath the surface of the large local refuse heaps which were left rotting for long periods, undisturbed. Unfortunately, it remains impossible to account for the presence and formation of the material; no research facilities were available to the author at this juncture, and the environmental conditions responsible must remain the object of speculation.

In the case of the peat bog, however, the state of scientific understanding differs radically. Evershed (1990) has examined bodies recovered from waterlogged peat bogs, and has demonstrated that certain tissues, such as muscle tissues, are preserved, a fact which stems from the presence of humic acid and tannin-like substances in the acid environment. Although causing rapid hydrolysis of the ester links in the triacylglycerols, this environment has not rendered soluble the lipid components, which, in consequence, have remained concentrated within the original tissue, only minimally leaking into the surroundings. In those circumstances where the, environment is less acidic, the shorter chain lipids, which are more water-soluble, are subject to increased vulnerability to microbial attack. (Parkes and Taylor, 1983; in Eglinton 1990).

4.13 Limitation of Molecular Damage during Fossilization.
There may be certain physical attributes of the deposited organic material, in addition to various chemical and mechanical processes, which may aid in the limitation of damage to the molecular structure of these biomolecules during the course of fossilization. The most significant of these, which are likely to have the strongest influence, are discussed below.

1. Gross Morphology.
Those hard parts of recovered plant materials, such as seed cases, may prevent the movement of biomolecules out of the seed, while simultaneously hindering the entrance to the interior of reactants. The softer tissues, by contrast, may have a high water-content, and possess, in consequence, an enhanced susceptibility to decomposition.

In a discussion of the charring process, Hillman has suggested that it may create a hard, protective layer around the seed which has a containing effect on the biomolecules. When the seeds in question were interred, moreover, the burial environment was dry and desiccating, which had the effect of further protecting the biomolecules, in this case from excessive water percolation. In structures such as teeth, bone and shells, biomolecules may be entrapped by the formation of the crystalline structure these items possess, and this structure may offer them additional protection.

2. Macromolecular Structure.
Cell organelles contain biomolecules such as proteins, carbohydrates and nucleic acids, which are so tightly stacked and folded that they are afforded some local resistance to attack.

3. Movement.
Physical structures and attributes may place limitations upon the actual motions of organisms such as worms, i.e. if the organism is unable to access the internal, molecular structure, there is a reduction in the potential for physical and mechanical damage; in addition, such features may also prevent or limit the exposure of samples to reactants and decomposing agents.

4. Water.
The presence of water can be highly problematic for attempts at preservation, as it may aid in the facilitation of numerous biochemical reactions and microbial activities. Even in solid materials, water may be present in tissues, and initiate hydrolysis reactions.

5. Environmental conditions: pH, eH, Temperature and Light.
The processes of oxidation and reduction in an environment specifically, may interact with the behaviours of microbes, that can greatly affect those many other natural processes which occur as a matter of course, including that of fossilization. Similarly with ambient temperature, increases may greatly enhance chemical reactions.

6. Ions as Nutrients.
The reaction of microbes within the fossilization environment is influenced by the presence of ions, as these function as microbial nutrients. They also function as electron receptors and as catalysts for the process of hydrolysis.

7. Microbes.
The action of microbes represents a major obstacle to the preservation of samples of organic materials, forming as it does the prime causal agent in processes of decay. This is especially the case when microbial action is considered in its interactions with other factors such as the presence of ions and electron receptors. There are, however, certain checks upon this action, and various compounds and processes exist within a number of specific environments that impede microbial decay; the peat bog, for example, is able to exercise a significant inhibiting effect thereon. The presence of humic and tannic acids in such environments acts as a preservative influence upon body tissues.

4.13.1 Factors Influencing Longer Term Survival:

8. Compaction and Lithification.
The increased compaction of fine-grained sediments can lead to the forced expulsion of water from the pores, and this reduces the movement of ions and molecules. Furthermore, the activity of microbes will be restricted when materials cease to be porous and permeable.

9. Radiation.
Radioactive ions such as potassium and uranium, which may be contained in clay, may emit radiation which results in damage at the molecular level.

10. Radicals.
Radicals created by radiation and exposure to light may become trapped within organic compounds, and these, in their turn, can prevent or inhibit degradation, however, they may also promote it.

11. Temperature.
The temperature of a burial environment may have considerable effect upon molecular debris. This effect is particularly acute in instances where the materials are buried deep beneath the surface. If the temperature is raised, thermolytic changes may occur with the passage

of time, leading eventually to the production of materials such as petroleum and natural gas. These hydrocarbons carry biomarkers which provide indications of their biological origins. (Eglinton, 1991).

4.14 Preservation

The preservation of organic matter depends heavily upon the condition of the burial environment in the period immediately following the death of the whole organism. This statement holds both at the level of gross organic debris, and at the molecular level. The organism is subject to relatively rapid decay in the early period of interment owing to the actions of gasses produced by its own physical breakdown. In circumstances in which the burial environment includes a physical barrier, the organism's chances of preservation are greatly enhanced; this is because such prophylaxis reduces the availability of nutrients for those microbes whose behaviour facilitates the process of decomposition. After the initial phase in the decaying process has exhausted itself, the relatively stable sedimentary milieu is capable of maintaining a steady state of preservation so long as the burial environment remains 'closed'. Effectively, such closure refers to the absence of gasses and solutions which would otherwise cause the complete mineralisation of the organic matter, breaking it down into its elements: carbon dioxide, water and so forth.

The term diagenesis refers to post-depositional processes of compaction, cementation and crystallization of sediments at or near the surface, which are subject to relatively low temperatures and pressures. The process of *compaction* occurs when the weight of overlying sediments increases, while that of *cementation* arises when individual particles are bonded together by materials precipitated by circulated fluids; these fluids may consist either of percolating ground water, the solution of part of the mineral matter of the sediment itself, or the squeezing of deposited organic matter. Most of the modifications entailed in diagenesis are controlled by the pH of the environment and the oxidizing potential (eH) of the circulating fluids.

When considering the process of decomposition of organic matter during the preliminary stages of diagenesis, it is necessary to look at both the availability and type of biomolecules within an organism. This is because different types of biomolecules deteriorate at differing rates, and may be afforded further protection by their given location within the molecular environment. This was indeed the case with respect to the preservation of those lipid components studied in the current research, which appeared to derive considerable protection from the outer covering of the seeds; this was itself strengthened by the charring to which the seeds had been subjected, and by the desiccating burial environment.

4.15 Decomposition: Microbial Activity.

Microbial activity seems to take place even in the least likely of environments. While it may be restricted in certain extreme contexts- high pH, salinity and temperatures- there may still be different groups of organisms which dominate the activity according to various specialized chemical conditions. Even when deprived of oxygen, for example, species of anaerobic organisms are specially adapted to, and able to flourish within, these conditions. Where pressure is an important factor in the advanced stages of diagenesis, decomposition may decrease due to compaction and the subsequent lack of availability of water dependent micro-organisms and chemical species.

4.16 Preservation in Terms of Environment.

While, a number of important biomolecules must be taken into account when speculating about the preservation and/or degradation of the sample materials, it is equally necessary to consider the influence of a range of environments, each which can impact significantly on the level of preservation. Lipid components, for example, are much more likely to be hydrolised and modified if buried within soft tissue as opposed to within a hard seed material whose covering offers it protection.

It is necessary to know the ways in which different classes of organic matter are susceptible to degradation, and the manner in which processes of molecular alteration differ according to the organism's timetable of fossilization and decay (Eglinton 1990). These issues pose, a set of fundamental questions concerning the events ensuing on the death of an organism, with specific reference to biological compounds: first, what is it that prevents or retards the post-mortem breakdown of organic material; secondly, what facilitates the long term survival of such material throughout the long period of interment: including the millions of years entailed when we invoke the passage of geological time.

The study of organic matter recovered from the fossil record has already yielded a large corpus of important information, and with the continuous advance of analytical techniques across the spectrum of scientific disciplines, there is every prospect that this will be expanded upon still further. It is my belief that the molecular level will feature prominently in this anticipated progress. In order to arrive at a full and adequate interpretation of the data gathered at this level it is necessary to develop, in tandem, an enhanced understanding of molecular taphonomy. This is the process of *decay*, and its study is the study of the ways in which processes of decay at the molecular level impact upon the fossil record.

4.17 Differences between modern and ancient seeds

The comparison of lipid biochemistry between ancient and modern samples requires particular care, as gross differences are likely to be present between the two sample populations. Processes which may affect fresh material, such as lipolytic activity, hydroxylation and the extraction of phenols, probably have no impact on archaeological material. Additionally, modern samples obtained from a cultivated source may have been sprayed with herbicides and pesticides during the growing period, as well as with a fungicide to prolong storage life. These additives may include a wide variety of substances, such as 2,4-D (2,4-dichlorophenoxyacetic acid [$C_8H_6Cl_2O_3$]) a growth regulator herbicide against broad leaved plants (MP 138°C), pyrethrum (a contact herbicide produced from esters of the chrysanthemum), carboxylic acid, or a fungicide (usually sulphur, polysulphide or sulphur containing chemicals such as dithiocarbonates $M^+(R_2NCS_2)^-$).

These chemicals have high melt points and may be retained on the seed coating, thereby introducing contaminants into the reflux process which may subsequently effect the IR result. This is especially problematic where pyrethrum compounds are concerned, as these are esters of carboxylic acids, and their major functional group structures may appear in the same IR regions as those of the analysed samples.

Ten thousand year-old charred archaeological material is very likely to contain only a small proportion of the original amount of lipids present. This may render the sample too small to be diagnostic with FTIR analysis. In addition, even where lipid content has been retained, it may have changed chemically and no longer be accessible using a normal solvent regime.

4.18 Infrared detection of triacylglycerols

Infrared analysis utilizes the vibrational region of the electromagnetic spectrum; molecules exposed to IR light become excited to a higher energy level, which returns to normal once the radiation source is removed, as there is no permanent change in the molecular ground state. The recognition of peak patterns is a product of the selective nature of frequencies absorbed by the bonds of functional groups, with vibrational frequencies that correspond to the infrared radiation source. The actual strengths of these absorbance bands will be affected by the relative amounts of the various triacylglycerols present. Once a molecule displays a spectral pattern from this energy absorption, it has produced a unique description of itself, as no two molecules of different structure will have exactly the same IR absorption pattern. However, some compounds may contain some of the same major functional groups and still produce patterns which are significantly different.

This type of information, gleaned via the bond absorption, is referred to as "structural information", as only specific bonds or functional groups can be identified with confidence, not distinct compounds. These known Sequences can be used as a fingerprint region for comparative analysis both visually and numerically, but prior knowledge regarding the sample origin is required if identification in terms of 'chemical species' is to be successful. To be absolutely certain of compound identification, a separation technique is required in which the individual constituents of the total analyte sample can be classified. The current project utilized the structural information to demonstrate the presence of the desired extract, namely the esters of fatty acids. Further work using HPLC or GCIMS would be required for an enhanced identification to be carried out.

4.18.1 Detection and Bond Frequencies

Technically, the identification requires not just the recognition of a point within a spectrum, but a number of complimentary ranges. These are produced by the energy absorption creating several modes of vibration on the bonds being investigated. The vibrational modes are termed "fundamental absorptions", which are expressed by stretching in either symmetric or asymmetric formation, or bending in scissoring or wagging in or out-of-plane formation. These inevitable bond actions yield complimentary identifiable peaks in different regions of a spectrum, without which a specific compound cannot be identified with any degree of conviction.

These action characteristics of bonds are part of a set of complex behavioural manoeuvres, which may combine with others to produce a spectrum with peaks in regions according to the material being analysed. These manoeuvres may be due to the presence of weak overtone, combination and difference bands. Overtones are generated by the excitation of bond activity from the ground state to higher energy levels, which correspond to integral multiples of the frequency of the fundamental $(\overline{\upsilon})$ [frequency in cm^{-1}]. This means that an absorption band at, for example $1740cm^{-1}$, may have an accompanying band of a lower intensity at $1100cm^{-1}$. The combination bands are the product of two vibrational frequencies coupling to give rise to a new frequency within the molecule ($\overline{\upsilon}$ comb = $\overline{\upsilon}_1 + \overline{\upsilon}_2$), whereas the difference bands arise from the difference between two interacting bonds ($\overline{\upsilon}_{diff}$ = $\overline{\upsilon}$ - $\overline{\upsilon}_2$). These three types of bands can be calculated by using direct manipulations of frequencies in wave numbers by simple multiplication, addition and subtraction respectively. (Solomons 1990) Despite the relative simplicity, these band and vibrational effects are some of the reasons why the utilization of FTIR, and/or an organics library, must be carried out with some degree of caution and a great deal of care. This is especially the case for the novice worker, and where ancient materials are concerned. Any lack of expertise in the fundamental organic chemistry of the materials being analysed may lead, not only to the incorrect identification of a sample, but more seriously still, to the establishment of an erroneous foundation or criterion, and subsequent flaw in all those samples interpreted in relation to it.

4.18.2 Detection and Functional Groups

The samples for this project were analysed at the mid infrared region with a range of $450\text{-}4000cm^{-1}$, allowing for the selection of the principal diagnostic regions for the identification of the esters of fatty acids, for both the known and unknown specimens. The peak locations for the spectra as visually displayed are as follows.

1. The **carbonyl C=0** bond has a large dipole moment, and therefore produces intense stretching absorptions around $1700cm^{-1}$. However, the exact frequency is determined by the specific functional groups around the carbonyl-containing molecule. Under normal conditions, the carbonyl bond for an ester is found in the $1735cm^{-1}$ region. However, certain chemical conditions such as *conjugation* render this ambiguous. The term conjugation refers to a situation where there is an alternation between a double and a single bond on the otherwise aliphatic chain. As a consequence, there is a lowering of the stretch frequency created by the partial π-bonding between the conjugated double bonds, in turn reducing the electron density of the carbonyl π -bond, thereby weakening it, and hence lowering its frequency, sometimes to as far to the right as $1680cm^{-1}$. When referring to the total structure of an ester: if the conjugation occurs in the R part, absorption moves to the right (down), if it occurs with the 0 in the R^1 part, absorption moves to the left (up).

Fig. 4.11 (Pavia, Lampan & Kriz 1979)

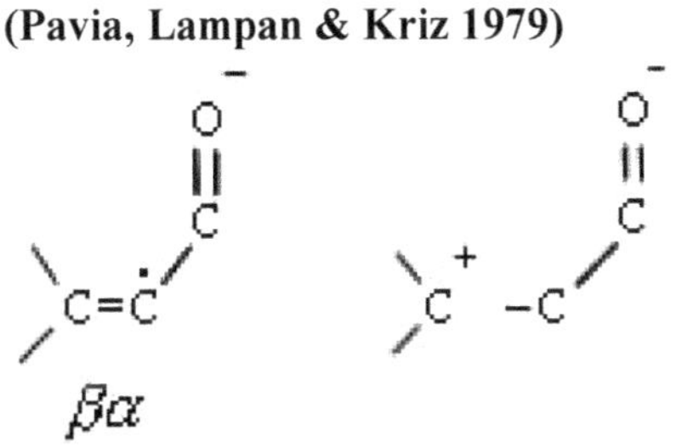

Conjugation Effect: the introduction of a C=C bond adjacent to a C=0 results in the delocalisation of the π electrons in the carbonyl and double bonds. This conjugation increases the single bond character of the C=0 bond and lowers its force constant, resulting in a lowering of the frequency of carbonyl absorption.

Deterioration of the ester, creating a peak shift in the diagnostic carbonyl group to within the $1710cm^{-1}$ region, may also occur as the possible result of a number of potential chemical changes. One such possibility would derive from the presence of free fatty acids containing the carboxylic acid functional group (COOH). This would be accompanied by a second characteristic peak in the$3000cm^{-1}$ region, which is a broad stretch frequency caused by the strong H-bonding in the COOH group. Another possibility may be that now only aldehydes are present; here the accompanying characteristic peaks are two low frequency stretch bands around the 2700-$2800cm^{-1}$ region, not produced by either fatty acids or ketones. Alternatively, ketones alone have been produced, where the carbonyl group remains in the $1710cm^{-1}$ region, but there are no accompanying distinctive peaks.

2. **Carbon-carbon** bond stretching regions have two levels of diagnostic determination; the single -C-C- chain which simply indicates that the analyte is an organic material. Characteristically they have small or zero dipole moments, and so, accordingly, produce extremely weak or totally absent non-diagnostic peaks. Whereas, by contrast, the double C=C alkene bond is essential for structural determination, as one of the most important structural characteristics is indicated by its presence: specifically, the unsaturated esters of fatty acids.

Fig. 4.12 (Pavia, Lampan & Kriz 1979)

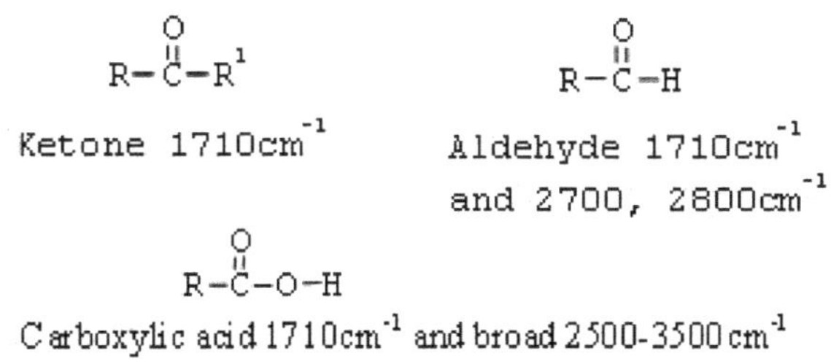

The specific frequency of double bond stretching depends on the previously mentioned conjugation effect. This leads to some π -bonding overlap between the two double bonds, which accordingly creates a reduction of the electron density in the bonds themselves, this reduces their strength and stiffness, and consequently causes a lower vibration frequency. The three examples below clearly demonstrate how the alkene peaks can move through the spectrum, even though there has been no deliberate chemical change to the molecule itself.

Fig. 4.13 (Pavia, Lampan & Kriz 1979)

C=C isolated $1640\text{-}1680cm^{-1}$
C=C conjugated $1620\text{-}1640cm^{-1}$

3. **C-H bond** stretching regions, (alkanes, alkenes and alkynes) will always be evident in aliphatic compounds, such as those under examination in this project. They are typically apparent at the far end of a spectrum, with frequencies at around $3000cm^{-1}$, but split into two, with an asymmetric peak at $2962cm^{-1}$ and a symmetric one at $2872cm^{-1}$. With a series of complimentary peaks at lower frequencies, the bending absorptions of the CH_2 at $1450cm^{-1}$ and CH_3 at $1375cm^{-1}$, which appears common with most of the strong bond absorption frequencies.

This region is characterized by the CH stretch, involving hybrid carbon atoms, those which absorb just below $3000cm^{-1}$ involve sp^3 hybrids, while those just above involve the sp^2 hybrids. The differences are due to the amount of [s] character in the carbon orbital used to form the bond. The s-orbital is closer to the nucleus than the p-orbitals, and therefore bonds formed with more [s] character will be both stronger and stiffer. An sp^3 orbital is $\frac{1}{4}$ [s] character, while an sp^2 orbital is $\frac{1}{3}$ [s] character; therefore a bond which utilizes sp^2 orbitals will be slightly stronger and have a higher vibration frequency. In essence, this allows for the distinction between -C-C-H which is an sp^3 orbital and C=C-H which is an sp^2.

4. C-0 bond is the single carbon-oxygen part of the total ester functional group. Although not as important as the double carbonyl, its presence is significant if the correct interpretation of the sample is to be made. These are stretch bonds covering a range of 1000-1300cm^{-1}, found in groups of two or three bands and varying in strength in terms of peak representation. The major significance of this region is that the bands are complimentary peaks for the C=0 part of the ester, therefore verifying the presence of an ester in the analyte. **(Ref. Points 1-4: Pavia, Lampan & Kriz 1979; Solomon 1990; Perkin Elmer laboratory Manual)**

4.19 Significance of functional groups

The regions discussed above were selected as the principal components of a spectrum, in order to identify the presence of an ester of a fatty acid, or the relevant deterioration products, indicating that one had existed previously. By looking for alkene groups, it was possible to infer the presence of unsaturated groups, which give further clues as to the actual ester present.

In order to ensure the reliability of results acquired, it is necessary to possess a good background knowledge of the materials under study, in addition to a solid grasp of the process of infrared analysis. Neither the mechanical reproduction of spectra, nor the generation of unmanageable quantities of raw data is sufficient to ensure the realization of the project goals. The ensuing chapters will provide an explanatory account of the manner in which the gathered data was handled. Meanwhile, the two methodological chapters will detail the precise criteria of selection for the compounds employed

CHAPTER 5

THE ORIGINS AND APPLICATION OF CHEMOMETRICS IN ARCHAEOLOGICAL CHEMISTRY

[Chemometrics is] chemical discipline that uses mathematical, statistical and other methods employing formal logic (a) to design or select optical measurement procedures and experiments and (b) to provide maximum relevant chemical information by analysing chemical data (Massert 1988:5).

5.0 Origins of Chemometrics

Chemometrics is a comparatively new discipline, emerging out of analytical chemistry in the early 1970s. An outgrowth of pattern recognition studies, its relative infancy has enabled chemometrics to benefit from a close marriage between statistical analysis and the development of increasingly powerful computational tools. This has facilitated the development of user- friendly computer software that leaves the laboratory-based chemist free to use data analysis without the need to become a computer expert. One of the leading proponents of the discipline is Richard Brereton, whose monograph (1990) has had a seminal influence on the field. Essentially the technique involves the application of statistical and mathematical methods as aids to the acquisition and interpretation of chemical data. Spanning the last three decades, chemometrics has been able to evolve in tandem with the information technology in which it has its roots. This in turn has allowed for the development of a new discipline within analytical chemistry.

Chemometrics consists of a combination of multivariate calibration and classificatory or discriminatory procedures and concepts that enhance the analysis of chemical experimentation. The discipline integrates signal processing and pattern recognition with experimental design in a form of bimodal experimental optimisation. The two modes can be described succinctly as (1) statistical techniques to develop precise experiment models, and, (2) further statistical methods which provide data trend analyses. As a consequence chemometrics is not merely a means of analysing experimental data, but also plays an integral part in the overall design and testing of experimental hypotheses. This is so because the optimisation of the overall experimental process entails the recognition that the method of data acquisition conditions analysis.

Chemometrics was chosen as the analytical underpinning of my research to introduce replicability and quantification to the use of FTIR in archaeology. To date, many of the IR spectroscopic analyses of ancient subsistents have relied primarily on a highly subjective visual interpretation of the spectra. The information content of spectroscopic data is potentially high, especially where multivariate data analyses are applied, regardless of whether the data are qualitative or quantitative. The traditional subjective approach is inherently flawed, then, for the following reasons:

1. Analyses may be biased by the observer's under-standing of biochemistry,
2. Noise and systemic errors may not be adequately recognized or controlled for with visual analyses, and
3. Manual (visual) analysis with large numbers of spectra can result in error through observer fatigue.

This latter point is especially problematic, since data sets may typically involve hundreds of spectra, and the possibility of data biasing with manual (as opposed to computerized) methods becomes a significant potential source of error. Compounding the problem, each point on the wavelength range of a single spectrum should be considered as a separate variable as well. In the project under discussion here, for example, the mid-infrared range was employed in the analysis of the targeted compounds. This resulted in the generation of very large data sets, some of which comprised as many as 3,500 data points.

5.1 Calibration and instrumental analysis

Although very large data sets are the norm with numerate IR analyses, various data compression routines can be used in order to enhance their manageability. These compression techniques include modifying the system resolution and using suitable statistical techniques. While the modification of the system-resolution is a trivial operation, the application of the appropriate statistical techniques requires stringent control. These prerequisites will be discussed in detail in the following sections.

A calibration normally consists of the analysis of known concentrations of the compounds of interest, in order to determine the instrumental response in the concentration range present in the samples, and to establish a benchmark. However, for this project, although the samples used to create the mathematical model were botanically identified, their actual ester concentration was not known. (see chaps. 4&7) Nor were the actual ester components known, as it was beyond the time constraints of this research to carry out a separation technique prior to infrared analysis. Nevertheless, a systematic type of calibration was carried out in order to determine the actual concentration range falling within the analytical capabilities of the FTIR. Beyond these limits, which were identified and delineated in the calibration process, the spectra produced became non-diagnostic in terms of the compounds required for this work. The compounds analysed were of Spectroscopic Gas-Chromatograph (GC) Assayed standards for linoleic, oleic and palmitic esters. Plant biochemistry literature (Galliard & Mercer 1975; Gunstone 1967; Harwood Russell 1984) suggested that these esters contained sufficient chemical information to allow this type of experiment to be carried out with a relative degree of accuracy, provided the whole system was standardized. These specific esters were also tested in various combinations, in order to identify any interactive effects between systemic components. Such concerns arise especially in circumstances where an unsaturated fatty acid ester containing an alkene group is involved. This combination (of ester and alkene group) may produce spectroscopic edge effects through shifting of the C=O and C=C peaks. Eventually the combined information would then suggest whether the peak ranges to be used for diagnostic interpretation, were adequate, and contained enough information about the material of interest, to enable the generation of a robust model from the identified knowns. Such a method was the only form of calibration that would operate as a system standardization, as the unknown materials were derived from many ancient sources, and could not stand up to a normal rigorous experimental control, as concentration and quality was not known. Also the apparent homogeneity of the ancient samples could not be relied upon, as the identification by morphology was not always definite.

The calibration set was used in order to model the chemical system of interest, for which the major functional groups were selected as diagnostic variables. These were the diagnostics peaks and regions discussed in the previous chapter (Chap. 4) as *bond absorption frequencies*, but in order for them to be utilized in chemometrics they had to be assigned via their spectral numerical value. A calibration training set must contain all of the likely sources of chemical variance which may be present in the unknowns. In this context, a calibration "training set" is the set of modern identified samples, which are used to "train" the statistical techniques to recognize the clustering tendencies of the sample population, the archaeological samples.

The term "training set" is one generally used in statistics when a model is constructed from some part of the overall data. These specific data have been subjected to rigorous analysis, and their properties are thoroughly known; the model is then used to analyse unknown data, and to manipulate them according to a defined statistical protocol.

Therefore the values of possible deterioration products had also to be included, if the system was to generate a robust model from which the unknowns were to be numerically selected. The diagnostic peaks and regions are effectively the major constituents, by which an ester is visually identified from a spectrum. They are also the column variables for the matrix model, ranging from columns two to thirteen for the spectral values, and column fourteen for alpha information.

The matrix model is the shape the matrix will take when variables and objects are combined; it can entail either (i) more objects than variables, in which case the shape is rectangular in the vertical, or (ii) more variables than objects, in which case the shape is rectangular in the horizontal plane.

These are 3400cm^{-1}, 2960cm^{-1}, 2872cm^{-1}, 1736cm^{-1}, 1712cm^{-1}, 1640cm^{-1}, 1452cm^{-1}, 1376cm^{-1}, 1252cm^{-1}, 1196cm^{-1}, 1152cm^{-1}, 676cm^{-1}, and the alpha column SAMPLE, respectively.

The alpha information is the column of the data which contains sample names. These identify the samples according to the solvent used in the extraction and the number allocated from the original batch.

Since quantitative analysis was primarily the aim of the purely analytical part of the research, it was necessary to create a system of calibration that would aid in quantifying the data. The calibration set comprised samples that were prepared to the levels considered likely to be the working range of concentrations in the unknowns. The information gathered from this was validated by the results of the assayed standards, and the random set (20) of the modern samples.(See sample preparation, chapt.6-sections 6.1-6.3) This basically required the sample being totally dried; as the original weight and that of the glass vial was already known, the weight of the extracted component could be calculated. To calculate the weight of the dried extract there were a number of steps that had to be carried out, which was the standard method for all the samples, both the modern and the archaeological seeds.

1. The clean, dried glass vial was weighed,
2. The cellulose thimble was weighed,
3. The weight of the thimble and the glass vial were zeroed on the scale, the seeds were then weighed to give their individual weight.
4. The extraction process was started, which proceeded for three days,
5. The subset of samples had the excess solvent evaporated off until dry,
6. The dried samples were cooled and weighed,
7. The weight of the individual vials were subtracted from each sample, leaving the weight of each separate extract,
8. The extract was finally converted into a percentage of the total sample weight.

Once all of the samples had been dried and weighed the percentages and concentrations were calculated, this allowed for the estimation of a working range for the esters of fatty acids, which should produce diagnostic spectra. Finally, all of the procedures were combined, to create a significantly robust numerical model, which was a major goal of the chemometric identification method.

The calibration contained two aspects, the internal standards and an external validation. Normally, only an external validation would be required, however, when dealing with unknown samples from an archaeological context, this was not adequate (C. James, personal communication 1995). This was due to the differences between the modern and archaeological samples, where morphological characteristics assigned species groupings to a number of archaeological samples, albeit tentatively, with reference to their modern counterparts, but the seed biochemistry could not concur. In other words, given the constraints of dealing with archaeological material, the bipartite calibration method enhances the robustness of the calibration.

Beer's Law would have been applied to the spectral data of both sets, to calculate the concentration of the extracts from the absorbance spectrum. However, due to the problems regarding calculating the concentration, this could not be applied, since the unity requirement of Beer's Law means that analyses of high resolution cannot be obtained from samples of low concentration, such as those found in both the modern and ancient samples.

Another aim of this research is to demonstrate the reliability of chemometric tools as a means of providing reliable replicable measures of botanical archaeological samples. Archaeological samples are substantially less easy to work with than their modern equivalents, owing to both the great age of the samples, and the taphonomic processes that would have continuously affected the seeds once gathered. These processes range from the simple charring to remove husks, to the long -term burial in soils, which vary substantially over time due to changes in the environmental conditions. This also extends to human influences through the archaeological recovery; these processes, either combined or on their own may have ultimately altered the chemistry of the seeds. Furthermore, while identification of species **may** be possible via microscopic analysis, genetic drift between a seed's collection in the distant past and a modern viable member of the same species can neither be ruled out nor necessarily quantified. However, work carried out by John Evans and his co-workers has demonstrated that the biochemical analysis of ancient subsistents to determine the genus of those foodstuffs does remain feasible.

5.1.1 Modelling

Once the calibration set has been built, chemometrics are applied to create a model of the system. It is imperative that the model be constructed from the data, rather than the reverse. The constructed model is then applied to both the experimental spectra and the validation set, in order to predict the properties that are of interest. This process represents the optimisation of experimental design, and requires the use of a number of statistical steps. It is these that are discussed individually in the following sub-sections of the present chapter.

Prior to any numerical analysis it was necessary to convert the spreadsheets to matrices, both covariance and correlation, and autoscale each variable in order to standardize the format. This does not alter the final results; as most SCAN applications are scale dependent, it is useful to apply this once a matrix has been constructed.

The SCAN applications are the statistical methods used by the computer programme of the same name. They are referred to as SCAN applications as this project utilizes those methods designed for the SCAN software and the accompanying manual.

Although this type of analysis has not been previously used in archaeological science, designing a project to utilize the chemometric methods allows for the testing of a quantitative numerical technique. Valuable information can be gained by trying out new techniques in hitherto untested areas, a fact which is demonstrated by the results obtained from the current work.

Autoscaling scales each variable to have a mean of 0 and a standard deviation of 1, where each observation is centered (subtract) by the column mean, m, and scaled (divide) by the column standard deviation, s. Thus a value x is scaled to $x^{!}=(x-m)/s$. This then allowed for the construction of a balanced numerical model, which was effectively the matrix of the modern chloroform extracted samples.

Once the model was made, it was then possible to work on the fit for validation and prediction process, which included looking for anomalies that may, inadvertently, have also been constructed. One such anomaly did arise, deriving from the application of the model to the GC assayed samples.

The anomaly in the GC assay was detected by the size of the peaks; due to the concentration of the sample, saturation peaks were created in the IR spectrum. These saturation peaks are over-sized peaks which have no clear point, as they literally do not fit on the spectrum page. In order to remedy this, the sample had to be diluted further, to a level similar to that of the modern samples. Once reduced to this level, all the major peaks of the assays fitted with absolute clarity on the spectrum page. (Further discussion of spectral interpretation can be found in chapter 6, section 6.10, followed by the interpretative explanation in chapter 7.)

Due to the purity of these samples, a significantly less diverse spectral range appeared when contrasted to samples extracted from a matrix which included those assays in combined form. Given the heterogeneous range types of samples employed, the anomalous phenomenon probably resulted from the complex mixture of unknown esters, which had the ability to produce any number of the previously mentioned bond-effects. An additional factor with a possible bearing was that the concentration of the total extract- and, hence, of the individual extracts, was at this juncture not known.

Consequently, the major diagnostic peaks for the modern samples had to be identified with the utmost caution, and the possibility of hidden chemical effects retained at the forefront of the mind. Proceeding accordingly, it was possible to achieve identifications with a high degree of reliability, even though the peaks did not correspond to the textbook standards as indicated for pristine compounds.

5.2 Interpretation of spectra — qualitative analysis
The qualitative interpretation of spectra relies on the fact that a relationship exists between the chemical structure of a compound and its molecular spectrum. A variety of statistical techniques, including discriminant analysis, cluster analysis and pattern recognition can be used to examine this relationship. For example, by using discriminant and pattern recognition techniques, sets of spectra representing different types of materials can be distinguished from each other.

Basically, materials of unknown chemical concentration (in this case the modern known species) are used as a training set, to "teach" the statistical algorithm the clustering tendencies of the data as it looks for patterns by evaluating similarities and dissimilarities between groups. The newly defined rules and parameters, thus derived, are subsequently applied to the unknown sample set (the archaeological spectra), thus utilizing a two step procedure for a single statistical operation. This system is known as Supervised Discriminant Analysis (SDA).

Unsupervised Discriminant Analysis (UDA), by contrast, can be considered as the chemometric equivalent of the visual overlay technique which has been applied to the analysis of FTIR spectra in other instances. Analysis with UDA does not use a training set; rather, it compares the individual spectra directly to determine clustering tendencies and cluster outliers. UDA can also be applied to the training set to upgrade a qualitative analysis to a quantitative analysis. In this case, UDA identifies the presence of outliers to clusters, since these outliers may represent a source of error in the calibration data as far as a quantitative analysis is concerned. The main difference between SDA and UDA is that SDA is only qualitative in nature, while UDA can, within certain constraints, be quantitative.

5.2.1 Exploratory Analysis: Hierarchical Cluster Analysis (HCA)
(HCA) (see Appendix pt. 2, page 121.)

The exploratory phase of data analysis uses statistical algorithms to reduce large and complex data sets into manageable views. It is effectively the graphical display and computation of the patterns of association of the raw, multivariate data.

HCA represents the initial step in data analysis, and examines distances between samples or variables. Where distances are small, this analysis suggests the likelihood of similarity or common origin (that is, belonging to the same species). Essentially, HCA groups data according to pre-defined criteria based on known trends derived from the training set. It is often advantageous to present the algorithm in terms of a dendrogramme, thereby enhancing the visualization of clustering tendencies by demonstrating the proximity of a sample to its nearest neighbour and the measurement of the correlation of similarity.

5.2.1.1 Mathematical Background of HCA

Equation -Distance Measures Fig. 5.1 (Miller & Miller 1993; SCAN manual)
The degree to which samples are not alike (their distance) is calculated using the following equation:

$$d_{ab} = \sum_{j=1}^{m} \left[(x_{aj} - x_{bj})^M \right]^{\frac{1}{M}}$$

where: d_{ab} = the distance between two samples measured at *m* variables.
M = the order of the distance.

In the analyses presented here (Chapter 5), the order of distance is assumed to be the Euclidean distance, which takes the value 2. This modifies the above equation as follows:

$$d_{ab} = \sum_{j=1}^{m} \left[(x_{aj} - x_{bj})^2 \right]^{\frac{1}{2}}$$

Equation –Similarity Fig.5.2 (Miller & Miller 1993; SCAN manual)
As the distances between samples often vary with the type and number of data present, it is advantageous to transform these distances into a more standard scale; is referred to as a scale of similarity.

$$S_{ab} = 1 - \frac{d_{ab}}{d_{\max}}$$

where d max= largest distance within the data
d_{ab} is defined as in 4.1

Identical samples will have a similarity equal to 1, while totally dissimilar samples will have a similarity of 0.

5.2.1.2 Clustering
Calculated similarities and distances between the samples indicate the manner in which the data group together (their clustering tendencies). These clustering tendencies are represented by a dendrogramme, (for a complete explanation of what a dendrogramme is and how it works see **Appendix ppg. 121-126**) where the similarities are represented as a vertical hierarchy and the differences are presented as horizontal branching. Each 'point' on the dendrogramme represents the eigenvector function of the similarity relative to the distance.

This process is iterative, with the final clustering tendencies representing what are essentially absolute eigenvalue differences within the whole sample population. The lengths of the vertical branches represent the 'goodness' of clustering (or measure of similarity), in that highly similar samples will have very little vertical displacement. Similarly, the differences between the clusters are represented by the horizontal displacement between the hierarchies. It is important to note that regardless of the kind of numerical analysis utilized, the results will not be pure, as these numbers are dependent on an instrumental measurement, which is not infallible- no mechanical system being entirely without error. When employing FTIR instrumentation, one specific problem generated is Type I experimental error. This is caused by signal-to-noise within the analytical equipment (instrumental noise), which is quantified to prevent data biasing via the Fast Fourier Transform algorithms built into the analytical equipment.

5.3 Spectral Interpretation: Quantitative Analysis

There is a mathematical prerequisite for spectrometric analysis:

Information derived from the spectrum of a sample is related in mathematical terms, to changes in the level(s) of an individual component, or several components, within the sample or a series of samples. That is the spectral response of an analyte can be related by a mathematical function to changes in concentration of the analyte (Coates 1990: 98 in Massart 1988).

By definition, data acquired via Infrared spectroscopy are not quantitative, despite being information rich. When IR spectra are first produced, the data is such that it cannot be used in a quantitative analysis without the addition of other types of data. This is because a peak on a spectrogram is not representative of one figure; it is actually in a position which is part of the range of that particular functional group.

This is particularly true when large numbers of well-resolved peaks (specific to different functional groups or structural features) are present. Conversion from qualitative to quantitative analysis is possible, but is complicated by the molecular environment found within the samples, since sub-molecular components may influence each other's peak positions, frequencies and strengths. As a consequence, the entire spectrum will not necessarily respond to changes in concentration in a quantitative manner. However, this problem is overcome through calibration, by running a series of concentrations to represent not just peak points but peak range as well.

In addition to enhancing replicability, quantitative analysis reduces the data set to a manageable size. This process begins with an analysis of the principle components.

When using numerical data collected from sources such as spectral analysis, it may be necessary to expose the data to several regression analyses before finding the most satisfactory set of results. Often, exploring and interpreting the data are not the only reasons for trying several methods- there may be problems such as collinearity or noise, which have to be considered when creating a model. Consequently, it may be necessary to compare the results from different methods and then analyse how and why they are different. This technique of 'method exploring' enables the experimenter to gain experience in both data handling and data observation, experience that can be applied to data processing. This is the preparation of the data so that it produces the best set of results, thereby reducing the prospects of both overfitting and underfitting. The data processing technique was applied in the present research with respect to the number of wavelengths. That is, only those that carried the best information (the ones with the high PLS-loading values or the highest *B*-coefficients) so as to render the model easier to interpret (fewer PLS-components) and reduce the prediction error. This is the result of avoiding wavelengths with much noise and irrelevant information, or non-linearities. Although PLS is able to handle full scan data, as it can extract the relevant information and disregard noise, using a full scan may also create problems. These problems arise from the risk of the noise and non-linearities being modelled in the earlier or later PLS-components, leading to overfitting and less accurate models. Consequently, the data in this project was subjected to filtering via the algorithms of the Fourier Transform, and by the above mentioned selection of the most data carrying wavelengths, which worked to handle most of the problems.

5.3.1 Principal Components Analysis (PCA)

(see Appendix pt. II pg. 112)

Although instrumental methods allow analyses which are beyond the scope of classical or 'wet chemistry', they are not problem-free. In particular, systematic errors may occur which must be identified and corrected to prevent erroneous interpretation. Typically, this form of error is controlled through the creation of mathematical models. PCA is the basis by which such models are created and tested.

PCA enhances the visibility of salient trends in complex data by constructing a new set of linear variables that combine the original (raw) variables. These variables form a new set of uncorrelated plotting axes, or *eigenvectors*, which are referred to as the Principal Components. Plotting axes are created according to the amount of variance within the data for which they can account. The First Principal Component describes more of the variance in the data than the second, and so on, in a descending scale. Due to the complexity of the chemistry of the samples, five Principal Components are used here. In effect, the spectra are transformed into these five components for subsequent analysis.

The nonlinear iterative partial least squares algorithm (NIPALS-page 112 Appendix) was utilized in order to compute the principal components, by calculating one variable at a time. As this algorithm is normally employed to calculate a few components with a large number of variables, the space and time requirements increase as the number of components increase. Meanwhile, the second option Singular Value Decomposition (SVD), (SVD pg.112 Appendix) calculates all the components together, utilizing a matrix with a large number of components and comparatively few variables, where the space and time requirements increase as the number of variables increases. Therefore it became a valuable comparative technique, as the data matrix in this project consisted of a larger number of observations by a considerably small number of variables. Given the choices available through SCAN, it is possible to apply the many different techniques or algorithms, in order to select that which performs the best for the data supplied. Examination of the end-result of each of the applied methods determines the selection of the appropriate algorithm or statistical technique.

As these methods are computer calculated, it is possible to inspect many results over a very short period of time. If the method is a step in a chain of numerical analyses, selecting the best method at each successive step should mean that the final result possesses a highly significant answer.

Intrinsic dimensionality within the data can be determined by PCA, even though only a few factors may influence the pattern of the data. This is achieved by extracting the diagnostic variables while eliminating all random variation and noise from the relevant signals. Data transformation **does not** change the relationships between samples, but it **does** enhance the order of importance of the variables. Effectively, it is a description of the variance contained within the data.

NIPALS and SVD provide extremely valuable views of the variability in a multivariate data set by indicating the natural clustering of the data, identifying outliers and indicating the factors which influence the observed patterns. The techniques make it possible to calculate the eigenvectors of either the covariance or correlation matrices, but via a different matrix format. They also determine the intrinsic dimensionality of the data; the variance retained in each principal component, and the contribution of the original measured variables derived directly from FTIR.

5.3.2 PCA goals

By applying the algorithms SVD or NIPALS to the original data on uncorrelated axes, it is possible to carry out the PCA or eigen analysis to find a new set of co-ordinate axes, onto which data can be projected. This procedure can also be used to determine new factors that represent independent information describing the data. The process by which this new data is acquired involves the application of matrix-decomposition techniques, and it results in the generation of two new matrices. These matrices are created by calculating the orthogonal linear combinations, **T**, of the autoscaled (see above) variable, **X**, which is based on the maximum variance criterion. The linear combinations are the principal component scores with the coefficients, **L**, which are also referred to as the loadings. The **T** matrix contains information about the variability in the system as a distance deviation from an ideal slope for the system, while the **L** matrix contains data about the correlation of each **T** variable to the axis of that ideal line. The three quantities have a relationship which can be represented in matrix notation by:

$\mathbf{T=XL}$ or $t_{im}=\Sigma_j x_{ij} l_{jm}$

with original variables being represented by: **Fig. 5.3 (Miller & Miller 1993; SCAN manual)**

$\mathbf{X=TL'}$ or $x_{ij} = \Sigma_m t_{im} l_{mj}$

As previously stated, the principal components are calculated according to the maximum variance, where each following component is an orthogonal linear combination of the original variables, to the point where it covers the maximum of the variance not accounted for by the former components. The principal components can be calculated from either the correlation or covariance matrix, **S**, of the original variables. These can be standardized, as using a correlation matrix, is the same as autoscaling then utilizing the covariance matrix. Consequently, all of the calculations can be described as calculations based on **S**. There are several methods used to calculate the eigenvectors, including the SVD and NIPALS algorithms referred to above. I will now proceed to outline these two algorithms.

SVD has two main steps; first, the symmetric matrix is transformed into tridiagonal form, and this is then diagonalised by the Q-R algorithm. (SCAN chp. 7). NIPALS starts with the autoscaled data matrix **X**, which then calculates the eigenvectors and PCs corresponding to the largest eigenvalue in an iterative algorithm. This iteration converges, the component is subtracted from the matrix **X**, and the calculation proceeds to the next or second principal component. This process continues until all the required data has been described. Thus, the product of these (transposed matrices) reproduces the data as linear combinations of the original variables with the following attributes:

1. The direction of the variability for each Principal Component in Cartesian space,
2. The magnitude of variability for each Principal Component in Cartesian space. This magnitude ranges as $-1 \geq$ loading ≤ 1.

A loading will only be 1 if it coincides totally with a principal component, but in the opposite direction.

Graphical representations of data sets may include any number of factor axes reduced by PCA to include as few factors as possible. The first factor is considered to be the regression line, which maximizes the sum of squares of the distances of each sample point to the line. As a consequence, this first factor's direction will be aligned with the largest spread in the data.

The second factor axis is perpendicular to the first axis, with a direction dictated by a regression line describing the maximum variation in the data. Each additional factor, as dictated by the original variables, is perpendicular to the preceding factors until all the data, including noise, have been described.

All of these factors describe some percentage of the variance in the original data, the sum of which will be equivalent to the total variance in the original data. The magnitude of the variance of each factor, the eigenvalue, does not correspond to information from the original variables; rather, the magnitude of each factor decreases with each subsequent Principal Component, since each represents smaller and smaller amounts of the total variance which remains undescribed.

Most of the relevant information in a data set is found within the first three factors generated by PCA. It is not unreasonable, therefore, to exclude some of the later factors, since doing so does not significantly reduce the ability of the factors to reproduce the original data.

5.3.3 Mathematical Background of Principal Component Analysis

Data are converted from peak patterns to a matrix of peak values of the form:

$$X = \begin{bmatrix} x_1 \\ x_2 \\ \cdots \\ x_n \end{bmatrix} = \begin{bmatrix} r_{11} & r_{12} & \cdots & r_{1m} \\ r_{21} & r_{22} & \cdots & r_{2m} \\ \cdots & \cdots & \cdots & \cdots \\ r_{n1} & r_{n2} & \cdots & r_{nm} \end{bmatrix}$$

Fig.5.4 (Miller & Miller 1993; SCAN manual)

where:

1. x represents the sample, running from 1, 2,…,n samples
2. r represents the principal components for each sample, running from 1, 2,…,m responses. Each matrix element is identified by a row and column subscript, in that order.

Thus, a single spectrum's principal components (eigenvectors) is represented a :

$x= [r_1\ r_2 \ldots r_m]$ **Fig. 5.5 (Miller & Miller 1993; SCAN manual)**

5.3.3.1 Matrix Operations

When working with vector and matrix rotation it is necessary to use the mathematical rules of linear algebra. Expressing a vector X graphically as a line in *m* dimensions, with an end point defined by distance x_1 along the first axis, by distance x_2 along the second axis, and so on through *m* axes. For example, a two dimensional vector takes the following form:

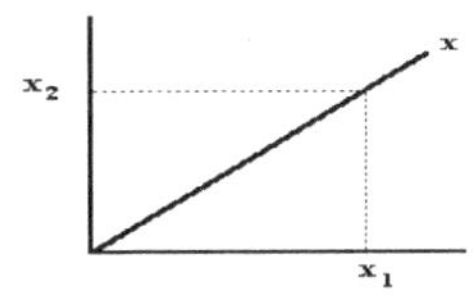

$$X\ (x_1\ x_2)$$

Fig. 5.6

This subsumes the orientation of the vector and its length follow Pythagorus' theorem.

5.4 Factor Analysis (see Appendix pt. 2, pg. 118)

To select the required components from the data it was necessary to include yet another step before the final regression analysis was decided upon. This next, invaluable step was factor analysis, which extracts the underlying factors of the data and performs orthogonal rotation to render the factors literally *more interpretable*. This is carried out by principal component analysis with specified variables as selected through SCAN.

The process models a set of variables as a linear combination of a small number of common factors, thereby obtaining a close and specific description of the observed or measured quantities. These are the common factors, which represent the previously mentioned underlying phenomena. These phenomena cannot be measured directly, requiring instead the application of appropriate statistical treatment to bring them into conceptual view.

5.4.1 Mathematical background to Factor Analysis

The factor analysis has a model which can be represented by matrix notation in the following format:

Fig. 5.7 (Malinowski 1991; SCAN manual)

$\mathbf{X=FL'+u}$ or $x_{jm}=\Sigma_m f_{im}\, l_{jm} + u_{ij}$ or I=1, n j=1,p m=1,M

or where each of the p observed variables x_{ij} is represented as a linear combination of M common factors f_{im} with coefficients l_{jm} and a unique factor u_{ij}. The coefficients, l_{jm} are called factors loadings. (SCAN pg. 7-26)

The common factors explain the correlation among the variables, while the so-called 'unique' factor covers the remaining variance, being referred to as the specific or unique variance. This unique factor is also called the residual error. The number of factors M, determines the complexity of the factor model. (SCAN pg. 7-26)

There are two steps in factor analysis: first comes the identification of the unobservable, underlying latent factors; subsequently, then, these extracted factors are rotated to obtain more meaningful, interpretable factors.

This factor-extraction, involves the determination of the number of factors and the calculation of their coefficients or factor loadings. (For further utilization and background to the algorithms see SCAN chp. 7.)

These processes have as their end-result the production of graphic representations of the calculated factors, in both rotated and unrotated form, for eigen value and score plots. The former gives a plot which represents the selected number of factors (x) against the eigen values (y), and the latter, a plot of first factor (x) against the second (y). The numerical outcome is presented in the form of two tables: the first contains the factor loadings, their variances and communalities, while the second contains the factor scores coefficients, with the rows corresponding to the variables, and the columns to the factors.

When everything is collated, it is possible to see what the ways in which the data is behaving, thereby giving the direction for the next step of analysis, if appropriate. As every step is, quite literally, exploring the data from a slightly different point of view, the cumulative effect provides a clear indication as to the desirability of further analysis. It answers the question, that is, of whether a continuation and refinement of the analysis will result in greater specificity and detail, or simply obscure those significant patterns within what is already an extremely large expansion in the original data-format.

Therefore the next step, albeit a small one, was vital for the determination of multivariate analysis, and in deciding as to whether or not it would be possible with the given data. That step was the matrix condition, which looks at the level of collinearity amongst the variables. The numerical analysis so derived will supply warnings regarding constant and highly collinear columns, which can create problems for regression analysis. In certain cases, regression analysis will be contraindicated.

The Matrix Condition provides a condition number in terms of a ratio of eigenvalues, as calculated from a correlation matrix. A number higher than 1,000 indicates high collinearity, which as a result will suggest the type of regression analysis best suited to the data. This analysis was carried out on the matrices of both the model from modern data, and the test data from the unknowns. (See SCAN chap.7)

5.5 Regression Analysis (see also Appendix pt. 2; PLS pg. 135, ACE pg. 140& NPLS pg. 141)

Chemometrics is actually a bimodal approach to the data: it involves both regression and calibration. PCA is a method by which data are converted into a suitable form for regression analysis and modelling.

Multivariate regression analysis requires an accurate calibration set, comprising a set of samples in which as many variables as possible are known, and which has been measured under the same conditions as the test samples, in this case the archaeological unidentified group.

The *sample set* is that which contains the modern extracts, whereas the *test set* contains the archaeological samples, which are the extracts under investigation. Both groups have been through the same preparation, extraction and FTIR analysis in order to determine their respective compositions. The calibration and validation data has also been added, so that the statistical analysis can be carried out.

In addition, when creating the regression model, only the relevant portion of the data must be included, since noise and other systemic errors may cause over fitting of the data and hence lower reliability in subsequent predictions. Similarly, reliability is lost if the model contains too little information.

Ultimately, the goal of any regression analysis is the development of a calibration and predictive model that correlates information according to some desired properties. This involves, as outlined above, analysing a range of measurements simultaneously and correlating the variables to build the calibration model.

Archaeological samples place additional constraints on the use of Chemometric modelling, as does the semi-quantitative nature of FTIR. As a consequence of these considerations, regression analysis in this research required external validation of the calibration results. External validation was actually intended to serve two purposes: both to verify the overall result, and to test the predictive capability of the calibration model. External validation was achieved by reference to the pure assayed triacylglycerols, which comprised a series of specific esters and ester combinations, thus establishing a benchmark of known properties against which the model was tested. This effectively ensured a measure of reliability for the prediction model and its data fitting power.

The two regression methods used in this analysis are Principal Component Regression (PCR) (see Appendix pt. II pg. 130) and Alternating Conditional Expectations Regression (ACE). These statistical techniques may be employed either in sequence or separately, depending on which method best suits the data in a given instance. In this case, sequence was considered the most appropriate option.

PCR is a three stage analysis. These stages are: 1. calibration; 2. validation and optimisation;. and 3. analysis of the unknown. The stages operate as follows:

1. Calibration: spectra are collected for the sample and test sets, the composition of which is measured by a reference method.

2. Validation and Optimisation: Validation entails the

analysis of samples either by two or more discrete instrumental methods (for example FTIR and GC), or by external testing of the model. As time constraints did not allow for two methods to be utilized, the external testing technique was opted for. This essentially tested the validity of the calibration set. As mentioned above, the process uses a second set of calibrated samples of known compositions...the validation 'set'. In addition, validation also optimises the PCR prediction model by ordering (ranking) the importance of the principal components in such a manner as to allow prediction of the quantified element. That is: the rank defines the number of principal components required to predict the element which constitutes the total number of important variables. The analysis-rank is determined by the comparison of the PCR predictions for a range of ranks and results in a set of known values containing: 1- the standard deviation; 2- the bias or mean deviation; and 3- the standard deviation corrected bias of the prediction. Thus, an optimal rank is defined as one with an associated minimal standard error, in subsequent analyses of unknown samples, when compared to the model comprising the identified modern specimens.

3. Analysis of unknown samples: Once the model has been validated and optimised, quantitative (or semi-quantitative) analysis of unknowns is possible.

PCR analysis also allows for direct comparison of spectral noise between the calibration spectra and the unknown spectra. This comparison is the 'F-ratio'. It is an important aspect of the analysis, as it can indicate similarity between the sample and the calibration set. If the 'F-ratio' of the sample is much higher than that observed during the cross validation, then there is a lack of similarity and the predicted results may not be accurate.

It was now necessary to construct a calibration graph in order to test the relationship between the sample concentration and the instrument response. This was only possible, and in fact only necessary, for the set of GC assayed samples. The graph was constructed in order to test for relationship direction; that is, whether the direction was linear or otherwise, as this determines the type of regression to be used as the final analysis. The graph construction was carried out using a simple Analysis of Variance (ANOVA) and polynomial quadratic function. As it was a standardized set, it was deemed reasonable to assume that the second calibration set (modern identified samples) would also behave in the same manner to the instrumental analysis. The test was also performed as a practical operation in order to assess the instrument detection limits for amount of triacylglycerol in the unknowns. This was primarily to determine the presence of diagnostic peaks, as without these it is not possible to utilize the peak numbers in the subsequent numerical analyses.

5.5.1 Outliers in Regression

A major problem in calibration experiments stems from the existence of outliers in the data. Although these are inevitable, they are not easy to deal with in regression analysis. The first problem is that, although the individual y_i values are assumed to be independent, the residuals $(y_1 - \hat{y}_1)$ are not, as their sum is always zero. Therefore the residuals cannot be used as a set of replicate measurements.

Outliers must, consequently, be identified in a calibration experiment. If there are peak values, but they are thought to be unreliable due to a questionable source, it might be possible to utilize the residuals, as they may provide the necessary diagnostic details with respect to either the samples or the instrument response. This can also provide a check on the correctness of the calibration set, because where the correct calibration has been used, the residuals should remain uniform in size and close to zero as y_i increases. The y residual may also be plotted against time where instrument shift or other time dependent effects are suspected.

5.6 Multivariate calibration

Any numerical analysis requires an hierarchical series of procedures to determine the most appropriate method for the data. In spectroscopic applications, not only are the data sets large, they are often affected by collinearity as well. Collinearity occurs where two predictors vary in the same manner, thereby masking the distinction as to which determines the observed relationship, and thus the effect of noise in the analyses.

As a consequence, multiple linear regression (MLR) cannot be used, since the technique cannot handle collinear variables; (i.e., it is unable to differentiate between dependent and independent variables.) If the x-variables are intercorrelated, the use of the MLR technique will result in prediction inaccuracies. MLR also requires a reduction in the number of variables in a collinear data matrix. For these reasons, it was judged inappropriate in the specific context of this research.

In essence a method is required that can handle large data sets in which there are more variables than objects, and which incorporates noise and collinearity. The method that was finally decided upon as the one best suited to the data was ACE, Alternating Conditional Expectation, to be preceded by PCR. The factors which influenced this selection are discussed below.

Multivariate calibration is versatile technique, it can; 1. aid in the predicting of abstract qualities of data set, 2. help to resolve areas of overlap or collinearity, 3. non-linearities and interferences create few problems, 4. minor instrument variation can be corrected.

5.6.1 Application of PCR

A multivariate mathematical model with parameters that

have been estimated from empirical data, is necessary if this type of analysis is to work. It is a form of calibration that is used to concentrate relevant information from many measured variables into a more manageable form. In essence, this provides a simplified map to the data that allows both prediction and chemical interpretation to take place.

Apart from pattern recognition, the two most effective techniques applied to this data are PCR and ACE , both of which achieve their goals in basically the same way. More specifically, PCR uses procedures which are similar to PCA. That is: with PCA, the data matrix of independent variables is decomposed into principal components. In addition, PCR goes on to reduce the principal components to regression vectors, which predict the value of the dependent variables for the test samples. ACE , on the other hand, calculates a nonlinear nonparametric regression model, where the response is estimated as the sum of functions of the predictors, which must be smooth. It is most applicable where there are sufficient degrees of freedom; that is, where the data set is well determined by containing more observations than predictors, thereby producing a model that assumes additive nonlinear predictor response, and identically and independently distributed errors, and which ultimately tests both the modelling capability (goodness of fit) and prediction capability (goodness of prediction).

It is not always obvious which of these two methods produces the most reliable calibration model. However, statistical theory suggests that each algorithm is advantageous in certain circumstances. This is most definitely the case where these two algorithms are concerned, as there are certain requirements regarding the overall data which optimises the individual methods. As previously mentioned singular value decomposition (SVD) functions at its best when utilized with data containing large numbers of components and only a few variables. Whereas non-iterative partial least squares (NIPALS) performs more effectively where the data comprises a small number of components, with a large number of variables.

Both methods are applied here as SCAN is able to carry out the numerical analysis with such speed and accuracy that certain types of trials can be performed without time loss. Although SVD best suits the data contained in the project, and was utilized to that end, NIPALS was also tried as a comparison. They were both used to extract the eigenvector from the data, from which the regression vector was to be calculated using the principal components regression (PCR) method, and the predictive procedures of the alternating conditional expectations (ACE) method.

While the fundamental concepts underpinning regression analysis (e.g. data compression, eigenvalues, eigenvectors and singular value decomposition [SVD] and [NIPALS]) have been elaborated in previous sections, their importance for data presentation and interpretation cannot not be overstated. When supplemented by a rigorous testing of its ability to predict, and an analysis of residuals, the technique of regression analysis provides a very high degree of reliability.

5.6.2 Principal Component Regression (PCR)

PCR is, in essence, the regression of Y dependent variables in a universe of X selected principal components. This method is generally employed in contexts where the X variables are expected to be collinear, as is the case with the data presented here. As a consequence of such collinearity, the **X** matrix show the dominating types of variability, those which carry most of the available information.

5.6.3 Mathematical background

PCR (Appendix pt. II pg. 130) calculates a parametric linear model for a single response, and calculates biased, nonleast squares coefficients. It is normally used when a data set is undetermined; i.e., where there are fewer observations than predictors and/or the predictors are highly correlated. Although the data set here was highly determined, PCR was utilized firstly to observe how it differs from ACE, and secondly because the data was highly correlated. PCR also assumes that the response is a linear function of the predictors, and that most of the information about the response relative to the variance of the predictors is in the first few principal components, which corresponds to the largest eigenvalues, and the initial PCA calculations.

The first step is to autoscale the data, then the PCR model is represented by:

$$\mathbf{y}=\mathbf{T}(m)\ \mathbf{a}+\mathbf{e}$$

Fig.5.8 (Miller & Miller 1993; SCAN manual)

where **T**(m) is the matrix for the first principal component scores, calculated from the matrix $\mathbf{X}^{!}\ \mathbf{X}$,**a** is the vector of coefficients, and **e** is the vector of errors. As the predictors are autoscaled, $\mathbf{X}^{!}\ \mathbf{X}$ is equal to the correlation matrix of predictors. The scores are calculated from **X** by:

$$\mathbf{T}(m) = \mathbf{XL}(m)$$

where **L**(m) is the matrix of the eigenvectors of $\mathbf{X}^{!}\ \mathbf{X}$, corresponding to first or largest m eigenvalues. The principal component scores are orthogonal vectors, and are linear combinations of the predictor variables.

The coefficients **a**, are found using OLS, with **y** as the response and **T**(m) as the predictors. The regression equation can then be expressed in terms of the predictors, **X**, as

$$\mathbf{y}=\mathbf{T}(m)\ \mathbf{a}+\mathbf{e} = [\mathbf{XL}(m)]\ \mathbf{a} + \mathbf{e} = \mathbf{X}[\mathbf{L}(m)\ \mathbf{a}] + \mathbf{e} = \mathbf{Xb} + \mathbf{e}$$

Therefore the coefficients for the principal components model with m components are given by:

b = **L**(m) **a**

Fig.5.9 (Miller & Miller 1993; SCAN manual)

There is a bias-variance trade off which is controlled by the number of principal components m. The more components, the larger the variance and the smaller the estimated regression coefficients. As the number of components increases, the PCR model converges to the OLS (Ordinary Least Squares) model. When m = p (number of predictors), PCR and OLS are the same. The optimal number of components is selected, based on a goodness-of-prediction criterion, e.g. by cross validation, (as utilized in ACE). PCR is a linear estimator, therefore the cross-validated residuals can be calculated from the ordinary residuals.

The sizes of these regression coefficients will only reflect the relative importance of the corresponding predictors if they are of equal variance; such is the case here as the data have been autoscaled. These standardized coefficients, such as those in the ACE model, have been multiplied by the standard deviation of the corresponding predictor, which demonstrates what a good measure of importance these coefficients are. Here, the predictor importance is calculated as the standard deviation of the corresponding nonlinear transformation.

$$s = \sqrt{\frac{\sum(\bar{y} - y_i^2)}{N-1}}$$ (general formula for standard deviation)

Fig. 5.10 (Miller & Miller 1993; SCAN manual)

The technique of OLS (Appendix pt. II pg. 126) will not be discussed in any depth here, as it was not a selected method of regression analysis. The primary reason for its rejection lay in its assumption of linearity, which was invalid in this case. Although closely related to the methods which were utilized, the nonlinear character of the data rendered OLS inappropriate, and ACE along with PCR were, accordingly, chosen as the techniques possessing the greatest utility.

5.7 Alternating Conditional Expectations (ACE appendix pt. II pg. 140)

ACE was considered to be the most productive of the available modes of analysis. This judgment was based upon indications present in the preceding analyses, the results of which confounded the experimenter's expectations. This analytic series had, furthermore, included the manually-calculated ANOVA and polynomial, which had indicated a nonlinear trend in the calibration assays. (see chap. 7)

As previously mentioned, ACE models goodness-of-fit and goodness-of-prediction; the former is via the residual sum of squares (Residual SS) and the R-squared coefficient, and the latter via the residual predictive sum of squares (resid PRESS) and the cross-validated R-squared coefficient (R2cv).

There are however, no exact formulae available in ACE with which to calculate the pre-eminent features in its analysis, (the cross-validated fits and residuals), of the type present in other forms of regression. Instead, these require calculation by means of the estimation-method of 'Leave-one-out', in which one observation at a time is omitted and the calculation repeated progressively until the entire matrix has been covered. Although time consuming and laborious, the results achieved in this way have proved to be of an extremely high level of validity for both the model and test samples. As there is no simple analytical form to ACE, the transformations can be analysed graphically by plotting each variable against its transformation function. See chaps. 7 & 8.

5.7.1 Mathematical Background

There is a basic mathematical theory that represents the ACE regression model.

$y = \Sigma_j t_j(x_j) + e$ **Fig. 5.11 (Miller & Miller 1993; SCAN manual)**

Where t_j are the transform functions, the smooth and relatively unrestricted functions of the predictor variables. In relation to the linear regression model, they are analogous to the regression coefficients. The response is modelled as the sum of transform functions. ACE functions are obtained by smoothers, which are themselves obtained by estimation by the least squares iterative algorithm. (See SCAN chp.9)

The variance function is represented by t_j, which indicates the importance of the predictor variable j in the model; i.e., the higher the variance, the more important the corresponding variable. In essence, the ACE model is defined as a set of point pairs:

$[x_{ij}, t_j(x_{ij})]$ $j = 1,p$

Fig. 5.12 (Miller & Miller 1993; SCAN manual)

Generally the ACE model has no simple analytical form, and the SCAN graphical and tabular output will provide the visual information required to see how the totality of the calculation has behaved, in terms of the data matrix. (see the results chapter for this discussion of these issues).

5.8 Conclusion

Having completed this complex series of calculations, it was necessary to evaluate the results in terms of the original aims of the project. The refined data included plotted matrices in the autoscaled format, supplemented by tables and graphs from all the techniques utilized; the latter were provided in order to enhance the visual detail. While the full and thorough discussion of these results will be resumed in chapters 7 and 8, it became clear at this juncture that they had fully justified the selection of chemometrics as the optimum form of analysis, and that, contrary to the pessimistic expectations to which the experimenter succumbed during difficult moments, what we were dealing with was a positive experimental outcome.

It is also necessary to point out that most of the statistical procedures utilized here can be found in any detailed statistical text, however credit is given to SCAN for all quoted, as they are utilized via SCAN software and hence via SCAN format.

CHAPTER 6

METHODOLOGY

6.0 Introduction

In order to ensure that any comparisons made between the ancient and the modern seed samples would be useful for further analysis, a standard method of handling them was developed. This standardization of treatment embraced all forms of preparation, the chemical extraction, and the FTIR analysis. Consequently, any assumptions made with regard to the fitness for analysis or contaminated status of a given sample could be extended to all the specimens, ancient and modern.

6.1 Sample acquisition

Each of the modern seed samples was identified by morphological examination. This entailed the observation of those external physical characteristics specific to particular species. Groups were established, each containing five seeds per sample. As actual concentration was not the main objective of the analysis, five seeds were selected for the modern extracts. Although the archaeological samples were confined to one seed, the set of five was used in order to obtain as much extract as possible. This was essential if the analysis was to produce clear and easily determined diagnostic peaks on the FTIR spectra.

The number was selected as the most likely to give the optimum extraction from the least number of seeds. To a certain extent, this selection was dictated by external constraints upon the experiment: as only one seed per sample set was available for the ancient specimen extraction, it was necessary to use as few seeds as possible for the modern equivalent. Under ideal circumstances, a series of experiments would have been performed, each employing a different number of seeds, and a decision arrived at accordingly. Since, however, unlimited time was not available for the research, it was considered appropriate to conserve it for those comparatively more substantial problems which would inevitably occur.

The next step in the preparation involved the individual peeling of each modern seed, which required the manual removal of the outer husks, the [2]lemma and [3]palea. This task needed careful handling, as many of the seeds were very small, and their husks adhered tightly to the seed kernel. In order to prevent oils from the hands from contaminating the peeled seeds, the wearing of surgical gloves was necessary whenever they were handled. Each sample set was kept separately to avoid any mixing of the species. It was then placed in a glass petrie dish under a binocular microscope, held with tweezers, and peeled carefully using a sharp scalpel. Once peeled, each set was placed in a clean sample tube, and marked with the relevant information transferred across from its original label. This operation was scrupulously performed in order to ensure the appropriate details were retained with each sample. All of the equipment used was then cleaned with acetone to preclude contamination of the next seed group to be peeled by oils left over from the previous group. This slow, painstaking procedure had to be carried out because: a) the archaeological seeds possess an outer covering, and b) to prevent possible lipid contamination. The high risk of this type of contamination was pointed out by Evans *et al* in 1990, (paper on identifying charred cereal grains using infrared spectroscopy), who had demonstrated in previous work that the husks contain their own lipid component. It was very important that these lipids be removed prior to instrumental analysis, as they would obviously be recognized by the FTIR and would distort the experimental data.

The comparison of ancient and modern sample sets being fundamental to the entire project, it was essential to omit from the outset any obvious differences between them. The major such difference was the unreliability of the morphological identification of the archeological seed remains, which were, accordingly, assigned only the most tentative of identifications based only on the most prominent of their physical characteristics. The archaeological seeds utilized in this research were subject to extensive charring, and were quite brittle; they had also been frequently studied, and were in a highly fragile state. This was a reference to the work carried out by Hillman and Nesbitt, in which great care had been taken to ensure the correct allocation of name to the specimen in question. Both of these researchers are highly experienced archaeobotanists, and are fully cognizant of the correct procedures for handling charred materials where these involve samples intended for use in chemical analyses. Furthermore, as Hillman and Nesbitt collected the seeds utilized in the project, they had also identified the samples and prepared them for storage. Nesbitt had himself made use of a number of these seed samples in the course of his PhD thesis, a work which included the accurate identification of the smaller seeded wild grasses, which had been subject to charring. This expert identification was of immense assistance to my research, enabling me to proceed with my work in the knowledge that a reliable identification had been previously undertaken, and, moreover, that it has been carried out by one who was keenly aware of the potential distortion of data which could result if contamination were to occur during handling. In summary, they required the most delicate treatment, and, notwithstanding this caution, their repeated handling inevitably led to minute pieces breaking off. In spite of these factors, however, it was possible for the experienced archaeobotanist to obtain a reasonable degree of morphological information, and, on the basis of the remaining features, to allocate some cautious classificatory identification.

[2] The lemma is defined as 'The lower of a pair of bracts beneath each floret in the florescence of a grass'. (Blackmore and Tootill 1984:208)
[3] The palea is defined as 'The upper of a pair of bracts beneath each floret in the florescence of a grass'. (Blackmore and Tootill 1984:257)

Due to the difficulty of making any firm identification of the ancient specimens, a precautionary procedure of one charred seed per archaeological sample set was decided upon, in order to enhance the prevention of cross-contamination by different species. As FTIR detects the presence of functional groups only, any lipids present in one seed should be detected as lower concentrations, but would appear in the same regions of the spectrum as the five seeds. This difference in seed numbers between the ancient and modern sample sets represented the only aspect of the analysis that featured a deliberately contrived variation between the two. This was, however, unavoidable if the archaeological extractions were to remain homogenous.

Before continuing with the description of the methodological processes entailed in the experimental work as such, it is necessary to note the status of the sample sets and the conditions under which they had been stored prior to their utilization in this project. The samples, both ancient and modern, were next removed from collection phials that had been stored in a dark, dry environment. All of the modern specimens had been collected some 20-25 years previously, and kept in special drying tubes plugged with cotton wool. This allows the continuation of the drying process during storage. It was considered safe, therefore, to conclude that the materials were thoroughly dry. A comparable time-period and set of storage conditions pertained to the archaeological samples, and with similar implications for their aridity. Any unusual circumstances or relevant treatments would have been clearly labelled as such by Hillman, e.g., if they had been subject to fungicidal or pesticidal applications. Finally, they were examined under microscope before use in the research, and any infected seeds were discarded.

These considerations are important because the dryness of the seeds was essential to that phase of the project in which a sub-set was selected for the calculation of the percentage of extract from a seed set. This undertaking involved: a) ascertaining the weight of each of twenty randomly chosen (dry) seed sets; b) running these through the extraction process, ensuring the identical treatment of each; c) driving off the excess solvent by distillation, leaving approx. 1ml.; d) placing the remaining 1ml under an IR lamp in order to completely dry off the sample.

This 1ml. was retained in its original glass vial when placed under the heat lamp, in order to prevent sample loss, which would occur if the samples were constantly transferred to other containers. Furthermore, it was very unlikely that any of the sample extracts would have been left in the distillation apparatus, as the extracts are heavier than the solvents; therefore, the only place they are going to be found is in the original glass flask which was part of the soxhlet extraction apparatus.

The weight of the extract versus seed weight loss was carried out for the sub-set of twenty, in order to ascertain the relative working range of extract concentration. The information received was utilized for experimental purposes, but the actual numbers were lost when the computer systems were changed. Unfortunately, before these numbers could be retrieved and stored on floppy discs the hard-drive on the old system broke-down. This rendered access impossible to any information left on the drive; happily, however, this was the only irreplaceable data left on the machine.

This procedure was carried out on a selected number of samples, rather than the entirety of them, for the following reasons: a) the time constraints placed on the research, which rendered impracticable the careful drying of every single sample; b) the risk of losing or destroying the sample by excessive heating, which may burn the extract and render it useless for further analysis; c) the time required to replace a sample if it is accidentally lost or damaged.

Given these uncertainties, it was decided that the optimum practical strategy involved the random selection of twenty samples, with which to demonstrate the likely percentage of extract available from each seed set.

Once the seeds were peeled, they were ready for the process of extraction. It was first necessary, however, to prepare the rest of the equipment, and to ensure it was clean and free from contamination. To minimize such contamination, surgical gloves were to be worn whenever handling the seeds and the cleaned glassware. The glassware was cleaned for each sample set by overnight soaking in a 10% solution of sulphuric acid in de-ionised water, then washing it for a further 2 hours in an ultrasonic bath, also with de-ionised water and a cleaning agent (decon 90). It was then dried beneath an IR lamp, assembled as a soxhlet system, and run for 24 hours with just the extracting solvents. The solvents were then reduced down, spotted onto both Kbr. discs and the sample cards, and run through the FTIR to see if spectra other than those representing the solvents were present, which may be indicative of unclean glassware. As this was not the case, it was considered safe to assume that the glassware was clean and ready to be utilized in the extraction of the lipids.

The systematic procedure for setting up the extraction process was as follows: a) weigh a 10ml. glass sample vial. b) weigh a 55mm cellulose thimble, in which the seeds are to be held. c) weigh the sample of seeds to be utilized. (These components were weighed separately as it was essential to ascertain the precise weight of each.)

The seeds were then placed in the thimble, using tweezers to avoid handling and to ensure that an identical number of seeds went into each thimble. Next, the seeds were introduced into the system via the thimble, which was itself held in a glass vessel called a soxhlet. The extraction process was then carried out by connecting a water-cooled condenser to the top of the soxhlet, and a glass flask containing the solvent to the bottom. This flask was placed in a heating mantle for thermostat-controlled slow heating, and having reached the required temperature, was maintained constantly thereat. This procedure would allow for the free-flow of the heated solvent, which was able to percolate through the sample without producing either excessive heat or loss of solvent. For adequate extraction from a sample of this size, some 20-25 mls. of solvent was required, to which clean anti-bumping granules had been added to prevent bumping or flaring as the cooled solvent returned to the system via the condenser.

Both hexane and chloroform were used in the extraction procedure; the hexane extraction was carried out first, followed by the chloroform. The two were not used in combination.

6.2 Sample Extraction
The lipid extraction was carried out using a standard soxhlet procedure; a series of solvents having an affinity for the lipids were employed, these being hexane and chloroform respectively, polysaccharides and phenols were avoided. Trial runs had demonstrated that, for optimum extraction, the system required to be run continuously for a period of three days. Upon completion, the extracts were distilled to remove excess solvent, retaining approximately 0.5 ml of solvent and extract (analyte) for the purposes of FTIR analysis. This method of retaining a minute amount of the solvent containing the sample was selected as it greatly reduces the chance of sample loss, which may occur if the sample is completely dried out under a heat source.

The subset of twenty samples was subjected to similar treatment. As these were to be employed, however, in the calculation of the weight of the extract, it was necessary to allow the whole of the solvent to evaporate, including the 0.5 ml. under the heat lamp. Once the samples had reached this point, they were permitted to cool, and then weighed again. Having already established the weight of the glass tube, what remained upon its subtraction from the total represented the weight of the extract. After ascertaining this weight, the sample was then reconstituted with the extraction solvent and prepared for FTIR analysis.

Complete lipid extraction requires a solvent system that has been devised specifically for the purpose. Insufficient attention to the details of the process may result in a variety of errors, stemming from either incomplete extraction or non-physiological alterations in the lipid structure. These considerations were of particular relevance to the modern extracts, as, although it was a relatively simple matter to obtain quantitative extractions, great care had to be taken to minimize lipase activity. Such activity may be observed where significant quantities of free fatty acids are present in the extracts due to hydrolysis of neutral lipids. The use of this system with hexane and chloroform, at temperatures which allowed for the extraction of the lipids without deterioration (i.e. 70 degrees C.) but destroyed enzyme activity, seemed sufficient to prevent undesired lypolitic activity. (Most enzymes work efficiently at temperatures around 40 degrees C.). (Gallirad & Mercer 1975)

6.3 Sample Preparation for FTIR Analysis
First, each sample was spotted onto a 3M (TM) disposable IR card, allowing for quick solvent loss and even sample distribution, without heat exposure. (The term ‘spotting’ refers to the slow application of an extract onto a card or disc, one drop at a time, thus allowing each drop to dry before the next is applied.)

The cards were chosen in preference to KBr. discs following a lengthy appraisal of their respective merits, as a result of which it was felt that they enabled superior retention and presentation of the samples. The decision was arrived at after experimental trials employing KBr. discs had demonstrated an unacceptable level of sample loss, deterioration due to heat, and a lack of sample distribution across the surface of the disc. In addition, the extended storage periods required led to an observable deterioration in the KBr. samples. As triacylglycerols are light sensitive, their being held on the disc-surface caused lipid breakdown and crystallization. Coupled with the opacity of the disc, (due to its hygroscopic nature), this had the effect of increasing beam-scattering, and caused a major shift in the functional group frequencies (C=O from carboxylic acids, aldehydes or ketones in the breakdown products), and also resulted in increased noise.

The original design for the research had included the use of the KBr. discs in an important role, and consequently the observed deterioration phenomena gave considerable cause for concern. This prompted a consultation with Dr. Raymond White of the National Gallery, whose acknowledged expertise in this field is based on an extensive and detailed study of the compounds concerned, and who confirmed my observations of deterioration products. Dr. White informed me that he too had witnessed an identical type of deterioration, which occurred in varying degrees according to exposure, apparently without regard to the origin of the oils.

A potential solution to the dilemma I was now facing was suggested by the manufacturer of the FTIR equipment utilized in the project, a company by the name of Perkin Elmer, whose staff are possessed of considerable skill and experience in the field. Following a discussion of the problem, they sent the disposable IR cards in the hope that these would provide a way forward.

Having observed the difficulties arising with the use of the KBr. discs, however, it was regarded as advisable to run a series of experiments with the cards before making any firm commitment to their use. It was essential to ensure that there would be no contamination of the samples from the card matrix, and no evidence of chromatographic separation by the aperture matrix of the spotted sample. The experiment involved the spotting of a number of modern seed extracts onto the cards, allowing the excess solvent to dry without heating. Upon close inspection, and following the procedure of slowly spotting onto the aperture one drop at a time, it was clear that there was no chromatographic separation of the specimen. The spotted sample was allowed to dry after each drop; finally, the totally dried sample was examined under the monocular microscope in order to look for any sign of separation. It was clearly evident that no such separation had occurred. The cards were then stored in a cool, dark place for a period of two weeks, after which time the samples were removed and analysed using FTIR.

The process entailed the removal of the card aperture, both for the hexane and the chloroform samples, and their immediate washing in the original extraction solvent, retaining approximately 0.5ml. These were then applied to clear KBr. discs, and left to dry without heating. Finally, they were analysed to ascertain whether peaks that were diagnostic of the card-matrix would be visible.

Tests were also carried out to ensure that the standard spectrum used for identifying the presence of the peaks for the card were constant. These tests required the random selection of a number of blank cards from a box of twenty-five, which were then analysed and compared, thus verifying the selected diagnostic positions.

Although the cards possess their own spectral region, it is possible to eliminate this as the type of FTIR equipment used allows for the ratioing out of any background spectrum. When combined with the chemometrics software, which formed the principal mode of spectral identification, this meant that any regions not required could be readily deselected from the data analysis.

6.3.1 Phenolic Contamination

The contamination of lipid extracts is less prevalent if the correct combination of solvents has been selected. The flavanoid compounds, the largest, most diverse range of plant phenolics, represent the most likely candidates for contaminating agents. They are not soluble in $CHCl_3$, the major solvent employed in this extraction procedure, but they do have an affinity for water and alcohol, deriving from the formation of esters of sugar via a linkage between the phenolic group and the sugar, whereby increasing the number of hydroxyl groups increases solubility. Consequently, if alcohol and water are used in the initial extraction process, phenols in the form of flavanoids may be present in the final lipid extract.

6.4 Investigating the Remains of Biomolecules

'Investigating the fossil remains of biomolecules requires expertise from several major branches of science'. This observation is made by both Eglinton and Curry the authors of Molecules Through Time. It is always an imaginative challenge to take account of the disciplines of others, which are often cast in theoretical forms and vocabularies foreign to us. Yet it is undoubtedly the case, that many significant advances have resulted from the cross-fertilization of and collaboration between specialists from the many and varied branches which go to make up the scientific endeavour in its totality. The numerous problems and challenges posed by the attempts of a specialist in a given field to construct a multi-disciplinary methodology were particularly acute in the case of the present dissertation. The researcher is, first and foremost, an organic chemist, and has therefore undertaken additional training (in a variety of analytical and statistical techniques, in chemometrics, and in archaeobotany) for the purposes of addressing the specific problems to be addressed within the terms of reference of her doctoral project.

Eglinton and Curry note that it is now a fairly widespread custom for the specialist to cross over into the territories of adjacent disciplines according to his or her theoretical and experimental requirements. They also comment, however, that the scope, complexity and extreme rapidity of development of contemporary scientific practice renders entirely unrealistic that mastery of multiple fields once accomplished by the fabled 'renaissance-man'. As a consequence of this, the pragmatic researcher, amongst whom the present author likes to consider herself included, must rely upon close co-operation with colleagues possessing expert knowledge in the appropriate fields, and draw fully upon their resources in pursuit of successful experimental work. An adroit usage of these resources can enable multi-disciplinary research to provide a comprehensive account and understanding of fossil remains of biomolecules.

In order to provide the necessary information with regard to those biomolecules investigated during the course of this project, a comprehensive understanding of living, related biomolecules was essential. As Eglinton and Curry point out, this data is not always readily available, and must be experimentally determined by the researcher.

Further practical problems related to the carrying out of the experimental aims were derived from the case with which the materials under investigation could become contaminated. These sources of contamination ranged from bacteria, light, water, and the touch of the human hand, to factors stemming from the original environment from which the matrix materials were recovered. In this case, the materials were extracted from seeds from the epipalaeolithic layer of an archaeological site. Eglinton suggests obtaining comprehensive information on the breakdown pathways of the biomolecules in geological situations; this data, however, is once more something which is not easily accessible, and it is, consequently, necessary to look highly critically at the materials under study, and perhaps to try to recreate some of the more obvious conditions. This is, no doubt, less problematic when dealing with less ancient samples, for example when utilizing materials from the relatively recent

Mesolithic hunter-gatherer societies. The most likely causes of damage to the biomolecules specific to the current research are those involved in the process of *charring*, with human handling coming in at a close second. At this juncture it might be useful to point out that the question of survival of biomolecules is much less of a problem when looking at materials relating to the epipalaeolithic or mesolithic periods, the difficulties actually arise when dealing with the much longer periods of the geological time scale.

In this context, it is necessary to make reference to basic research which helps to establish the essential conditions of possibility for the current work, carried out by the author in 1992 (Cave 1992). This work involved conventional dispersive infrared analysis of a number of charred samples of wild barley. The research was undertaken under the auspices of an undergraduate's third-year dissertation at the University of London and laid the foundations upon which a more complex doctoral thesis was built. This incorporated, in addition to my basic training as a chemist and the specific techniques learned under the supervision of Evans, the archaeobotanical work undertaken with Hillman. It also drew upon related studies at Evans' laboratory that were being undertaken by colleagues Sue Wales and Frances McLaren.

6.5 Relationship between Modern and Ancient Seeds

Owing to their extreme age and to the extensive charring to which they had been subjected, the ancient seeds had lost many of their diagnostic features, and this fact made their morphological comparison with modern seeds specimens difficult. The charring process renders the seeds very brittle, with the consequence that even a careful handling may result in the breaking off of tiny but potentially diagnostic fragments. Charring can also cause the seeds to become swollen, and normally rather flat specimens may be distorted and assume shapes and sizes which can mislead the morphological examiner. While morphological characteristics may be modified by the process of charring, however, Evans *et al* (1990) have demonstrated that the lipid biochemistry is not subjected to such drastic alteration. This finding suggests that, if it can be successfully utilized, the lipid composition may offer an alternative modality of identification.

6.5.1 Charring

As mentioned above, the archaeologically derived seeds utilized in the research are charred. More specifically, we can state with certainty that they have been exposed in some way to the actions of fire, perhaps in the course of a partial cooking, etc., but that they have not been totally burned and carbonised. Although subject to external dehydration, they have to a certain extent retained their original physical features. Furthermore, while the charring process may have destroyed enzyme activity, and perhaps also the D.N.A. structure, (which, if it were intact, would represent another potential modality of identification), it seems to have had a minimal effect upon the lipid content.

As the extraction of triacylglycerols has been successfully achieved, even where the smallest of wild grasses have been targeted, it does appear that lipids are maintained within the seed matrix. It has been suggested by Hillman (1993: MSc. Archaeobotany lecture on charred seeds) that the charring process may produce a casing around the seed, thereby protecting the lipids and possibly causing them to migrate inwards, away from the external heat source. The suggestion was made in the context of a personal communication from Gordon Hillman which included a discussion of the possible reasons for the long term persistence of certain biomolecules, such as lipids, within the matrix of archaeologically recovered seeds. Hillman was drawing upon his great experience in the archaeobotanical field, which was supplemented by collaboration with numerous colleagues; his judgment was that this was indeed a plausible explanation. The hypothesis has been supported by subsequent experimental work which includes that of the present author.

6.6 Experimental charring

Utilizing those projects having a direct bearing on the development of my own research enabled precious time and resources to be efficiently deployed, and to avoid the duplication of costly scientific labour. With reference to the effects of the charring process upon the lipid content of seed samples, for example- this being of potentially decisive interest in developing an elective mode of chemometric identification- I was able to draw on the extensive experimental work of Evans and McLaren. This work, which is due to be published upon the completion of McLaren's PhD thesis when combined with previous studies on this topic undertaken by Evans, supplied data in which I had every confidence, and rendered quite unnecessary its replication by myself. McLaren covers, in a thorough and extensive manner, the potential effects of charring on the lipid content of seeds. To give a sketched summary of the work: fresh seeds were subjected to charring under a variety of conditions judged relevant to the deterioration of lipids. A sample population of seeds was divided into two groups, with one (the control) being subjected to the extraction process without charring and its lipid content subsequently being examined by means of infrared analysis. The second group was subdivided, its constituent groups being subjected to a series of varying heat treatments in order to determine the charring effects according to factors including the duration of exposure, mode of application, the atmosphere of the heating-chamber, and the condition of the seeds.
In essence, the experimenters were looking at:

1. Rate of heat rise; i.e. the exposure of the seeds either to instantly high temperatures or a gradual increase from room-temperature.
2. Duration of heating; i.e. how long seeds were exposed to a given temperature.
3. Range of temperature; i.e. the precise temperatures

required to produce degrees of charring.
4. Condition of seeds; i.e. whether the level of moisture present in the seeds would affect the lipids, by way of interaction with the charring process.
5. Oxygen availability; i.e. the seeds were exposed in both aerobic and anaerobic conditions during heating for purposes of comparison.

This set of primary criteria was rearranged in a multiplicity of combinations. Subsequently, the treated seed-groupings were solvent-extracted for lipid content, and the extracts were analysed by means of infrared spectroscopy. **The experimenters arrived at the conclusion that there was no significant difference between the fresh and the charred extracted lipids.**

6.7 Taphanomic Processes

Among those numerous environmental influences to which the ancient seed samples have been exposed, it is likely that the arid climatic conditions have subjected them to processes of desiccation. However, once again successful extractions have demonstrated that these processes appear to have had little significant impact upon lipid presence. The most marked effect on the seed seems, rather, to have been the physical damage produced by abrasions arising from the burial-matrix and the recovery processes.

6.7.1 Recovery Techniques

When carrying out any form of scientific analysis on archaeological materials, it is essential to consider the handling techniques employed in its recovery, as these may impact upon and adversely affect the overall results. This is particularly important where chemical analysis is concerned, as extracted material is usually recovered in trace amounts only.

The most successful method of recovering charred materials from a soil matrix involves the technique known as *flotation*. At its most fundamental, this requires a source of running water and a holding sieve, and its application will ostensibly have no effect on the hydrophobic chemical extract. Although the seed biochemistry remains unaffected by this process, the flotsam may have been subjected to other treatments. For example, bulk soil samples may often have been examined for other materials such as mollusc, arthropod and pollen remains. When carried out separately, such examinations are time-consuming and expensive, and, as a consequence of these factors, the individual components will often be separated out by means of the application of certain chemicals. It is necessary to mention these procedures in this context because, although the possibility of the separation techniques and sources of contamination associated with them affecting the results of the analysis is minimal, it is important to be aware of them and to remain vigilant.

In order to remove mollusc remains from soil samples, a 30% hydrogen peroxide (H_2O_2) is added to the soil/water mixture. This produces bubbling, which releases the required material from the matrix and allows it to float to the surface. In those cases in which a high clay-content prevents the soil from easily releasing the samples, sodium hexmetaphosphate $(NaPO_3)_6$ may be added to assist in breaking it up. This is a cleansing agent which will affect the oil extracts if the residues are not washed away. It can cause saponification - the alkaline hydrolysis of an ester to an alcohol- effectively breaking down the oils.

For pollen analysis, octanol $[CH_3(CH_2)_5CHMeOH]$, a heavy secondary alcohol, is used as a foaming agent with which to induce the release of minute pollen spores from the matrix. The principal problem with the employment of this alcohol lies in the fact that is prepared by distilling castor oil with concentrated NaOH, and is, in consequence, not very volatile. If, however, the seed samples are placed in the reflux system in a dry state, and are completely solvent-free, this factor should not affect them.

In summary, knowledge of the pre-handling techniques and procedures utilized is of the utmost importance, as there are many sources of contamination that may distort the chemical data obtained. Even those oils exuded naturally by the human skin may be exported into the dry seeds.

6.8 Spectra Acquisition

The equipment used for the FTIR analysis was a Perkin Elmer 200 series Fourier Transform Infrared, single beam system, with a sealed nitrogen chamber purge, and an infrared data manager control terminal.

Prior to the actual FTIR analysis of the sample extracts, it was necessary to prepare the instrumentation in order to obtain the optimum performance from it. This involved purging the sample chamber for fifteen minutes in order to remove excess water-vapour and carbon-dioxide, both of which can produce interference in spectral results. Subsequently, it was necessary to set the instrument parameters to match the type of materials to be analysed. These parameters were obtained from the appropriate chemistry and biochemistry texts, in which was contained the information regarding the spectral range for these types of organic compounds. (George and Willis, 1990); (Galliard and Mercer, 1975); as well as trials using similar types of extracts, and prior work carried out by Cave, 1992.

The general format for the analysis was as follows:

a). A data interval of 4. This is the abscissa spacing of data points in a spectrum or interferogram at optimum resolution.
This setting for optimum resolution was selected, as previous trials of short test runs on other levels showed it to be the most appropriate for the optimum clarity to a spectrum.

b). A spectral range of 450-4000cm^{-1}, which was demonstrated as ideal for this type of organic extract.
c). Setting the [4]J-stop or aperture at 1.63mm with a gain of 1, helping to reduce noise further with information-loss occurring.
d). Setting the [5]B-stop at 21.2mm, which represents the maximum output to the detector, in order to allow the greatest possible energy to reach it without inducing overload.

These parameters being set, a background scan was made, which would then act as a ratio for the chamber environment and the card. Combined with the Fourier Transform, this permitted the filtering of any unwanted spectral detail. Each batch analysis was preceded by the random selection of a card and its separate and single analysis, in order to check both its condition and to ensure that a blank card was analysed under the same set of conditions as the extract itself.

As it was necessary to obtain the feasible optimum quantity of information, the project was to generate a large number of extracts, which translated, in turn, into a vast array of spectra. As a consequence, a decision had to be arrived at concerning the most viable of two available analytic techniques, which consisted in the following:
a) In which all the extracts were produced, stored, and then run concurrently through the IR analysis.
b) In which the extracts were produced in batches according to the number of soxhlets, and each batch run immediately upon extraction.

In the light of the problems of deterioration referred to in the preceding discussion, it was decided to adopt the latter of the two methods, and to use the extracts immediately after their production. This required a standardization of the analytic procedures in order to ensure that, each time a batch was examined, identical criteria would be employed. This was, in fact, the case for all the spectra gathered and utilized in the research.

The spectra were generated according to the following protocol:

1) Each extract was spotted (i.e. pippetted onto a card one drop at a time) onto a card immediately following extraction.
2) Each prepared card was left to stand in a fume-cupboard to allow the solvent to evaporate.
3) The sample was placed into the holding slot of the FTIR chamber.
4) The chamber was closed and purged with Nitrogen for 3 minutes.
5) The FTIR analysis was carried out.
6) The spectrum on the screen was checked to ensure the instrument was functioning correctly.
7) The spectrum was named, and then filed with the appropriate sample and solvent details onto the hard-drive of the computer.

6.9 IR Detection of Triacylglycerols

The inclusion within the methodology of the two sections dealing with Spectral Interpretation and IR Detection of Triacylglycerols is intended to make clear the manner in which this information was gained from the IR analysis, and the way it was then used in the determination of precisely which of the components included in the information were salient in subsequent analyses. Although these considerations do not belong to the methodological domain as customarily conceptualized for the purposes of 'wet chemistry', they do form an essential methodological component in the current project. In order to facilitate the replication of the work by future researchers, each step in the methodology must be described, even where these entail processes which do not resemble the practicalities of the traditional 'wet chemistry' experiment. For example, it was important to describe exactly how the extracts were utilized in the instrumental analysis, how they were affected by this utilization, and what type of information they ultimately generated. Only when equipped with such knowledge is it possible to understand the derivation of the diagnostic features, and by what criteria they were so defined.

It is the vibrational region of the Infrared spectrum which is used in chemical analysis. The molecules of an exposed sample become excited and attain a higher energy state as they absorb IR radiation. This state of excitation does not represent a permanent modification of ground state; upon removal of the radiation-source, the molecules return to their normal energy-level. Furthermore, not all of the molecules will absorb IR energy, even where the vibrational frequencies are matched with the bond motion, because the transformation of the energy is dependent upon the presence of an electrical dipole which changes at the same frequency as that of the incoming radiation. Those bonds which are most affected by this process, and which are, consequently, of primary relevance to this research, are the *symmetric alkenes*. (C=C). A detailed discussion of these issues is found in Chapter 4.

As each bond type possesses its own specific natural frequency of vibration, no two molecules of differing structure will have an identical pattern of IR absorption. While some of the absorbed frequencies may be the same, the different molecules will produce different spectra. In the case of pure compounds, these spectra function in a manner equivalent to *fingerprints*, i.e., as indicators of a unique molecular identity. In those cases involving an analyte comprising a mixture of similar

[4] The J-stop (Jacquinot stop) is the aperture which determines the beam divergence through the interferometer, the size of which is set to the maximum acceptable for the required resolution. (PE manual pp.5-6, 6-8)
[5] The B-stop is located between the source and the interferometer, this defines the area of the beam passing through the interferometer and can act as a variable attenuator. Reducing the size of the B-aperture limits the energy from the source that passes into the interferometer, and thus protects the detector from being overloaded. (PE help series on instrument set-up)

compounds, however, matters are more complex.

In the current project, IR analysis was employed to provide structural information, to indicate the presence of specific functional groups and their location in specific regions of the spectrum, and to demonstrate whether the extracts contained the compounds in sufficient concentrations, which would in turn provide clear, diagnostic peaks. It was hoped, in addition, that the use of IR analysis would determine whether or not the retained samples had suffered deterioration in the course of time, and whether or not the newly obtained extract had survived its exposure to the environment.

6.10 Spectral Interpretation

Although the task of spectral interpretation has been previously covered in detail, a brief reiteration of peak origins may be of assistance in understanding the process of generating the diagnostic regions, and of the data upon which that process depended. The selection of the peaks has been included in the methodology owing to its fundamental influence upon subsequent methodological strategy.

Each spectrum represents the regions in which the functional groups of fatty-acid esters are identified. The recognition of peak-patterns for specific functional groups is made possible by the selectivity of frequencies which may be absorbed by a molecule. Only those frequencies of infrared radiation which correspond to the natural vibrational frequencies of the molecule in question will be absorbed, effectively increasing the amplitude of natural vibrations of the bonds within the molecule. A visible spectrum is thereby generated, which appears on the [6]V.D.U./ computer screen, and a more precise interpretation is facilitated, the peaks having been expanded- and their visibility correspondingly enhanced- by the amplification of the naturally occurring bond vibrations. This permits even the most minute of samples to be detected and specified.

The simple recognition of visually displayed patterns is, therefore, the principal business entailed in FTIR identification; i.e., the observation of the major spectral regions into which the diagnostic bond structures fall. The principal regions selected for diagnostic recognition are those which correspond to the main molecular structures: the ester functional group; bonds indicating unsaturation and the presence of the aliphatic hydrocarbon-chain; and the bond-regions which correspond to the major deterioration-products (i.e., carboxylic acids, aldehyde and ketones).

Making use of these regions enabled the identification of the extracts as esters of fatty-acids, although the individual components of these esters were not themselves specified. The primary aim of the analysis, which consisted in the identification of the fatty-acid esters, was therefore achieved. These esters constituted the regions defined as the principal variables to be utilized in the numerical analysis.

6.11 Constructing a Spectral Library

The first step in the project entailed the construction of a comprehensive spectral library. This comprised the whole of the spectra from both solvent extracts, in addition to the GC-assayed standards: a total of some 400 spectra, each containing 3500 data-points. This was, in numerical terms, a large database, and required a process of refinement in order to separate out the useful information from that which was merely incidental. The initial spectra, accumulated directly from the FTIR analysis, were stored in the form of raw data on both the hard-drive of the computer and on paper. This storage of raw data, undertaken upon the completion of each batch of extracts, insured against accidental loss of data, while at the same time allowing for the first analytic step, which consisted in the visual inspection of each set of spectra to check that every run had produced a viable spectrum. In instances in which the data gathered were insufficient, the result perhaps of sampling or machine error, the sample was run again. If it remained ambiguous, a replacement sample was made. If on the final attempt the spectrum continued to produce small or non-diagnostic peaks, it was considered evident that this was the best result the particular extract was likely to yield.

When carrying out an experiment of this nature, in which the primary aim is the production of quantified data, it is of the utmost importance that all procedures are rigorously standardized. Although the FTIR analysis is fundamentally a qualitative technique, nonetheless a certain level of systematisation can be achieved, allowing for the creation of a calibration set which may be used to test the data. From the outset, therefore, (as mentioned above), it was necessary to treat all of the samples in an identical fashion in order to preclude the influence of extraneous variables. Accordingly, each sample was exposed to the same environment in the laboratory fume-cupboard; each contained an equal quantity of solution pipetted onto the card; and, when ready for analysis, each was placed in the chamber under comparable conditions of nitrogen purge. Consequentially, it was possible to make certain assumptions in regarding the behaviour and relationship of the results when these were ultimately translated into numerical form.

The spectral library is made up of spectra of extracts from modern, identified plant-species, stored within allocated directories which may be accessed for purposes of visual comparison. Such a system forms a resource which may be drawn upon by future researchers wishing to check and compare known samples, and to look for possible matches for new or unidentified species. Although this is a useful method, and is both functional and convenient when small numbers of spectra are to be compared, it was deemed unsuitable for a project such as this one, in which large numbers of unidentified samples required

[6] V.D.U. visual display unit

comparison (in excess of 50). As these would require to be made the subject of comparison with 135 library standards, the application of numerical analyses was considered to be the more feasible methodological strategy.

6.12 Constructing a Numerical Library

In order to analyse the data in a more systematic way, the totality of the spectra were converted to numerical form, each individual data-point having a value of thirteen decimal places. In spite of the fact that these numbers contain very large amounts of information, modern computer software is able to utilize them according to programmed statistical analyses. At this juncture, in addition, the numbers had already been made subject to the initial reductive technique, by means of the setting of the resolution at 4 during analysis. This resulted in the allocation of 1000 data-points per spectrum, by contrast with an original 3,500. Such a large reduction notwithstanding, the spectra still contained a host of inessential information, and required further sifting and filtering in order to sort out the relevant data essential to the final analytic format. This process of refinement required the selection of the principal components, represented by individual diagnostic peaks.

6.13 Data Analysis

The interpretation of raw chemometric data required the standardization of the data-formats. As [7]SCAN and the infrared software were not compatible, an initial conversion into Lotus 1-2-3 was necessary in order to translate IRDM (infrared data manager) formatted data into a form in which it could be converted into Excel, and subsequently into SCAN. The incompatibility of packages meant that extensive works of translation had to be undertaken before the various components could communicate effectively. This translation was, however, successfully achieved.

It was deemed inappropriate to attempt the data analysis for the entire spectrum of each single sample, as this would generate a matrix of immense proportions, much of which would comprise information of no practical utility. The principal peaks provided the variables (or columns) for the matrix, with each sample representing a row, giving a 13x133 matrix for the modern samples and a 13x58 matrix for the archaeological samples.

The data is accumulated according to the following steps:

1: Calculation of the matrix type.
2: Calculation of variability and collinearity, using the highest and lowest eigenvalues to determine the 'condition number'.
3: Determination of the most appropriate regression analysis for the data, based on the degree of collinearity.

6.14 SCAN Software for Chemometric Analysis/Explanation.

Scan is essentially a commercial chemometrics software package, designed for the analysis of large sets of data within Windows. It was, accordingly, pre-eminently suited to the handling of spectral information. Its employment in this project was selected on account of its objectivity in the interpretation of the spectra, which was adjudged to be superior to that entailed in the overlay pattern-recognition technique. The use of SCAN permitted, in addition to the above, the standardization of the pre-processing of the data, in terms of both the modern and the archaeological sets. The actual interpretation of this technique, and the manner in which it aided in the understanding of the data as a whole, will be discussed in the following chapter.

Notes:

Several sections have had explanations added to them in order to clarify various concepts and procedures employed within the work. Although not strictly belonging to the domain of methodology, often these explanations aid in putting the specific techniques into perspective.

[7] SCAN (from minitab) is essentially a commercial chemometrics package designed for the analysis of large data-sets within Windows.

CHAPTER 7

INTERPRETIVE METHODOLOGY

7.0 Introduction

This second methodological chapter comprises a detailed account of the procedures employed within the research, with special reference to the preparation of samples, the acquisition of data, and the ways in which this data was utilized to obtain the final results. A detailed and clearly recorded methodology plays an integral role in the overall project, providing as it does the condition of possibility of accurate interpretation and experimental replicability upon which the validity of the conclusions must in turn depend.

7.1 Interpretation and analysis

It is a defining characteristic of the scientific method that the optimum degree of objectivity is brought to bear upon each and every stage in the process of investigation. Experimental results and the conclusions drawn from them are, nonetheless, liable to the influence of subjective interpretation in a number of ways: the researcher does not, in the course of his or her experimental work, cease to be a human being. This should not be taken to imply any conscious dishonesty or bad faith on the part of the scientist; much more insidious is the natural enough human proclivity to come up with a positive result from one's labours. It requires a constant vigilance to avoid such tendencies. Perhaps the greatest scope for the influence of subjective factors of this order arises in those circumstances where final results remain ambiguous, and consequently allow for the entry into the equation of one's own personal assumptions and preferences. Perhaps the best way to guard against these factors is to record clear and precise results, and to present methodology in such a way as to facilitate the replicability of the work. Clearly, all of these prerequisites are best met if the experimental results can be represented in quantitative terms which allow for the assigning of a fixed meaning by any independent and objective observer. It was this imperative which determined the selection of the chemometric technique, which, as previously discussed, was adjudged to provide a greater objectivity in the identification of samples.

7.2 Interpretation of spectra (qualitative)

Fundamental to the carrying out of this research was the interpretation of infrared spectra, which entailed the recognition of diagnostic peaks representing the principal compounds present in chemical extracts. These representations appeared in the form of graphic images on the visual display unit of the computer; a certain degree of previous experience in the use of FTIR was required for their accurate interpretation, as was the possession of background knowledge of the biochemical composition of the compounds. Peaks which occurred in the graphical representations were thus utilized in the *diagnosis* or identification of functional groups within the compounds, and in order to ascertain their status as targeted extracts, contaminants, or extraneous noise. Experience was, in addition, necessary in order to recognize the presence or otherwise of a *diagnostic spectrum*, a key analytic term whose nature is described below.

There are numerous factors which can influence the constitution of diagnostic spectra, the most significant amongst which derive from the nature of organic chemistry itself. This can affect the areas of diagnostic value by generating spectra which, while belonging to the same chemical species, are possessed of distinct similarities or differences, and may mislead the untrained eye. Prime examples of these factors include *bond absorption*, *bond effect* and *conjugation*, which may produce the 'spectral shift' phenomena discussed in chapter 4.

The initial form of analysis, however, entailed the relatively simple task of pattern-recognition, the essential first step by means of which the quality of the spectra are examined. This qualitative analysis was the prerequisite for any further analyses, and was undertaken in order to ascertain whether the extracts did in fact contain the relevant information.

7.2.1. What constitutes a spectrum?

A spectrum is essentially a series of peaks and troughs produced by specific bonds absorbing infrared energy. The pattern of absorption is recorded and represented by means of a graphic display appearing on the computer V.D.U. This representation takes the form of a graph, with the co-ordinates of %T on the y-axis and cm-1 on the x-axis. The required compounds are esters of fatty acids, and previous experimentation undertaken by Cave (Cave 1992) had demonstrated the optimum parameters within which to analyse these extracts. The most important amongst these parameters consists in the frequency at which the scans are run, which lies in the range of 450-4000cm^{-1}. This range represents the mid-infrared region, i.e., the area of vibrational energy within the electromagnetic spectrum which is employed in the analysis of organic compounds. The resulting spectra are then subjected to visual analysis in order to identify the diagnostic peaks, which are determined according to the biochemistry of the compounds.

7.2.2. Relevance of Diagnostic Peaks

The diagnostic peaks were those which were utilized in detecting the presence of the required substance within the analysed extract. Further peaks may be present in a spectrum which do not derive from the analyte, and may be representative of contaminants, deterioration products, or artifacts of the instrumentation (e.g., noise from the ambient environment of the sample chamber- specifically the airborne impurities carbon-dioxide and water-vapour).

Two basic options presented themselves for utilizing the diagnostic regions within the data-analysis: a) they

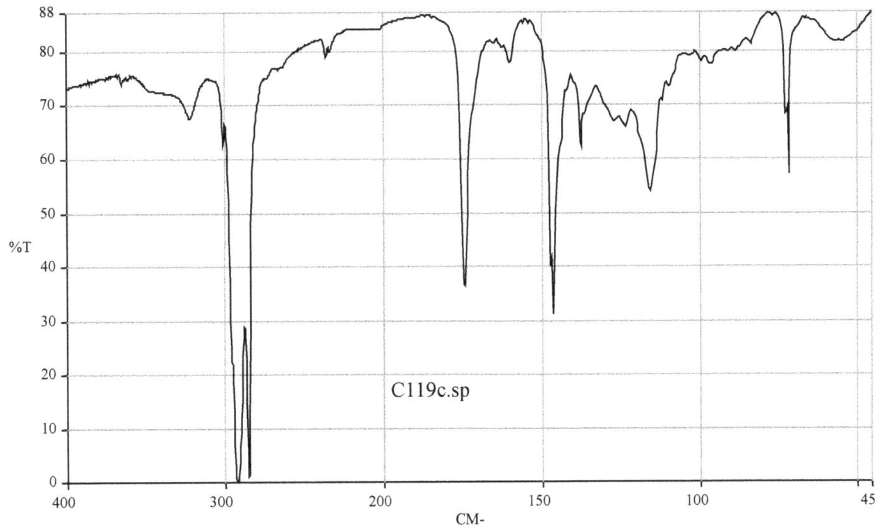

Clearly defined diagnostic peaks.

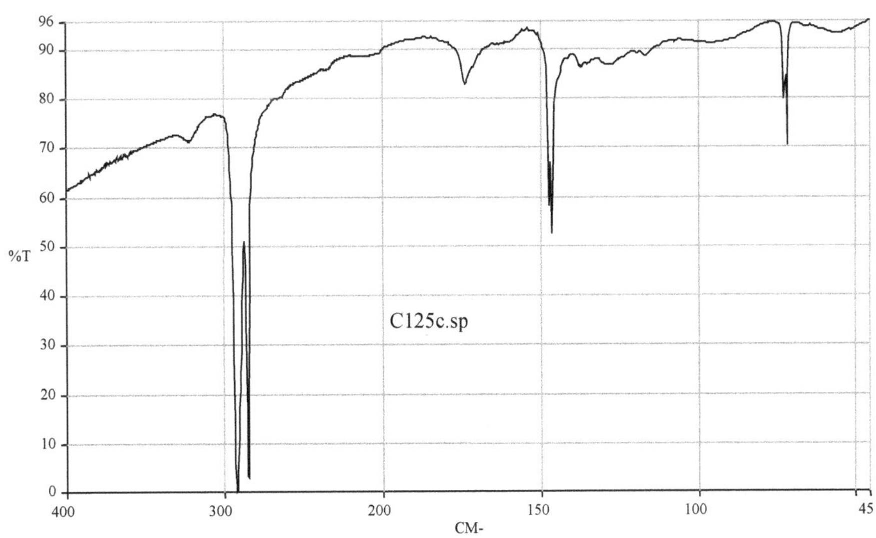

Indicating regions lacking diagnostic peaks.

Fig.7.1 Spectrum with defined diagnostic peaks (top);

Fig.7.2 Spectrum lacking diagnostic peaks (bottom).

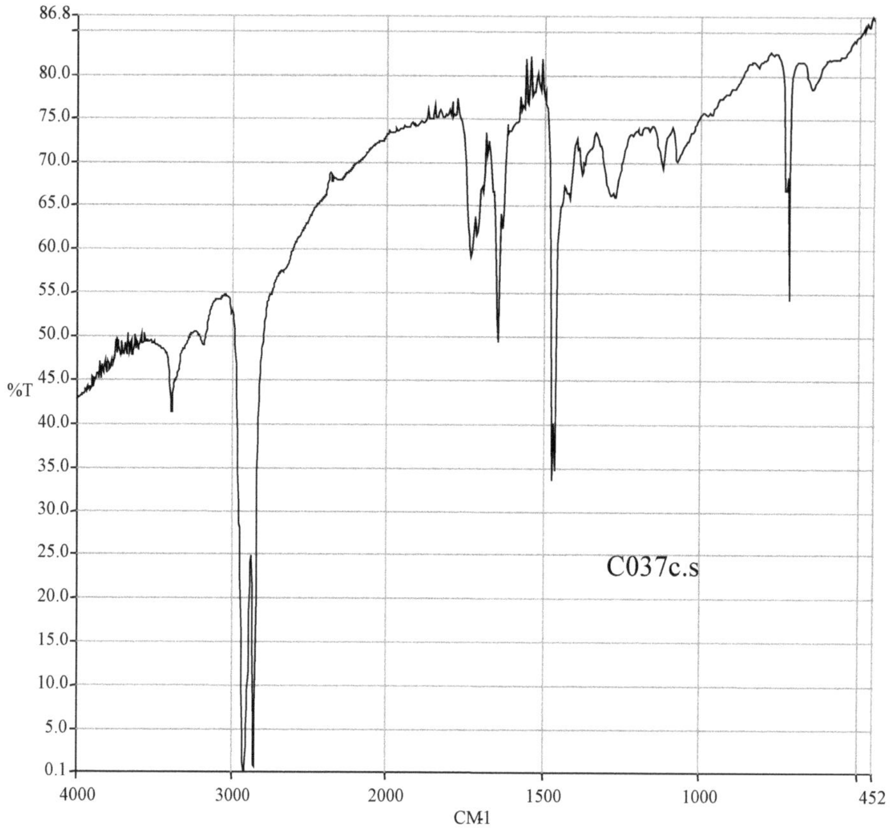

Fig.7.3 C=O peak at 1711cm^{-1} and OH, indicating peak shift due to presence of COOH in extract.

might be separated out from the extraneous peaks and modeled as specific, isolated variables; b) they might remain included within the spectrum as a whole, the totality of which would thereby constitute the manipulated data. After consideration of the two alternatives, it was decided that the latter would result in the generation of extremely large data sets; these may have been too vast for accurate numerical analysis, and produced, as a consequence, systemic errors in the results. It was primarily in order to avoid this possibility that the first option was selected as the most viable, and the numerical analysis was carried out using only the diagnostic peaks. These contained the relevant information for the major characteristics of the esters of fatty acids. It was also necessary to include the major structural combinations of the principal deterioration products, such as carboxylic acid.

7.2.3. Targetted chemical species

In the pre-existing biochemical information, the functional group structure was defined in terms of the infrared vibrations according to the regions of absorption. When the project samples were created by extraction-procedure utilizing the correct solvent regime, each of the specific functional groups or bonds which went to make up the compound were represented by a spectrum in which various peaks defined their structure. Assuming, therefore, that the correct criteria for the selection of chemical species had been employed, these peaks would represent the targeted chemical species. These species could therefore be identified with a high degree of confidence.

7.2.4. Diagnostic Peaks

While a background knowledge of biochemistry provided the necessary information for the selection of diagnostic peaks, it also made it possible to identify those regions whose diagnostic function would be inadequate. This represented a type of judgment which, potentially at least, posed a greater problem, because it is possible for infrared peaks to actually lose their diagnostic information. Such is the case most especially in those instances where there is no other region to support the peak's presence for a particular functional group in the targeted chemical species.

By way of example, consider the carbonyl group: if one peak is present in the 1740cm^{-1} region, and none elsewhere, it is quite unsafe to make any unequivocal assumption that this represents part of an ester of a fatty acid.

A spectrum can be identified as an ester of a fatty acid if it contains the specific peaks illustrated in Fig. 7.1. (Page 57) Here is an example of a modern sample extract, with the diagnostic peaks clearly defined in the transmission mode. By contrast, Fig. 7.2., on the same page, demonstrates the uncertainty of identification which stems from a lack of peaks present within certain regions.

7.2.5. Specified spectra similarities and differences

As discussed above, it is the presence of certain clearly defined peaks which defines the diagnostic value of spectra. However, the specific *location* of peaks may also perform this function. This fact derives from chemical effects arising within the structure of a compound, where specific bond types that are adjacent may affect the region to which they are assigned. For example, if a double bond is alternated by a single bond, the phenomenon of conjugation occurs, whereby a lower frequency is assigned to the diagnostic peak. This process does not modify the ester structure, only the location of one of the major identifying features.

Again, these considerations demonstrate the advantage conferred by a prior knowledge of chemistry; in the absence of such knowledge, it is doubtful whether a firm identification can be achieved. These factors also point to the underlying need for the re-running of samples in order to ensure experimental validity.

The phenomenon of spectral shift is evident in cases where the sample has deteriorated into carboxylic acid: the major diagnostic peaks will now be in the region of $1710cm^{-1}$, indicating the presence of a carbonyl bond. This, however, is accompanied by an OH bond in the $3000cm^{-1}$ region, which suggests that this shift is due to deterioration rather than bond-effect. (Fig.7.3.)

7.3 Data analysis

An essential prerequisite for carrying out a reliable data analysis was the establishment of a series of internal standards to which the unknown sample extracts could be compared and contrasted chemically. The removal of a compound from its natural matrix (in this case, the seed) may result in the modification of its spectral signature. Against such a background, it is imperative that a range of anticipated frequencies be devised, a range which is able to function as a benchmark. Such a standardized set may be derived from pure assays, and is employed in the determination of diagnostic regions for both the knowns and the archaeological samples.

7.3.1 Internal standards

These standards possessed a supplementary function, aiding in the determination of the degree of sensitivity and precision of the FTIR in terms of ester identification. To this end, the structural information supplied by the peaks was utilized. The set took the form of a graded series of dilutions of the esters of three major fatty- acids found in plant extracts, and ranged from 100% pure to minute, trace amounts. They were held in the same solvents which had been employed in the soxhlet extraction procedure.

The three pure extracts selected comprised the esters of palmitic, linoleic and oleic acids. These were chosen as substances most likely to be representative of the entire spread of acids found in this particular plant seed material. (Gailiard & Mercer 1975)

The standards were made up with these extracts into the two solvent groups according to the following proportions:

1. A weighed quantity (1 ml.) of each 100% was added to 10mls of each of the solvents, hexane and chloroform, thus forming the 1-10 dilution.
2. 1ml. of this solution was then taken and added to a further 1ml. of solvent, producing the 1-20 dilution.
3. This procedure was continued until all the required dilutions had been prepared, up to and including a 1-160 dilution. This was an extremely weak solution, whose function would be to test the absolute sensitivity of the FTIR.

7.3.2 Calibration

These dilutions formed a part of the calibration data set, by means of which the operational parameters of the experiment were defined. They represented a range of concentrations which approximated, as near as was practicable, to that of the experimental data, within which the known samples were included. The dilutions demonstrated, in addition, the relationship between the instrument response and the levels of analyte concentration. In that relationship, the former term initially rises in increments relative to the increase in the latter. There is, however, a certain critical point in the level of concentration, beyond which a quite dramatic change takes place in the relationship: the instrument response begins to take on an almost exponential aspect, with the measurements showing a sharp upturn and, apparently, ceasing to follow the single incremental trend observed in the earlier stage. The observed phenomena are demonstrated in graphical form in the calibration-graph included in chapter 8. (see fig. 8.1 pg. 65)

7.3.3 Problems with calibration, the linear relationship and instrumental analysis

Having completed the chemical components of the calibration process, it was necessary to undertake an FTIR analysis of the extracts. To exclude the action of extraneous variables, each extract required analysing under identical conditions. It may be observed from the printed results that the ratio of concentration to instrument-response appeared to constitute a relation of linearity. The graphical peaks were translated into numerical terms and plotted as a calibration-graph. (See Chapter 8)

The semi-qualitative character of infrared analysis has been elaborated upon in previous chapters. In view of this partially quantitative status, it is perhaps not surprising that the use of calibration and techniques of numerical data-transformation represents a relatively unfamiliar application of FTIR. The most frequently employed form of IR interpretation is probably that which involves the

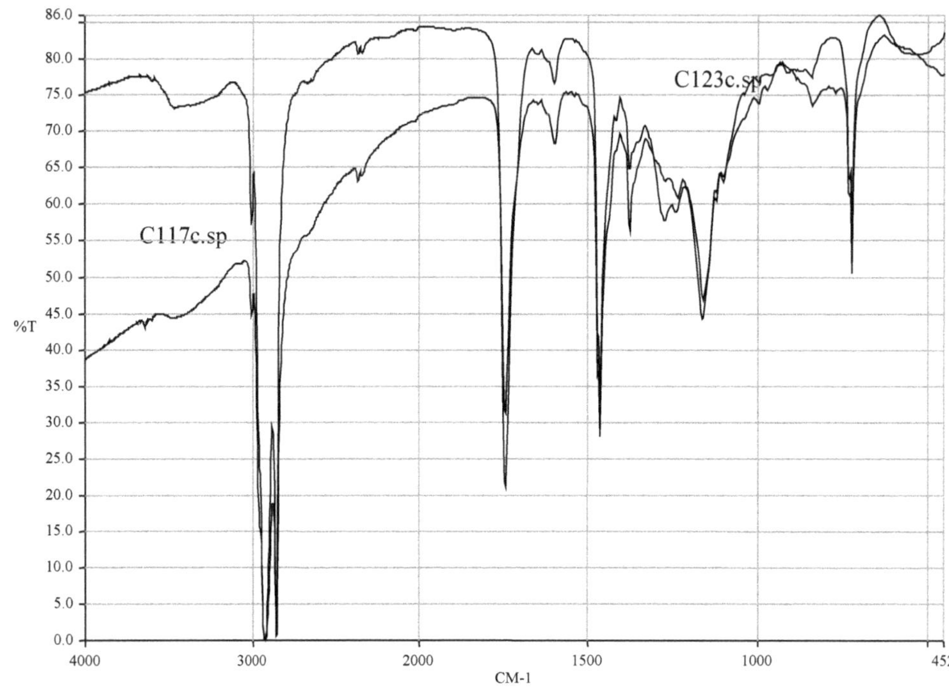

Two spectra showing masking of diagnostic peaks

Fig. 7.4 Spectra overlay for peak comparison.

making of visual comparisons between sets of peaks and regions, achieved by the physical overlaying and superimposition of one spectrum upon another.

While the visual overlay technique has proved itself to be of considerable utility, its reliability and validity are attenuated in certain circumstances; specifically, in cases where mixtures of unknown compounds are present, such as in the current project. The collated data remain, however, highly valuable when the method is combined with other, complementary techniques.

The selection of thirteen 'channels' was not made arbitrarily; these specific peaks contain all of the relevant data needed to numerically analyse chemical extracts of this nature. Even though the spectra themselves, in the raw form, contain around 3,500 data points, much of this information is not required. When a PCA was performed on the raw data, the significant information derived was that contained in the thirteen. When this data was subjected to further clustering, many single outliers and few significant groupings were found in a position beyond these thirteen.

(The actual library of raw data is available on the computer in the Wolfston Laboratory at the Institute of Archaeology. It was not possible to include this spectral information here, as it is an enormous data-base and would require many floppy discs to store it (there was no CD-write available), if they are to be kept with the thesis.)

Contemporary FTIR systems, such as the Perkin Elmer 2000 series used in conjunction with interactive software, have facilitated the semi-quantification of the spectral data. In this research, both the calibration data and the samples known and unknown, were converted into this semi-quantitative format. The process permitted numerical comparisons to be made, which were more refined and closer to the less 'visible'[8] data, by contrast with the visual overlay technique. As a consequence, it was able to demonstrate those flaws which are inherent in that technique, flaws which derive directly from its dependence on raw human observation, which is often unable to detect minute peaks which may be present in spectra. Only those data that are visible to the human eye are captured.

The axes on the spectrum, and indeed all of the spectra in visual format are [%T] percentage Transmission on the vertical axis, which was the mode the FTIR machine was set at for scanning. On the horizontal axis, the measurement unit was [cm^{-1}] reciprocal centimetres, i.e. the measurement between data points. The FTIR can also be set in absorption mode [%A]; all spectra can be converted to either mode via the software. This conversion is necessary when the spectral data is being used in numerical analysis, as was the case with the data utilized in this project; all data were changed to %A for the ANOVA and the chemometric analysis.

7.4 Calibration graphs and instrument analysis

The construction of the calibration graph proceeded according to complex and rigorous criteria. The optimum

[8] When the term 'visible' is used to describe IR data, it denotes a spectrum displayed on a computer screen (and not, of course, the IR energy itself!)

reduction of error was imperative at this juncture, as the graph formed the standard against which all subsequently gathered information would be compared. (Table 8.1 chapter. 8 pg. 66)

The problems and considerations itemised below required attention if the calibration graph was to be utilized in a reliable manner.

1. Is the graph linear or curved?
2. Due to the fact that anyone point on this graph is subject to error, the drawn line has to be the best fit.
3. If the plot is linear, then what are the estimated errors and confidence limits for the slope and intercept of the line? As demonstrated by the ANOVA for the calibration graph where the slope (b) and intercept (a) are represented by the following:

$$\frac{\sum_i \{(x_i - \bar{x})(y_i - \bar{y})\}}{\sum_i (x_i - \bar{x})^2} \qquad a = \bar{y} - b\bar{x}$$

4. When the plots are used for the test samples what are the errors and confidence limits for specified concentration.
5. What is the least concentration of the analyte that can be detected with a predetermined level of confidence?

The calibration of an experiment of this nature is a process which demands a constant vigilance in order to avoid reliance upon certain unsafe assumptions, of which the two major examples would be: 1. That errors in calibration had occurred solely in the y values, i.e., in the instrumentation. 2. That the magnitude of errors in y was independent of the analyte concentration.

These premises are, in fact, incorrect, as relative errors in measurement are constant, and absolute errors will increase as the analyte-concentration increases. These facts are clearly demonstrated in the calibration-graph, where manual calculation was employed.

7.5 Chemometrics and data analysis

The complexity of spectral data, in addition to the sheer volume of spectra generated, meant that the familiar method of overlay would not, on its own, prove adequate to the demands placed upon it by this project.

The comparison of so many variables, drawn from such a range and multitude of individual samples, would place an untenable strain upon the resources of visual interpretation. This strain would not be confined to problems of precision and definition stemming from fatigue, but might also tend towards the failure of objectivity, and the perception of the objects of one's desire where no such objects in reality exist.

The aim was, therefore, to devise a procedure which possessed the ability to produce results of enhanced accuracy; a procedure which was not intended to replace established methods or to render the observer redundant, but rather to supplement them by expanding the yield of interpretable information. The greater the data, the better the chance of an interpretation of it that is both accurate and productive.

7.5.1 Computer analysis

To reiterate: the choice of computer analysis rested upon its ability to handle such a large accumulation of data, an ability superior to that of the human eye. The SCAN software is able to both treat the data highly accurately and to reproduce it in the form of clearly definable graphics and numerical tables. In essence, it can present the data in a format that is both manageable and attractive, thereby allowing for a more accurate type of visual analysis and comparison.

7.5.2 Data analysis (quantitative)

The primary data on which the chemometric analysis was carried out was qualitative in nature; once the calibration system was applied, however, the information was rendered in semi-quantitative terms. This quantification of the data defined the parameters within which the experiment operated, as the calibration treats pre-defined information regarding the chemistry of the material under analysis. As a consequence of the procedure it was possible to utilize a higher form of statistical treatment, the application of which enhances, in its turn, the interpretation of the chemical data.

7.5.3 Regression analysis

SCAN utilizes the principal components of the raw data entered into it. It does not alter the original calculations; rather, it modifies the form in which they are presented, with the consequence that it becomes possible to employ a much wider range of analyses. These include **Exploratory data analysis**, an example of which is the principal components analysis, detailed in chapter 5. PCA employs specific algorithms according to the data under investigation (e.g. NIPALS and SVD, which may be supplemented by matrix condition, factor analysis and hierarchical clustering.)

These tools are able to generate highly descriptive results, and may be sufficient to supply the required information. If further analysis is desired, however, the next step is that of **classification**. This is a discriminant analysis, and includes 'soft independent modelling of class analogy' (SIMCA).

Following this, we reach the higher statistical treatments such as **regression (calibration) analysis**. Regression analysis is the final, most illuminating and information-rich form of statistical interpretation. It incorporates ordinary least squares (OLS); partial least squares (PLS); nonlinear partial least squares (NPLS); and the nonlinear, nonparametric alternating conditional expectations (ACE). These techniques are discussed fully in both the chemometrics and the results chapters, and also the **Appendix pt. II**, and need no further elaboration here.

7.6 Numerical analytical series

Once all of the data had been exported to the SCAN system, the total analytical series could be carried out. This entailed firstly the chemometric analysis of the spectral data; and secondly, and of equal importance, the exploration of the first set of results. This initial step involved the conversion of the spreadsheet into matrix format, which was a simple matter of utilizing the default settings in the pre-processing section of SCAN. In order to render them fully commensurable, all matrices were converted to covariance with autoscaled data, (a procedure which continued the standardization of the experimental data).

The second step, vital in ensuring the selection of the appropriate type of regression analysis, consisted in the examination of all matrices in terms of the matrix condition. This quantifies collinearity amongst the variables, where all x-variables are intercollated (approximately linearly dependent); $x1 = f(x2, x3,..,xn)$.

Where a high collinearity obtains between x1 and x2, a greater degree of variability is found along the solid line in the x1 and x2 plane than on the lateral plane. This latter direction, with only a small variation in x, then creates problems in the estimation of the variability of y. This is especially the case where the direction of x is important for the prediction of y; and, in such circumstances, the collinearity of the data can present a number of difficulties. Fortunately, it does not generally apply when dealing with spectroscopic data. The cardinal rule in these matters is to select the appropriate type of statistical tool with which to treat the specific data in hand; the choice of techniques able to treat highly collinear data will guard against the introduction of error.

As spectroscopic data is usually characterized by a high degree of collinearity, it was essential to subject it to rigorous testing. In order to assess the quantity of collinearity, the ratio of the highest to the lowest eigenvalue was calculated. This calculation yielded a figure of 1000 or above, indicating that the data were highly collinear, and it was therefore necessary to employ regression analysis, a treatment which has the ability to deal specifically with collinearity, and to produce significant results.

This preprocessing of the data represented a refinement that would ensure that the subsequent analytical procedures could be carried out with a high degree of confidence. The first such technique consisted of a treatment known as Hierarchical Cluster Analysis, an exploratory method which assists in the identification of clusters of objects (represented by the rows of the matrix) in dimensional space, based on inter-object distances. (See chapter 5). This system effectively produced a hierarchy of partitions of objects, represented graphically in the form of a dendrogramme.

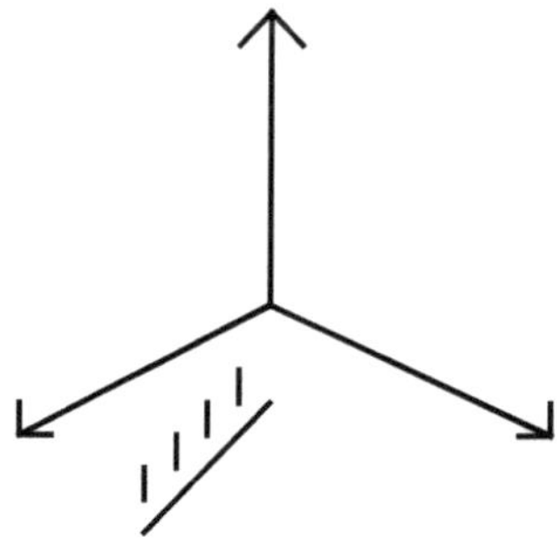

Fig.7.5 Collinearity: reproduced from pg.2 Multivariate Calibration in Practice. Suzanne Schonkopf, Dominique Guyot 1995

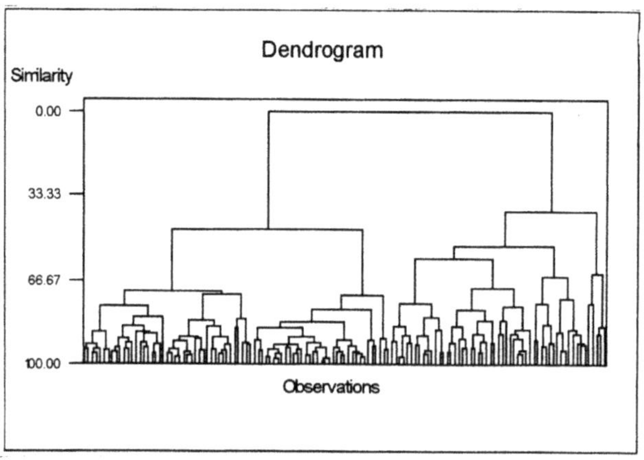

Fig.7.6 A Dendrogramme as calculated from the HCA matrix

The second technique entailed another form of cluster analysis, which was, once again, performed in order to test the behaviour of the given data. Going by the name of the Jarvis-Patrick technique, this was a non-hierarchical cluster analysis, the function of which is to reveal natural groupings of objects in high dimensional measurement space. The technique deals exclusively with large data sets, and calculates the relationship using Euclidean distance, which is expressed by the following formula:

$$d_{ik} = \sqrt{\sum_j \left[x_{ij} - x_{kj}\right]^2}$$

Fig. 7.7 (Miller & Miller 1993: Perkin Elmer manual)

Having carried out these preliminary steps of data exploration, the serial analysis proceeded to its ensuing phase, in which Principal Components Analysis (PCA) and the closely related Factor Analysis (FA) were undertaken. This stage involved the selection of the major components of the data, those that contributed to the bulk of the final results. PCA utilized the two algorithms entitled 'Singular Value Decomposition' (SVD) and 'Nonlinear Iterative Partial Least Squares' (NIPALS), which were subjected to detailed discussion in Chapter 5, in order to calculate the eigenvectors and the principal component scores of the matrices. Both were employed in the investigation of all aspects of the data, and in the

C1	C2	C3	C4	C5	C6	C7	C8	C9	C10
Sample	3400	2960	2872	1736	1712	1640	1542	1376	1252
C001	-0.29751	0.09535	0.00512	0.77153	0.98791	-0.46778	0.28838	0.55510	0.65076
C002	-0.56369	-1.4462	0.20030	0.12110	-0.04656	-0.80232	0.03079	0.15313	0.16362

Table 7.1 An example of the autoscaled matrix

determination of the most effective formula, which was achieved by the use of the diagnostic peaks from spectra as the predictor variables, and the sample constituents as the observations. In the tables below, **Table 7.1**, the former are represented by the columns, the latter by the rows.

The subsequent stage of the numerical analysis involved the use of factor analysis. The application of this technique allowed the extraction of the underlying factors, and performed orthogonal rotation in order to render the factors more interpretable. A two-step procedure was utilized, consisting of: 1) Factor extraction, and 2) Factor rotation. The factors were calculated by means of Principal Component Analysis, which modelled the set of variables, (as described above), in the form of a linear combination of a small number of common factors. By this method, an exact or highly proximate description of the observed or measured quantities was obtained, assuming that these observed common factors represented the underlying, but not directly measurable, phenomena.

Having established distance-relationship and grouping, the data may be classified. Classification consisted of a discriminant analysis, in which objects were assigned to one of several existing classes according to the classification rule of supervised pattern recognition. The intention here was to calculate class models and class boundaries, and to define a classification rule based upon a training set derived from objects of known class; this rule would then be applied to the test set containing objects whose class was undetermined.

The procedure consists, essentially, in a collection of parametric classification methods which model each class by its centroid and its covariance matrix, and then assign an object to its class of nearest fit. While classification may, under certain conditions, represent a tool of high diagnostic utility, the responses generated were in this instance of low significance, in regard to both the archaeological and the modern samples. This technique was, consequently, omitted from the analytical series. It was decided that its inclusion would represent analysis for its own sake, producing copious amounts of low-value information that would clutter and obscure the experiment.

In order to obtain the optimum method of numerical analysis and obtain the requisite level of refinement, it was necessary to run a series of techniques which ranged from the coarse groupings of cluster analysis to the precise goodness-of-fit and prediction of principal components regression. Regression entails a collection of statistical treatments which model the relationship among measured and observed quantities, aiming ultimately at two major procedures: modelling and prediction. The technique describes the predictor-response relationship in algebraic form, and calculates unknown responses. Each step represents a further refinement of the data: enhancing the results and yielding additional information, it indicates the subsequent step to be taken in order to arrive at the ultimate experimental destination.

Each statistical technique was carried out using the same matrix format, and the results tabulated and recorded in graphical form in order to present every stage as a discrete, self-contained component within the totality. As a consequence, the printed output was highly descriptive, but somewhat bulky; and it was decided, in order to avoid overwhelming the main text with an unwieldy mass of raw data, to present the entirety of the data in the form of an appendix, which may be read in conjunction with the text, or consulted as necessary. **(See Appendix part II.)**

These statistical treatments were utilized as tools in a manner analogous to the various techniques employed in the course of the wet chemistry. In the following chapter, the application of this cross-disciplinary methodology is elaborated. The methodological strategy employed was to detail each facet in turn, and then to draw the various strands together in order to demonstrate the comprehensive realization of the experimental aim: the application of an innovative multidisciplinary methodology to provide positive and objective responses to a set of problems posed from within archaeology.

CHAPTER 8

FINAL RESULTS

8.0 Introduction

In this, the culminating phase of the project, each of the constituent methodological strands was drawn together in order to assess their efficacy in terms of realizing the aims and core objectives of the research. In a multidisciplinary programme of this type, each component must be considered in its relation to the whole. The primary aim here was to explore the use of chemical criteria in the identification of charred remains of small-grained grasses retrieved from archaeological sites. The results would, therefore, provide the basis for a decision as to the general viability of employing chemical criteria in such a context, and of its merits relative to those of the more orthodox techniques. They would, furthermore, constitute a means of assessing the effectiveness of specific research tools utilized in this project, notably the chemometric method upon which heavy reliance was placed throughout.

8.1 Complete Results

Upon the project's completion, the results were assessed in order to determine its validity: i.e., whether or not the experiment had in fact provided answers to the questions posed, and in order to evaluate the contribution of the individual components to the overall design. The interrogation of these individual experimental components proceeded as follows:

1. **Processing of botanical samples.** Were the modern seeds adequately prepared? Does removing the lemma and palea prior to extraction allow them to be employed as templates for the ancient materials? Did the arbitrary groupings for the ancient samples assist in the final identification?
2. **Wet chemistry.** Did the actual extraction procedures for the seeds (both ancient and modern) provide quantities sufficient to allow for further analysis?
3. **Instrumental analysis**. Did the technique selected supply the results required for diagnostic interpretation?
4. **Numerical analysis**. Was chemometric analysis appropriate for the type of data generated in this project?

8.1.1 Botanical Materials.

The identity of each of the contemporary samples was checked carefully by Hillman, and subsequently placed under a binocular microscope in order to ensure that methodical cleaning had removed all outer covering. This made certain that only the tissues of the archaeological specimens remained, allowing these clean modern seeds to be employed as templates. This tentative identification enabled the inclusion of the optimum number of species of each genus found in the designated geographical area.

8.1.2 Chemistry

The chosen solvent regime, in combination with a three-day extraction-period, proved adequate in providing sufficient quantities of sample for further analysis, the modern samples in general generating spectra with relevant diagnostic peaks. The actual range of concentrations was shown by randomly selected subset of twenty samples, which were dried and weighed to ascertain the quantity extracted. These were then reconstituted with their extraction solvent (0.5), and analysed with the FTIR in order to determine the diagnostic qualities of the resultant scans. This process served the additional purpose of testing the suitability of the instrumental analyses: if the machine possessed the ability to recognize such tiny quantities, and produce a viable result, the selection of analytical procedure (both extraction and instrumental) would be justified. In the case of the archaeological samples, some were unable to generate diagnostic peaks regardless of the conditions under which they were obtained. As the extraction and analytic format appeared to be optimal for all the other specimens, this implied that additional variables were involved, probably stemming from poor preservation. Overall, the fact that any of these extremely ancient materials yielded any organic extracts was in any case remarkable, having an average age of 10,000 years and a prolonged exposure to the environment in which they were buried.

8.1.3. Instrumental Analysis.

As it is capable of producing precisely the type of results required, the technique of Fourier Transform Infrared was central to the project's experimental design. Indeed, the application of this Fourier Transform method represented an important aspect of the research, and was attempted in order to ascertain whether its semi-quantitative procedure might provide an advance on a purely morphological identification of seed-remains.

The positive character of the results obtained indicates strongly that the requisite information can be accessed via the implementation of this technique, and that the core objectives of the research would be reached. However, before elaborating fully upon these conclusions and considering the key results in terms of groupings and identifications, it is necessary to discuss the overall effects of applying numerical analyses and of calibrating raw spectral data.

The graphs and tables of the results are contained **in part II of the Appendix starting at page 105**; each section has a heading according to the specific analysis used. This section starts with the very first, and most basic numerical procedure, the matrix condition, progressing to the final and more precise regression analysis that consists of several individual components. The spectra which were used as the initial raw data are also contained in the appendix, these are in part III, starting at page 110 for the modern spectra, and page 196 for the archaeological.

8.1.4 The results of numerical analysis

The positive results generated by the construction of a

numerical model, and its capacity to provide an accurate identification of the ancient specimen, indicated that chemometric analysis was indeed an appropriate format with which to treat the data. Although there remains scope for further modification and improvement in the design, the underlying theory and concepts are able to supply results which can be interpreted with a high degree of confidence. As with all forms of numerical analysis, however, it is essential to carry out a series of simple tests before committing oneself to any given technique. These tests include calibration, optimisation and validation, and are undertaken in order to assess the accuracy of the methods selected for the principal series. The results of steps are outlined below.

8.2 Results of calibration

To reiterate: one of the project's major intentions was to construct, via the application of a series of numerical techniques, a model that would function as a template for the archaeological samples. Although the results appear quite straightforward once they have been obtained, the route taken sometimes appeared convoluted; each step was, however, a necessary component in the construction of a good, reliable set which was simple to interpret and to replicate.

The initial results were gleaned from the set of GC assayed standards, to which reference has been made above; this entailed the use of ANOVA and a polynomial quadratic equation in order to calculate instrument response against analyte concentration. The procedure generated a slightly curved graph, a characteristic response that became more evident as the concentration increased. nonetheless; as the data was not totally skewed it was still possible to proceed with the preparation for the regression analysis. This preparation involved principally the observation of differences between a plot best fitted by a straight line and one best fitted by a curve: that is, a line of best fit was required, which was then drawn across the diagonal. This process demonstrated clearly the respective linearity/non-linearity of the points. As may be observed in **Fig. 8.1**, the curvature occurs over only a part of the plot, and creates, therefore, relatively few problems for the calibration and attainment of high r values.

The r value represents the relationship between the analyte and the instrument response; where this is calculated with a high value, the relationship is good and likely to be linear, or a close approximation thereto. This signifies that the instrument is responding in a positive way according to the concentration of the sample under analysis. As a consequence, the system of analysis may be considered valid and useful in the generation of the requisite order of result.

Although the curvature of the line was slight, it remained significant insofar as this data was utilized as one of the calibration sets, and a method of testing its effect was therefore necessary. Given the parameters, it is possible to apply a set of tests for curvature which are fairly routine and entail no extensive procedures of calculation. The tests in question are based upon the use of the *y* residuals on the calibration plot, and employ in the first instance unweighted linear regression calibration to all of the data, (slope, intercept, etc.).

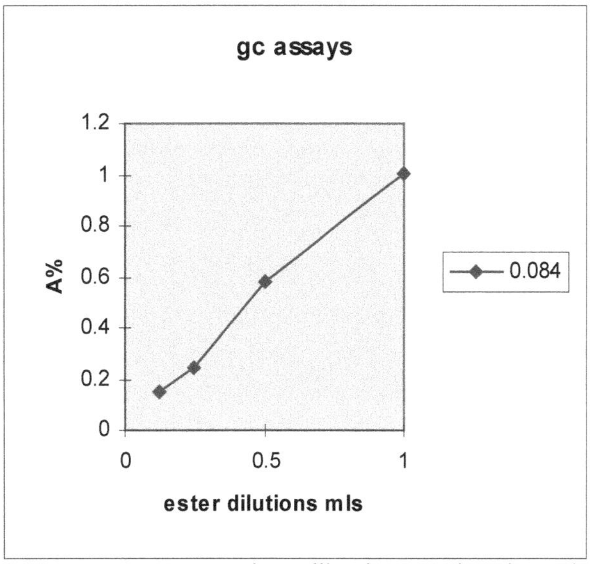

Figure 8.1 represents the calibration graph, where the main points have been plotted, through which a line of best fit has been drawn. The x-axis represents the dilution of the esters in chloroform, and the y-axis represents the instrument response in absorption. **(see Notes pg. 216 Appendix part III)**

As applied to straight line plotting, least squares criteria can be used in order to decide the quality of fit, while aiming to minimize the sum of the squares *y* residuals. This procedure enables the utilization of ANOVA calculations to evaluate the quality of successive polynomials where the assumption that only y direction random errors occur.

For example if the calibration points have values y_1, y_2...yi...y_n and mean value $\bar{y}$, then $\sum[y_i - \bar{y}]^2$ is the sum of squares about $\bar{y}$ (Miller and Miller: 133). This can be shown as equal to **1:** the sum of squares due to regression $\sum[\hat{y}_i - \bar{y}]^2$ and **2:** the sum of squares about regression $\left(\sum[y_i - \hat{y}]\right)^2$. The "sum of squares due to regression" is that part of $\sum[y_i - \bar{y}]^2$ which is due to the successful fitting of an equation to the data and should, as a consequence, be as large as possible, whereas the "sum of squares, about regression" is equal to the sum of squares of the *y* residuals, which should be as small as possible if the curve is to be a good fit.

R^2- the coefficient of determination or multiple correlation coefficient. **(Fig. 8.2):**

$$R^2 \frac{\text{SS due to regression}}{\text{Total SS}} = 1 = \frac{\text{Residual SS}}{\text{Total SS}}$$

Fig.8.2

The analysis of variance (ANOVA) calculations provided for curve fitting, by dividing the total sum of squares of y_i about $\bar{y}$ into two, whose individual contributions to the total will vary according to the goodness of fit, given by $\mathbf{R^2}$. From this it appears that all that is required is to calculate $\mathbf{R^2}$ values until $\mathbf{R^2}$ no longer increases. However when another term is added to the polynomial, $\mathbf{R^2}$ will increase. The chemometrics programme offers a further or adjusted statistic:

$$R'^2 = 1 - \frac{ResidualMS}{TotalMS}$$

Fig. 8.3

This accounts for the number of residual degrees of freedom in the polynomial regression that changes as the order of the polynomial changes. However R'^2 is always smaller than $\mathbf{R^2}$. Polynomial regression is given by [n-k-11 where k is the number of terms in the regression equation containing a function of x, represented by the number of degrees of freedom.

The above described method of analysis of variance was used on the GC assayed samples in order to test the linearity, while making it possible to extrapolate the way the bulk of the samples behave under the same conditions. This was essential prior to totally committing all of the data to numerical analysis, due to the complexity and time constraints related to the project as a whole.

8.2.1 ANOVA on GC Assays

All of the spectra are analysed in the transmission mode. However, to carry out any type of numerical analysis, such as the ANOVA, the numbers have to be converted to percentage absorption (A%). The following example demonstrates the use of these converted spectral data in the calibration calculation, the primary numerical working as standard for any type of chemical analysis, if quantification is to be attempted.

Table 8.1 ANOVA (Pt. 1)

x_i	y_i	$x_i - \bar{x}$	$(x_i - \bar{x})^2$	$y_i - \bar{y}$	$(y_i - \bar{y})^2$	$(x_i - \bar{x})(y_i - \bar{y})$
.0625	**0.084**	**-0.324**	**0.104**	**-0.327**	**0.106**	**0.105**
.125	**0.152**	**-0.262**	**0.068**	**-0.259**	**0.067**	**0.067**
.25	**0.240**	**-0.137**	**0.018**	**-0.171**	**0.029**	**0.023**
.5	**0.577**	**0.113**	**0.012**	**0.166**	**0.027**	**0.003**
1.0	**1.003**	**0.613**	**0.375**	**0.592**	**0.350**	**0.214**
$\sum = 1.9375$	$\sum = 2.056$	$\sum = 0$	$\sum = 0.577$	$\sum = 0$	$\sum = 0.579$	$\sum = 0.412$

$\bar{x}$ **= 1.9375/5 = 0.387**

$\bar{y}$ **= 2.056/5 = 0.411**

r = 0.412/(0.577x.579)$^{1/2}$
= 0.412/.577
= 0.714

b = 0.412/0.577 = 0.714
a = 0.411-(.714x.387) = 0.411-0.276 = 0.135

Table 8.2 ANOVA (Pt.2)

$\hat{y}$	y **- residual**	y **- residual**^{2}	$[\hat{y}_i - \hat{y}]^2$
.135	**-0.051**	**.002**	**0.076**
.849	**-0.697**	**.485**	**0.191**
1.563	**-1.323**	**1.750**	**1.327**
2.277	**-1.7**	**2.89**	**3.481**
2.991	**-1.988**	**3.952**	**6.656**
		9.079	**11.73**

$$\mathbf{R^2} = \frac{ss - due - to - regression}{total - ss}$$

$$\mathbf{R^2} = \frac{\text{Re}\,sidual - ss}{total - ss}$$

$$\therefore \mathbf{R^2} = \frac{11.73}{20.81} = \mathbf{0.563}$$ **= 56% relationship between x & y**

$$\mathbf{R'^2} = \mathbf{1 -} \frac{\text{Re}\,sidual - MS}{total - MS}$$

$$\mathbf{R'^2 = (1-[2.26/4.162]) = 0.456}$$

Fig.8.4

Table 8.3 ANOVA (Pt.3)

	SS	df	MS
Regression	11.73	1	11.73
Residual	9.079	4	2.26
Total	20.81	5	4.162

Sum of square $\sum\left[y_i - \hat{y}\right]^2 = 0.579$

Sum of squares due to regression $\left(\sum\left[\hat{y}_i - \bar{y}\right]^2\right) = 11.73$

Sum of squares about regression $\sum\left[y_i - \hat{y}\right]^2 = 9.079$

Fig.8.5

$$r = \frac{\sum_i\{(x_i - \bar{x})(y_i - \bar{y})\}}{\left\{\left[\sum_i (x_i - \bar{x})^2\right]\left[\sum_i (y_i - \bar{y})^2\right]\right\}^{1/2}} \qquad b = \frac{\left\{\left(x_i - \bar{x}\right)\left(y_i - \bar{y}\right)\right\}}{\sum_i\left(x_i - \bar{x}\right)^2}$$

slope

$$a = \bar{y} - b\bar{x}$$

intercept

Fig 8.6

The outcome of the ANOVA, which included the residuals to look at curvature, was a poor one, this indicated that its effects on the overall results are negligible, if at all significant. It also indicates that further numerical analysis can be carried out with a high degree of confidence, particularly the regression analysis, as the data are showing a positive relationship between sample concentration and instrument response.

The results of these calibration calculations are important in order to demonstrate the relationship between the instrument and the analyte concentration. This is fundamental to the whole analytical procedure; it is essential to show that the instrument is performing in accordance with the sample being analysed, and the specifications required for that compound. It enables the assumption that the analytical system is a totally standardized scheme, given the constraints of the instrument being utilized. Consequently a sample can be analysed and thence identified with confidence, as the worker knows that the instrument is displaying the results clearly and precisely and will do so for the most dilute of solutions, as specified by the calibration data. The analytical procedure required primarily looking for the r-value utilizing the formulae in (**Fig. 8.6**), followed by the slope and intercept previously cited in (chap. 7: 7.4), but are restated here as they are necessary when demonstrating the route via which this information was obtained.

The r-value is calculated at 0.71, which implies a 71 % relationship between the x and y variables, and the slope (b) and intercept (a) respectively are 0.714 and 0.135. The slope is, by definition, the steepness of the line, while the intercept is where the y-co-ordinate crosses the y-axis, which in practice helps solve problems regarding the identification of the unknown samples, since where absorbance is known for a band or sample, concentration can be calculated from the calibration line, or where concentration is known absorbance can be calculated, as this defines whether a peak is diagnostic or not. This can be demonstrated by using an ideal situation, where the component of interest is represented by a single absorbance band, for which the concentration is known, with no noise or any other extraneous interferences. **Figures 8.7** and **8.8** diagrammatically represent the ideal situation, where **Fig.8.7** demonstrates how a calibration graph is used, from which concentration, slope and intercept are calculated. Where light and absorption are part of this calibration equation, it can be expressed quantitatively by the relationship known as Beer's Law **(Appendix part III pg.216). Fig. 8.8** represents a single absorbance band of the carbonyl (C=O) peak from the ester functional group of the triacylglycerol, containing three concentrations at around 1740cm^{-1}, with the increased absorbance measurement as the concentration increased. Both of these illustrate the relationship between concentration and instrument response, and shows how the ability to identify diagnostic peaks and hence samples relies on this relationship. By showing that a concentration increase initiates a complementary instrument response, means that all the information generated is integral, such that a peak and/or total spectrum can be a used as a chemical map of the known and unknown samples.

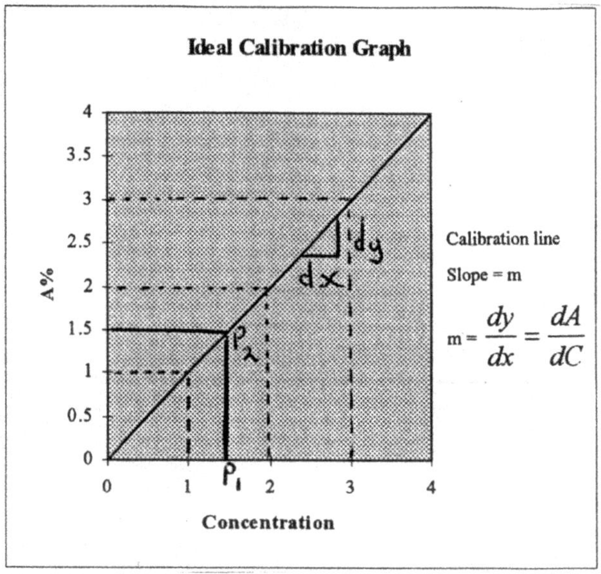

Fig. 8.7 Ideal Calibration Graph

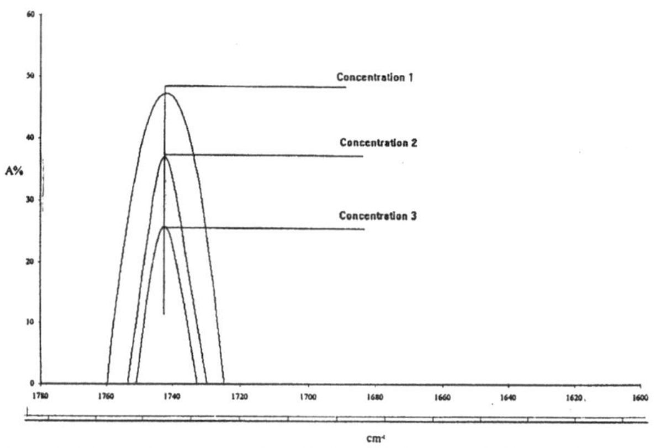

Fig. 8.8 Ideal Spectral Peak: The C=0 peak from the ester functional group of the triacylglycerol. Idealized: no noise, 3 different concentrations showing one absorbance band.

This systematic form of calculation, where each step is verified before proceeding is used because the analysis as a whole relies heavily upon the measurements of extracts from natural products, which by virtue of what they are, create problems with the building of a calibration set. Once these problems are overcome, as far as the experimental design allows, then the procedure has to be applied to the experiment as a whole if reliable results are to be obtained from the use of chemometric analysis. In other words, for a positive outcome on the final analysis, it is imperative that a robust calibration system be created.

To illustrate more clearly, an idealized situation is used, with strict adherence to Beer's Law which assumes unity,

PALMITIC	RESPONSE	LINOLEIC	RESPONSE	OLEIC	RESPONSE
1-10	Diagnostic	1-10	Diagnostic	1-10	Diagnostic
1-20	Diagnostic	1-20	Diagnostic	1-20	Diagnostic
1-40	Diagnostic	1-40	Partially Diag.	1-40	Partially Diag.
1-80	Partially Diag.	1-80	Partially Diag.	1-80	Partially Diag.
1-100	non-diagnostic	1-100	non-diagnostic	1-100	non-diagnostic
1-160	non-diagnostic	1-160	non-diagnostic	1-160	non-diagnostic

Table 8.4 Diagnostic response for calibration data

where the height of the absorbance peak is proportional to the concentration of the constituent. On the calibration graph a straight or calibration line is drawn on the diagonal, from zero to the maximum x and y co-ordinates, as specified by the actual concentration levels of the calibration data. (**Figs. 8.7 & 8.8**))

It is possible to examine an unknown sample by measuring its absorbance at P_1, then projecting the reading to the line at P_2, and then projecting the point of intersection to the concentration axis. In the real situation, both concentration in terms of instrument response for the calibration data, and the direction of linearity were calculated. This was carried out in order to assess how the bulk of the data may behave given the same circumstances.

The culmination of all the steps of the calibration, and the sequential validation, clearly manifested the desired positive results. Although the actual calibration graph did appear to be curved, when a line is drawn through each individual data point, a line of best fit clearly indicates that the nonlinearity is negligible. If a non-linear result is produced, and continues to be so even after obvious checks have been made, (such as the influence of extreme outliers, the lack of consideration of collinearity, or errors in the calculation of the calibration- all of which could lead to the incorrect choice of statistical analysis) then another type of statistical analysis would be required, according to the extent of the non-linearity. This is extremely important as it means the ensuing procedures can be used with a high level of confidence: as long as the system is functioning accurately, then the final results will be (and are) as correct as the optimised experimental parameters will allow.

This table was included to show which levels of dilution gave a "visually" diagnostic response. It is a simple table, with little or no quantitative value, it is merely a representation of what can be "seen" on the respective spectra.

8.3 Results of SCAN Analysis

To enable any kind of data production and analysis, it was necessary to include a numerical technique as an integral part of the overall project design. Consequently a number of commercial chemometric packages were investigated, as they are designed to deal with the large sets of data produced by spectroscopic analysis. Given the statistical criteria which are required to produce both a positive and a reproducible experiment, SCAN was selected as the most appropriate technique for describing statistically the type of data accumulated in the chemistry laboratory. SCAN is a comprehensive software package for chemometric analysis. It includes a wide range of methods, including exploratory data analysis, classification, regression (calibration), modern validation and diagnostic techniques, and data pre-processing, all of which are displayed via detailed graphics.

Prior to any formal numerical analysis in the series, it was necessary to examine the quality of the matrices being used, and determine whether they can be used in multivariate analysis. This was carried out by using the SCAN data exploration technique of Matrix Condition, which actually quantifies the collinearity among the variables and warns about constant and highly collinear columns. This was an important preparatory step, as regular high collinearity may have rendered some classification and regression results arbitrary, or even impossible.

[To reiterate, all data used in this analysis were in correlation matrix form, and any covariance matrices were autoscaled at the beginning, in order to standardize all the matrix information output (see SCAN 7-13).]

The theory regarding the matrix condition concerns situations in which the condition number is the ratio of the largest and smallest eigenvalues, calculated from the correlation matrix. When the condition number is larger than 1,000, then the matrix indicates high collinearity, which occurred in the calculations for both matrices (see chemo. appendix). However, there were no warnings regarding constant or collinear columns. Therefore it was possible to proceed with the more advanced forms of analysis, but ensuring that the methods used were able to deal with highly collinear data.

8.3.1 The results of Hierarchical Cluster Analysis (HCA)-Appendix pt.II pg.121

As the results indicate, each step of the project produced a positive result, from the experimental design to the final regression analysis. This can be demonstrated by looking at the results of each consecutive part of the numerical analysis, which influenced that which succeeded it. The analytical series began with HCA, which creates groups according to similarity and differences, with reference to their distance. **Fig. 8.9** is a dendrogramme and illustrates how these groups are represented, with each sample number at the beginning of each 'arm'. The similarities are represented on the vertical axis (hierarchy), while the differences are the horizontal branching. Each point represents the function of similarity relative to distance (eigenvector), while the final clustering represents the absolute difference within the whole sample population (eigenvalue). The lengths of the vertical branches represent 'goodness' of clustering or the measure of similarity: very similar samples will be displaced very little vertically, whereas the differences are represented by the horizontal displacement between the hierarchies, between the clusters.

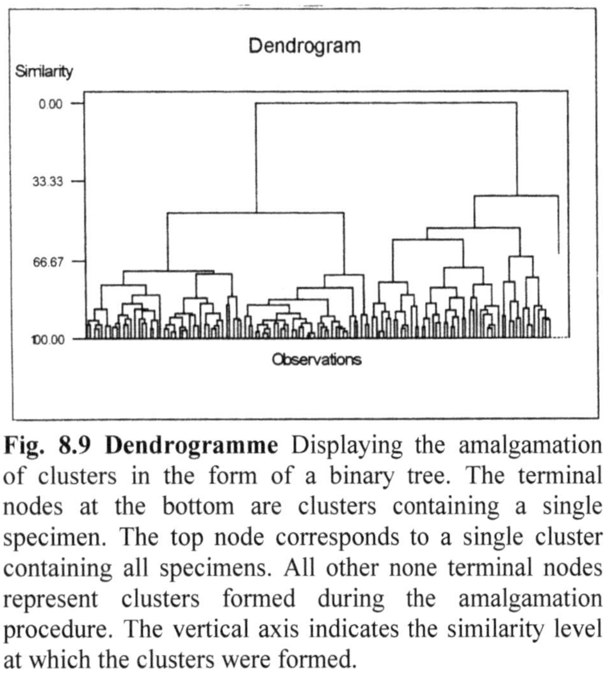

Fig. 8.9 Dendrogramme Displaying the amalgamation of clusters in the form of a binary tree. The terminal nodes at the bottom are clusters containing a single specimen. The top node corresponds to a single cluster containing all specimens. All other none terminal nodes represent clusters formed during the amalgamation procedure. The vertical axis indicates the similarity level at which the clusters were formed.

To produce the final picture SCAN was set for complete linkage, showing the relationship from the closest to the least similar, utilizing the correlation matrix. There were five major clusters, as beyond this the dendrogramme becomes unmanageable with autoscaling which effects the standardization of all the measurement types from the spreadsheet to the matrix. Finally, the measurement used to demonstrate the distance in terms of similarities was the Euclidean Distance, which is written as:

$$d_{ik} - \sqrt{\sum_j \left[x_{ij} - x_{ki}\right]}^2 \quad \textbf{(Fig. 8.10)}$$

The dendrogramme clearly illustrates clustering, but the groups are much too unrefined to demonstrate distinct species or genus differentiation. As can be seen from the cluster tables, cluster 1. contains 43 observations, while cluster 5. has only 5, which indicates that despite the presence of the divisions, they are much too crude to give accurate descriptions of the samples. Therefore more refined analyses were required, with clearly defined data that could be used to create a model, in which the differences and similarities were not so loosely described.

8.3.2 Results of Principal Components Analysis (PCA)-Appendix pt.II pg.112

The next step in this analytical series was PCA, which is also part of the SCAN data exploration, but much more detailed in data description. As with all of the techniques used, the full working procedure is covered in the chemometrics chapter (Chap. 5), therefore only brief, relevant, details are given here to describe the results. PCA calculates the eigenvalues, eigenvectors and principal component scores of the matrix, using the two previously defined algorithms SVD and NIPALS. (pg. 121 Appendix part II)

Essentially the PCA gives an overview of the dominant patterns of information in the data. For this project it explores the relationships between the spectra and the relationship that exists between the frequencies. This is carried out mathematically by using a lower number of variables, the principal components, than those present in the original data matrix. This effectively maximises the amount of information while reducing the number of variables, thereby enhancing the understanding and manipulation of the data.

Primarily PCA looks at the percentage of variance which is the eigenvalue for each component calculated, the proportion of the total variance covered by that component, and the cumulative proportion of variance covered up to that component. The correlation matrix for the modern samples gave optimum results for 4 PCA's, which indicated that at least four were needed to account for more than 90% of the total variance. This can be seen on the eigenvalue scree plot, where the variance is illustrated by the plateau that is created as the variance decreases beyond 4 PCA's. Consequently no more are required to describe the situation, as they would contribute nothing of any significance. The value for each of these relevant principals components in terms of the total amount of variance taken by each, can be seen on the data output table as eigenvalues, which also indicates the percentage proportion for each value. Finally, how these values divide into groups of either distinction or ambiguity, can be seen on the PCA scores and loading plots. If it is the latter, then the analysis needs further refinement. **(See Chemo. Appendix part II for plots and tables pg. 121; and explanations of all table and data headings)**

Principal Components Analysis Data Table Calculated from Covariance Matrix by SVD

Eigenvalue	9.7232	1.2397	0.4172	0.1972
Proportion	0.810	0.103	0.035	0.016
Cumulative	0.810	0.914	0.948	0.965

Table 8.5 Eigenvectors

Variable	PC1	PC2	PC3	PC4
v3400	**-0.176**	**-0.656**	**-0.569**	**0.285**
v2960	**-0.310**	**-0.053**	**-0.078**	**-0.509**
v2872	**-0.302**	**-0.196**	**-0.179**	**-0.476**
v1736	**-0.293**	**0.286**	**-0.250**	**0.092**
v1712	**-0.300**	**0.128**	**-0.002**	**0.419**
v1640	**-0.222**	**-0.545**	**0.487**	**0.017**
v1452	**-0.312**	**0.061**	**0.157**	**-0.275**
v1376	**-0.313**	**0.093**	**0.031**	**0.022**
v1252	**-0.303**	**0.122**	**-0.153**	**0.300**
v1196	**-0.310**	**0.184**	**-0.024**	**0.094**
v1152	**-0.301**	**0.252**	**-0.048**	**-0.071**
v676	**-0.287**	**-0.097**	**0.535**	**0.261**

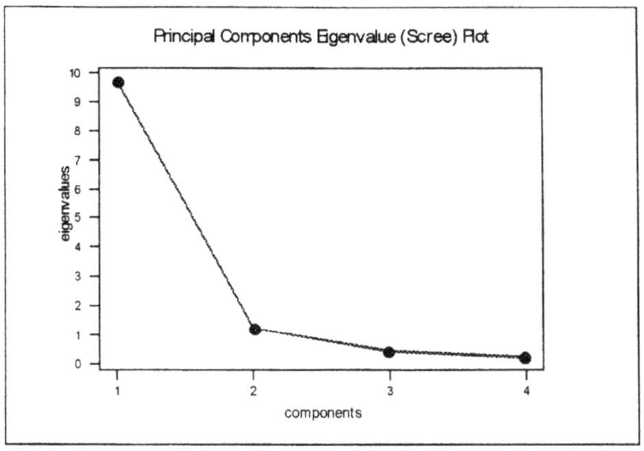

Fig. 8.11 The eigenvalue or scree plot, is a scatter plot of eigenvalues against the component ids. The points are connected forming a decreasing curve. This plot helps determine the number of significant components that need to be used.

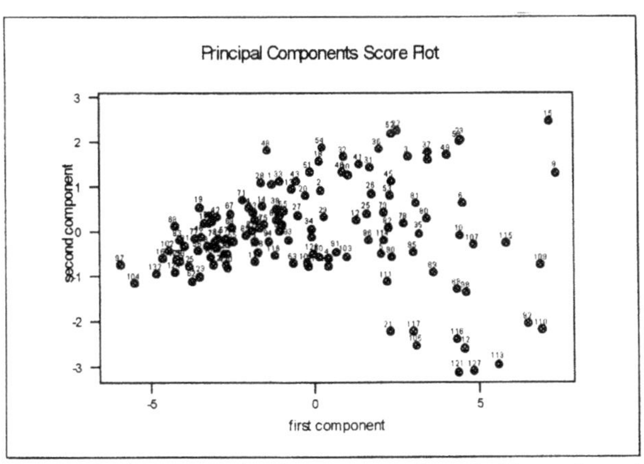

Fig. 8.12 The score plot is a scatter plot of objects (rows) projected onto the plane of the first two components, i.e., the first principal component scores are plotted against the second P.C. scores. When the first two components cover most of the total variance, the configuration of the points on this plot closely reflects the original multidimensional configuration. Outliers and clusters can easily be spotted on this graph.

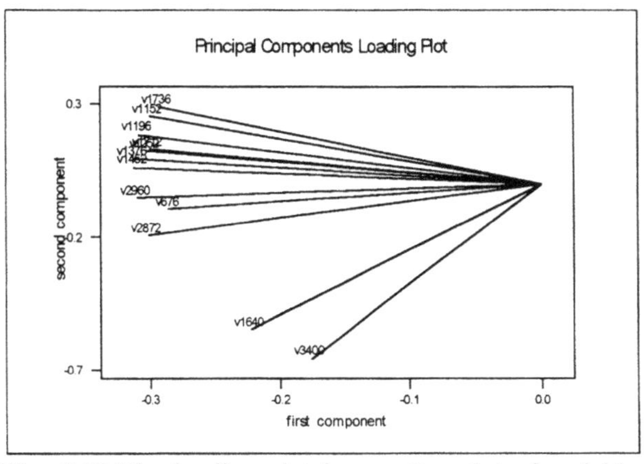

Fig. 8.13 The loading plot is a scatter plot of variables projected onto the plane of the first two Components, i.e., the first component loadings are plotted against the second component loadings. The variable points are connected to the (0,0) point. The angles of the lines reflect the correlations among the variables.

All told, the eigenvalues represent a portion of the total variance contained within the data, while the corresponding eigenvectors provide the coefficients for the PCAs and the directions of maximum variation through the data. It is effectively a numerical data distillation and not a data reduction process.

8.3.3 Results of Factor Analysis (FA)-Appendix pt.II pg.118

Each successive step of the chemometric analysis is actually another level of understanding of how the data are numerically expressing the chemical information, and of the relationships between and within the data groups. Consequently a further analysis within data exploration was carded out, as an additional refinement towards regression.

This was the Factor Analysis (FA), which, via principal components, aims to reduce a large data matrix to one of a more manageable size. It indicates both the number of factors influencing the data set and the nature of significant parameters. It utilizes the same number of principal components as the initial PCA, but has a rotation factor which aims to find common factors that are composed of only a few variables. It is effectively an expansion on the PCA; however, this is a two-step process, (See Chap. 5) with PCA inclusive, but the goals are the same. The results for the modern samples confirmed the data optimisation at 4 principal components, while the archaeological data confirmed at 5 principal components. Both showed that the inclusion of a rotation factor did not significantly enhance the results. They did, however, indicate that the manageable views created by PCA and FA aided the understanding of the complex matrices of both the modern and archaeological data, and what would be the best way to deal with them, in order to extract the highest quality results.

The results are available in both tabular and graphic form. Table one contains factor loadings, their variances, and their communalities, and table two contains the factor score coefficients. The graphs, which present the data in a visually more easily interpreted format, represent the eigenvalue or scree plots, which show both the rotated and unrotated factors as scatter plots, with eigenvalues plotted against factor IDs, which make it possible to separate out the significant factors. The second important graph is the score plot, also containing the rotated and unrotated factors, but in this case its axes are the factors, with the first being factor one plotted against factor two, indicating where most of the variance lies, individual significant clusters, and outliers. Both the graphs and the tables enable a clear interpretation of the data results. **(The latter can be seen in their entirety in the Appendix part II pg. 118, arranged according to their sequence in the numerical analysis).**

Factor Analysis
Principal Components from Covariance Matrix. Unrotated Factor Loadings and Communalities

Table 8.6

Variable	Factor1	Factor2	Factor3	Factor4	Commnlty
v3400	-0.550	-0.730	-0.367	0.127	0.986
v2960	-0.968	-0.059	-0.050	-0.226	0.993
v2872	-0.940	-0.218	-0.116	-0.212	0.990
v1736	-0.915	0.318	-0.161	0.041	0.966
v1712	-0.937	0.142	-0.001	0.186	0.933
v1640	-0.694	-0.607	0.315	0.007	0.949
v1452	-0.974	0.068	0.101	-0.122	0.978
v1376	-0.975	0.104	0.020	0.010	0.961
v1252	-0.944	0.136	-0.099	0.133	0.937
v1196	-0.966	0.205	-0.016	0.042	0.977
v1152	-0.939	0.280	-0.031	-0.031	0.962
v676	-0.895	-0.108	0.346	0.116	0.946

Table 8.8

Factor Score Coefficients				
Variable	Factor1	Factor2	Factor3	Factor4
v3400	-0.057	-0.589	-0.880	0.642
v2960	-0.100	-0.047	-0.120	-1.147
v2872	-0.097	-0.176	-0.278	-1.073
v1736	-0.094	0.257	-0.387	0.207
v1712	-0.096	0.115	-0.002	0.944
v1640	-0.071	-0.490	0.754	0.038
v1452	-0.100	0.055	0.243	-0.618
v1376	-0.100	0.084	0.047	0.050
v1252	-0.097	0.109	-0.237	0.675
v1152	-0.097	0.226	-0.074	-0.159
v676	-0.092	-0.087	0.829	0.588

Table 8.7

Variance	9.7232	1.2397	0.4172	0.1972	11.5773
% Var	0.810	0.103	0.035	0.016	0.965

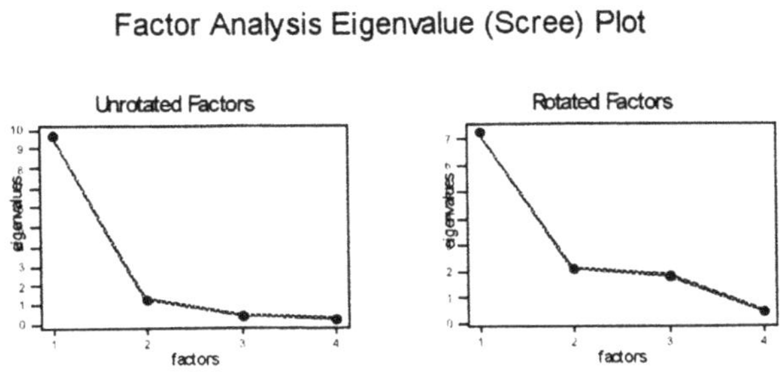

Fig. 8.14 The eigenvalue or scree plot is a scatter plot of the eigenvalues against the factor ids. the points are connected forming a decreasing curve. This helps to determine the number of significant factors.

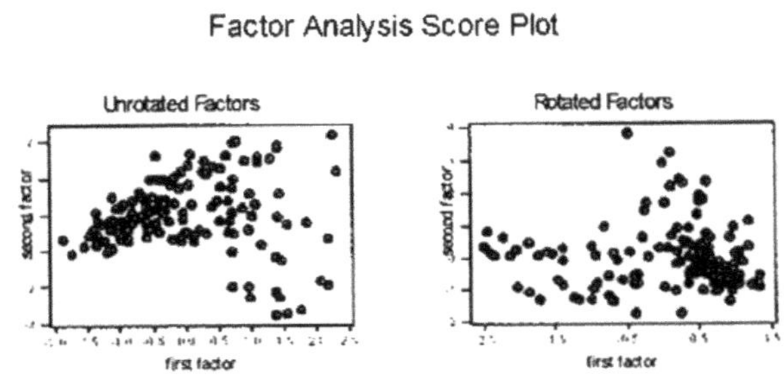

Fig. 8.15 Score plot is scatter plot of objects (rows) projected onto the plane of the first two factors, i.e. the first factor scores are plotted against the second factor scores. When the first two factors cover most of the total variance, the configuration of the point, on this plot closely reflects the original multidimensional configuration. Outliers and clusters can be spotted.

8.3.4 Results of Classification

Initially the next step was to carry out a classification of the data, which is essentially a supervised pattern recognition analysis. This type of analysis aims at calculating class models, class boundaries and class rule; this rule is based on a training set that contains objects of a known class. The ensuing rule is then applied to a test set, containing objects of unknown classes, by assigning objects to the class with the largest class density function. All told, there was little significant response to this analytical procedure, from both the modern and the archaeological data. It was therefore, considered better to proceed without this step, as regression is not dependent on it.

8.3.5 Results of Regression Analysis

The final step in the analytical procedure was regression analysis. This demonstrates the level of quantification achieved by calibration and validation, and indicates whether the chemometrics has produced results significantly better than those achieved by visual analysis alone.

All the information collected prior to this indicated which statistical method would best suit the data. However as optimisation was a primary objective, several were included, namely Ordinary Least Squares (OLS), Principal Components Regression (PCR), Partial Least Squares (PLS) and Alternating Conditional Expectations (ACE). As previously mentioned in Chapter 5, the chemical data indicated a non-linear relationship, although both the calibration and the validation demonstrated the effects of this on the overall results to be negligible. It was nevertheless thought prudent to carryout a series of tests, in order to either verify or negate this question of linearity.

8.3.5.1 Ordinary Least Squares (OLS) **–Appendix pt.II pg.126**

As regression has been discussed quite extensively in chapter 5, it is necessary only to give an overview of what each process does, and whether there are any differences between them that may significantly affect the results. The series began with OLS, which calculates unbiased, least squares coefficients, and would normally be used only on a well determined data set, (i.e. where there are more observations than predictors, which are not highly correlated). This was followed by PCR, which calculates biased, non-least- squares coefficients, on a data set considered to be undetermined, (i.e. where there are fewer observations than predictors and/or the predictors are highly correlated.) Both methods calculate a parametric linear model for a single response, assuming the response is a linear function of the predictors.

The numerical output is in the form of tables. The first shows the goodness of fit and goodness of prediction, and contains the model validation parameters of how well the calculated model fits the training set data, and how well it would predict future data. The goodness of fit or modelling capability is measured by the residual sum of squares (Residual SS) and by the R^2 coefficient (R2), whereas the goodness of prediction or predicting capability is measured by the residual predictive sum of squares (Resid PRESS) and by the cross-validated R^2 coefficients (R2cv).

The results of the modern data matrix for OLS are 69% for the goodness of fit, and 62% for the goodness of prediction. Although there is not a great deal of difference between the two, neither are high enough to justify the use of the procedure for the model. Where the matrix for the archaeological data is concerned, both goodness of fit and prediction are considerably higher, with 87% for the former and 76% the latter. However, a high value for the test data is not the decisive factor for the adoption of a numerical technique.

Ordinary Least Squares Regression (OLS)
Response Variable: sample
Goodness-of-Fit and Goodness-of-Prediction

Table 8.9

Residual SS	R2	Resid PRESS	R2cv
63432	0.6903	78065	0.6189

Table 8.10 Regression Coefficients, Predictor Importances

Intercept	v3400	v2960	v2872	v1736	v1712	v1640
68.14	-3.783	26.13	35.02	-13.04	-16.80	19.32
Importance	-3.783	26.13	35.02	-13.04	-16.80	19.32

Table 8.11

v1452	v1376	v1252	v1196	v1152	v676
-107.4	41.29	-11.76	41.70	-11.44	7.957
-107.4	41.29	-11.76	41.70	-11.44	7.957

Analysis of Variance

Source	Df	SS	MS	F	p
Model	12.00	141400	11783	22.29	0.0000
Residual	120.00	63432	528.6		
Total	132.00	204832			

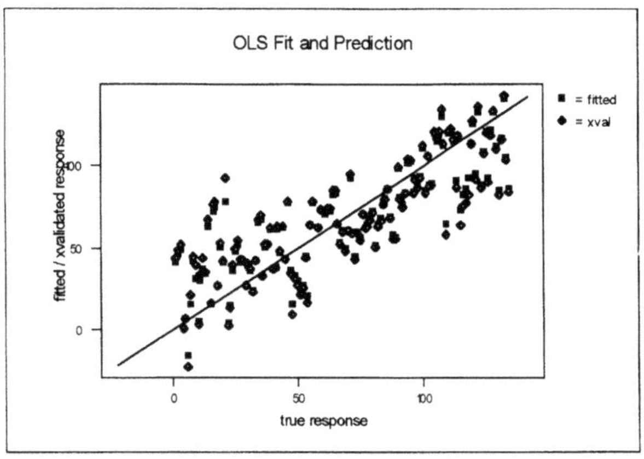

Fig. 8.16 Response plot- a scatter plot of the calculated response versus the true response. Two calculated response values are plotted: the ordinary fitted response (squares) and the cross-validated predicted response (diamonds). Observations (rows or points on the table) where the cross-validated value differs significantly from the ordinary fit may be leverage points. (see Appendix OLS pg. 126) Ideally this plot is close to the straight line; calculated=true. This line is also on the plot.

As well as the tabular presentation of the data, there are also a number of graphs, which illustrate the positions of the data response. One of the most important of these is the OLS fit and prediction graph, which shows a line of best fit running diagonally through the components, with axes x the true response and y the fitted/cross validated response. Basically it is a scatter-plot of the calculated versus the true response, which ideally should be close to the straight line, where calculated = true i.e. where the Principal Components Regression (PCR) response is the numerically calculated predicted answer from the OLS functions compared with the true or actual plotted observations. It also contains the F-value (F) and the corresponding probability (P), calculated from the model and residual mean square, which indicate the significance of the model; a large F-value and a small p value, indicates a high level of significance for the model.

8.3.5.2 Principal Components Regression (PCR). Appendix pt. II pg. 130

The **Table 8.11** indicates a high level of significance for the constituted model from the modern samples, with F being 22.29 and p being 0.000. This is an important result, as it was to this end that OLS was carried out. However, to enhance these results, another similar regression analysis is used, the Principal Components Regression (PCR), in order to examine the R2, R2cv, F-value and P. Although the procedures are very close, as OLS is a sub-case of PCR, there are slight differences in how the regression coefficients are calculated; in particular PCR calculates biased non least squares coefficients.

Table 8.14 shows the outcome of the PCR analysis using the same number of components as used in OLS. The R2 or goodness of fit (measured by the sum of squares and the R^2 coefficients) was 87%, and the R2cv or goodness of prediction (measured by residual predictive sum of squares and the R^2 cross validated coefficients) was 76%. Both of these are significantly higher than those calculated by OLS alone, indicating that PCR has refined the results further. As was the case when the model

Response Variable: sample
Optimal Number of Components: 10

Table 8.12 Goodness-of-Fit and Goodness-of-Prediction

Comp SS	Df	Residual	R2	Resid PRESS	R2cv
1	131.00	203605	0.0060	210421	-0.0273
2	130.00	122577	0.4016	127487	0.3776
3	129.00	121921	0.4048	129067	0.3699
4	128.00	121193	0.4083	130535	0.3627
5	127.00	97097	0.5260	105774	0.4836
6	126.00	91406	0.5537	103082	0.4967
7	125.00	90985	0.5558	104230	0.4911
8	124.00	88105	0.5699	102592	0.4991
9	123.00	73463	0.6414	86368	0.5783
10	122.00	64952	0.6829	77082	0.6237
11	121.00	64942	0.6830	78451	0.6170
12	120.00	63432	0.6903	78065	0.6189

Table 8.13 Regression Coefficients, Predictor Importances

Intercept	v3400	v2960	v2872	v1736	v1712	v1640
68.14	-4.032	24.29	34.28	-32.63	-11.82	18.00
Importance	-4.032	24.29	34.28	-32.63	-11.82	18.00

v1452	v1376	v1252	v1196	v1152	v676
-95.31	31.10	17.08	1.749	18.92	5.446
-95.31	31.10	17.08	1.749	18.92	5.446

Table 8.14 Analysis of Variance

Source	Df	SS	MS	F	p
Model	10.00	139881	13988	26.27	0.0000
Residual	122.00	64952	532.4		
Total	132.00	204832			

significance calculations were examined; the F-value had increased, while p remained zero, suggesting that the significance level of the model was improved upon, as before.

The results for both OLS and PCR can also be seen graphically, in order to see how each compares, and to see where the PCR analysis confirmed and enhanced the initial OLS results. Due to the quantity of the graphic data, it is not possible to include all of them in the text Correspondingly, in order to maintain sequential understanding, all chemometric graphics appear under the appropriate heading in the Appendix part II, page 126 for OLS, and page 130 for PCR.

8.3.5.3 Results of Partial Least Squares analysis. Appendix pt. II pg.135

Utilizing a system of analysis as sophisticated as SCAN enabled the comparison of a number of closely related methods that, prior to the advent of high speed PCs, would have been too time consuming to contemplate. This allowed for the isolation of a procedure that optimised the data, without sacrificing valuable experimental time. Consequently, PLS was performed on the data, even though it was considered not to be the most suitable, as it did not fit all the specified criteria. However, it enabled the close comparison of the results with the other forms of regression analysis, so as to demonstrate where each method either excelled or performed poorly.

Partial Least Squares (PLS) calculates a parametric, biased, linear regression model, which may contain one or several responses. It aims to produce a more predictive model than OLS, as it assumes the information about the responses is in a subspace of high variance and of high correlation with the response(s), (i.e. in the first few PLS components). However, there are no exact formulae open to PLS for the calculation of cross validated and residuals, so they have to be estimated by what is actually referred to as the "leave-one-out" method. This simply requires the omission of observations, and the subsequent repeating of the model calculations for a specified number of times. This produces the best predictive PLS

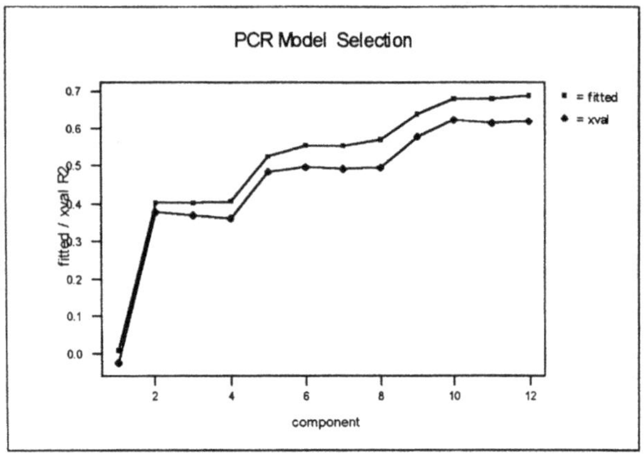

Fig. 8.17 PCR Model Selection: displays a scatter plot of the ordinary fitted R-squared (squares) and the cross-validated R-squared (diamonds), versus the number of components. This plot shows the differences among potential models in terms of modelling and predictive capability.

model, if the number of components are optimised, but decreases the predictive capability if the component number increases. The optimal results for the PLS method for the modern samples are 67% for the goodness of fit and 57% for predictive capability, which is somewhat less than those for the OLS model, but nowhere near as good as that calculated by the PCR.

The results of the PLS calculations are also illustrated by both tables and graphs; the actual numerical output consists of two tables for each of the responses, the first being the table for the goodness of fit and prediction which contains the model validation parameters. This is information about how well the calculated model fits the training set data, and how well it could be expected to predict future data. To reiterate, the training set is that information which is identified, known chemical data, as used to set the parameters for evaluating the similarities and dissimilarities between groups. It is used to define a set of rules that can be applied to the unknown, unidentified chemical data, which comprises the test set, This table is represented graphically by the model selection profile, which shows it to reach a plateau at 5 components for both the predictive capability curve (the line with diamonds), and the modelling capability curve (the line with squares).

The second table presents the regression coefficients and predictor importance, and includes the regression coefficients and the calculated importance of each predictor in the model. This is represented graphically by the regression coefficient plot, which shows which of the variables has the most significant effect, and how important it is compared to the others.

There are a number of graphs for the PLS results, including the latent variable plots which suggest linearity or non-linearity for each of the components; e.g. if curvature is evident then there maybe a case for a non-linear PLS. The normal residual plots can indicate error: where the lines are relatively straight, the error is within acceptable limits. The ordinary residual plots indicate which observations are the most likely to be outliers on the ordinary residual and fitted response plot. Lastly I calculated, the OLS fit and prediction plot, which shows where the observations lie in relation to the true line, from the true response on the x axis and the cross validated/fitted response on the y axis. The furthest from the predicted = true line, is the observation(s) likely to have the largest residual. All of these tables and graphs are contained in the chemometrics appendix, where they can be viewed in full sequence.

Partial Least Squares Regression (PLS)

Number of Cross-validation Groups: 133
Optimal Number of Components: 9

Response Variable: sample
Goodness-of-Fit and Goodness-of-Prediction

Table 8.15

Comp	Residual SS	R2	Resid PRESS	R2cv
1	146940	0.2826	158227	0.2275
2	112581	0.4504	119733	0.4155
3	85109	0.5845	97256	0.5252
4	79748	0.6107	93660	0.5427
5	71014	0.6533	85788	0.5812
6	67408	0.6709	80131	0.6088
7	66527	0.6752	80459	0.6072
8	64557	0.6848	77308	0.6226
9	63894	0.6881	77205	0.6231
10	63770	0.6887	77576	0.6213
11	63433	0.6903	78316	0.6177
12	63432	0.6903	78065	0.6189

Table 8.16 Regression Coefficients, Predictor Importances

Intercept	v3400	v2960	v2872	v1736	v1712	v1640
68.14	-3.814	26.03	33.88	-25.39	-13.48	18.71
Importance	-0.09682	0.6608	0.8600	-0.6447	-0.3422	0.4750
v1452	v1376	v1252	v1196	v1152	v676	
-102.6	38.54	2.225	22.08	5.138	5.663	
-2.604	0.9783	0.05647	0.5605	0.1304	0.1438	

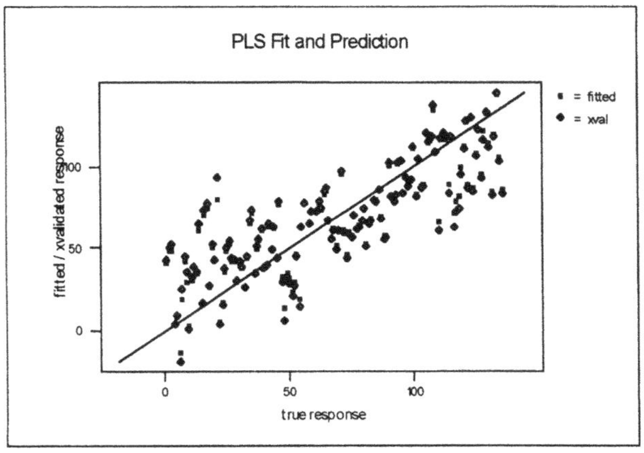

Fig. 8.18 Scatter plot of the calculated responses versus the true response. Two types of calculated response values are plotted ordinary fitted response (squares) and cross-validated predicted response (diamonds). Observations where the cross-validated value differs significantly from the ordinary fit are likely to be leverage points. Ideally the plot is close to the straight line (calculated = true).

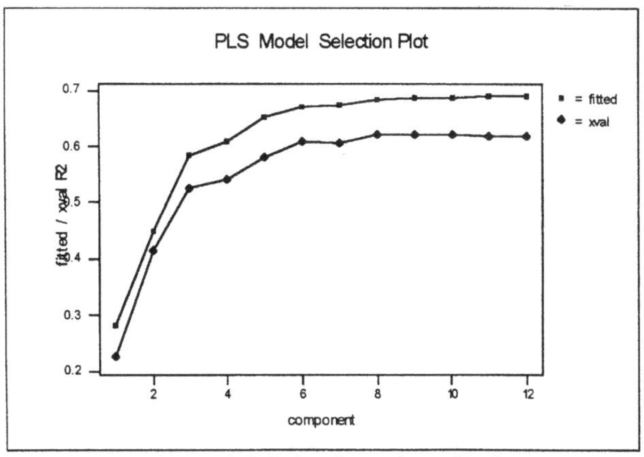

Fig. 8.19 PLS Model Selection Plot: displays a scatter plot of the ordinary fitted R-squared (squares) and the cross-validated R-squared (diamonds), versus the number of components. When there are several responses, four plots are displayed in one window, each for a different response variable. This plot shows the differences among potential models in terms of modelling and predictive capability.

Looking at these results all of these analyses demonstrates the requirement of more than one numerical analysis, as, theoretically, PLS should be an advancement on the OLS result, but in reality it failed to meet the level of the initial regression step. By carrying out the full series, incorrect judgements may be avoided, as each procedure can be determined according to its own merits or misgivings. Consequently the onus related to errors, will not fall solely on the data and the overall experimental procedures.

Due to the poor performance of PLS, it was evident that another type of regression analysis was required for a final comparative method. As previous formulae all contained a linear function, it seemed apt that a non-linear method be utilized this time, primarily because the initial calibration data created a curve, but also, for comparative reasons, to check the non-linear effect on the results. Given the dimensions of the modern data matrix, the Alternating Conditional Expectations (ACE) regression was considered to be the most appropriate where improvement on linear models are required.

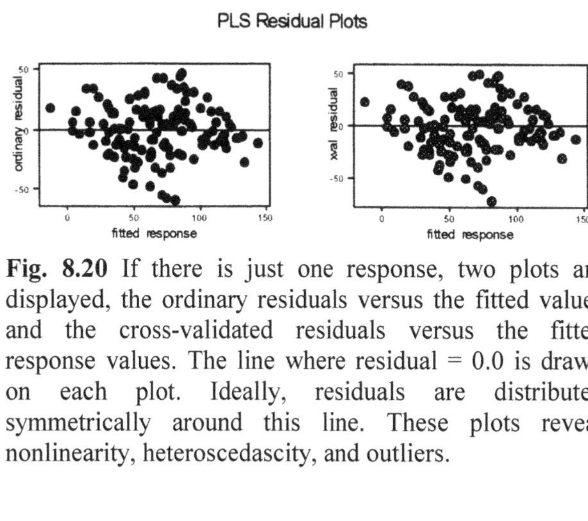

Fig. 8.20 If there is just one response, two plots are displayed, the ordinary residuals versus the fitted values and the cross-validated residuals versus the fitted response values. The line where residual = 0.0 is drawn on each plot. Ideally, residuals are distributed symmetrically around this line. These plots reveal nonlinearity, heteroscedascity, and outliers.

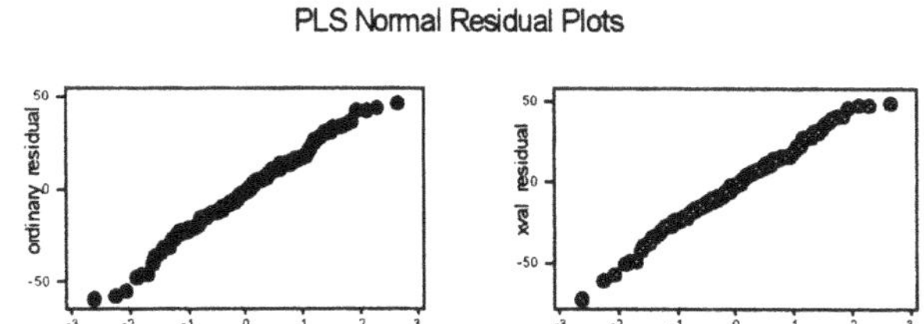

Fig. 8.21 If there is just one response, two plots are displayed, ordinary residuals versus normal scores (quantiles from standard normal distribution), and cross-validated residual versus normal scores. This plot is especially useful for checking the normality assumption on the residuals.

Alternating Conditional Expectations Regression (ACE)

Response Variable: sample
Smoothing Parameter (SPAN): 0%
Number of Cross-validation Groups: 133

Table 8.17 Goodness-of-Fit and Goodness-of-Prediction

Residual SS	R2	Resid PRESS	R2cv
37960	0.8147	99748	0.5130

Table 8.18 Predictor Variable Importances

v3400	v2960	v2872	v1736	v1712	v1640	v1452
0.1438	0.6811	0.8846	0.3398	0.3853	0.5268	2.766
1376	v1252	v1196	v1152	v676		
1.054	0.3085	1.022	0.2928	0.2386		

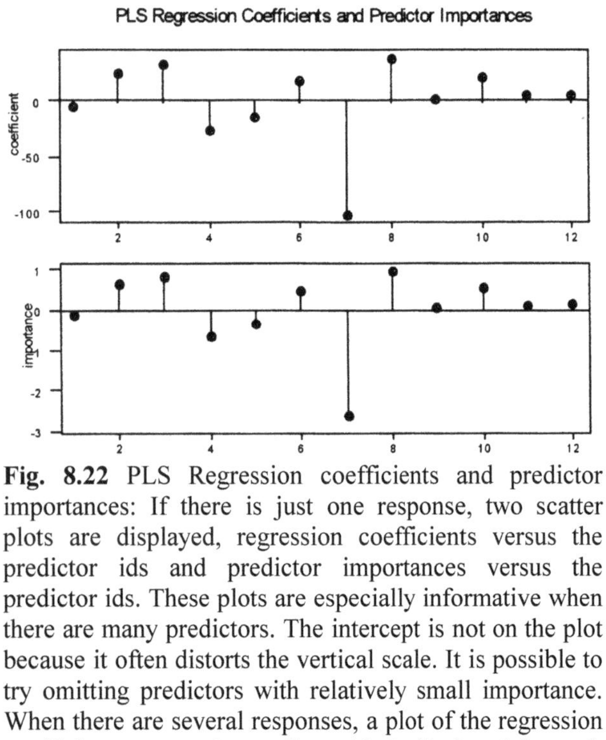

Fig. 8.22 PLS Regression coefficients and predictor importances: If there is just one response, two scatter plots are displayed, regression coefficients versus the predictor ids and predictor importances versus the predictor ids. These plots are especially informative when there are many predictors. The intercept is not on the plot because it often distorts the vertical scale. It is possible to try omitting predictors with relatively small importance. When there are several responses, a plot of the regression coefficients versus the predictor ids is displayed for each response. Plots of the predictor importances are not produced.

8.3.5.4 Alternating Conditional Expectation (ACE) Appendix pt.II pg.140

Similarly with ACE, it is not necessary to cover the method in depth as this detailed in chapter 5, however a basic overview is appropriate. ACE calculates a nonlinear, nonparametric model, where the response is estimated as a sum of functions of the predictors, and should be utilized only when OLS performs poorly due to nonlinearity. It can be used only on well-determined data, where there are more observations than predictors. This is a major aspect of my modern data matrix. As with PLS, this technique does not have an exact formula to calculate cross-validated fit and residuals. ACE also requires a "leave-one-out" (sequential elimination) and repeat-the-model calculation method. For the model data a full cross validation, using the "leave-one-out" method was undertaken, and this repeated the model calculation *n* times (*n* = number of observations). Although it is the most time consuming of techniques, the outcome produces the most accurate estimate of the predictive capability attainable. The results can be seen in the tables for the goodness of fit and goodness of prediction, which contain the model validation parameters of how well the calculated model fits the training set and how well it would predict future data.

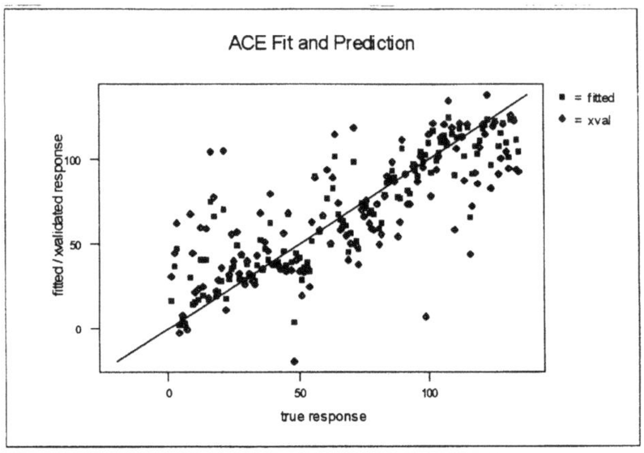

Fig. 8.23 ACE fit and prediction plot: displays a scatter plot of the calculated response versus the true response. Two calculated response values are plotted: ordinary fitted response (squares) and cross-validated predicted response (diamonds). If no cross-validation was done, only the fitted response is plotted. Observations where the cross-validated value differs significantly from the ordinary fit may be leverage points. Ideally this plot is close to the straight line, calculated=true. The diagonal line is also on the plot diagram.

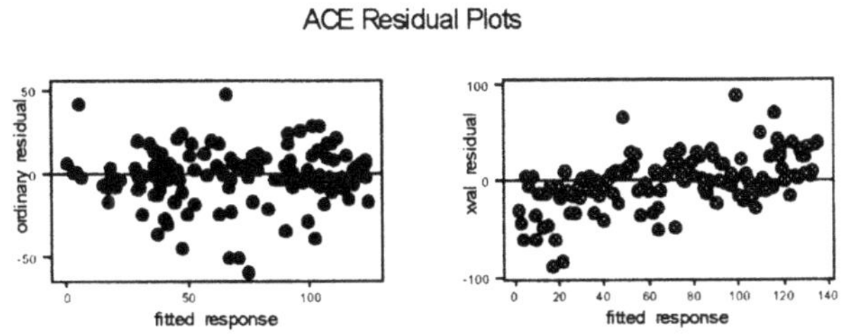

Fig. 8.24 ACE residual plots: display two plots, ordinary residuals versus the fitted values and the cross-validated residuals versus the fitted values. If no cross-validation was done, the only plot of the ordinary residuals is displayed. These plots reveal heteroscedasticity and outliers. The horizontal line where residual = 0.0 is displayed on each plot. Ideally, residuals are distributed symmetrically around this line.

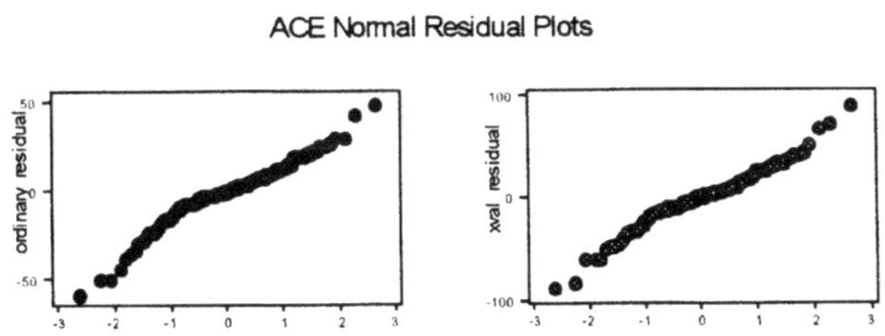

Fig. 8.25 Normal residual plots: displays two plots, ordinary residuals versus the normal scores (quantiles from standard normal distribution), and cross-validated residuals versus the normal scores. If no cross-validation was carried out, only the plot of the ordinary residuals is displayed. These plots are especially useful for checking the normality assumption of the residuals.

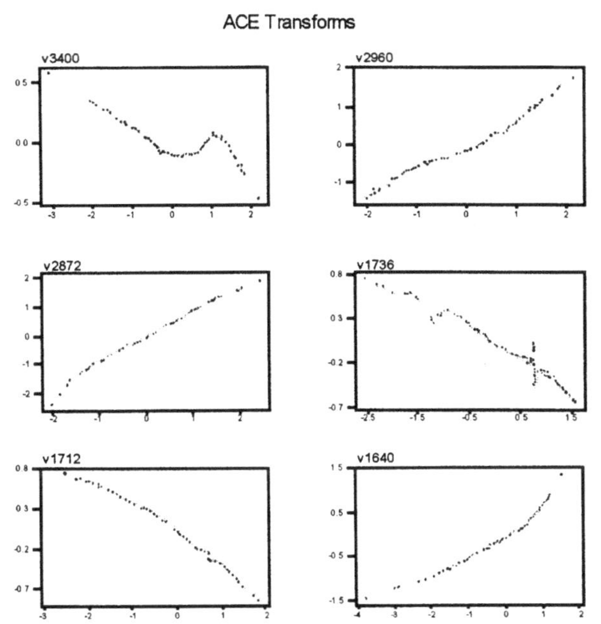

Fig. 8.26 ACE transforms: displays the transformation function for each predictor variable. In each plot, the transformed values are plotted against the original predictor values. Up to four transformation plots are displayed in one window. These are the most important plots in ACE.

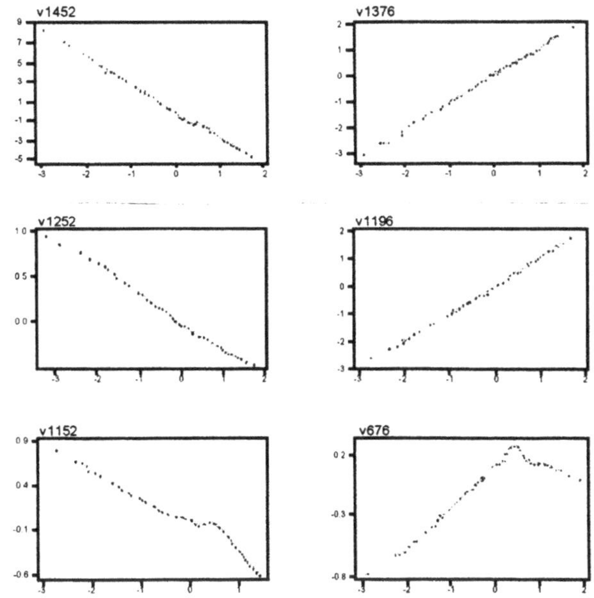

Fig. 8.27 ACE Transforms continued

The goodness-of-fit or modelling capability is measured by the residual sum of squares (Residual SS), and the R^2 coefficient, while the goodness of prediction or prediction capability, is measured by the residual predictive sum of squares (Resid PRESS) and by the cross validated R^2 coefficient R2cv. **Table 8.19** gives the modelling capability as 81% and the predictive capability as 52%. Although this is a significant improvement on the OLS calculations for modelling, it is dismally lower on the predictive scale. To be precise, ACE produced a lower predictive result than any of the other comparative techniques, which may be due to the highly collinear nature of the data.

Table comparing each method: Table 8.19

	R^2	R2cv
OLS	**69%**	**62%**
PCR	**87%**	**76%**
PLS	**67%**	**57%**
ACE	**81%**	**52%**

Overall, each regression technique provided valuable information about the data concerning the significance, modelling and predictive capability, which is essential when a model is being designed and built as a comparative and selective tool for unknown data.

8.3.6 Validation

The information produced by the analytical series was very positive, and the model from the identified modern samples gave a high goodness-of-fit. However, due to the ambiguous nature of archaeological material it was necessary to validate these results to ensure the data was behaving correctly for a prospective model. This was carried out by exposing the calibration set and the subset of twenty to the same sort of analyses as the modern, identified samples, in every format, but most specifically the regression analysis OLS, PCR, PLS and ACE. Once completed, these results were compared and the initial model data adjusted accordingly. In this case, the adjustment was minimal, and the goodness-of-fit was still significant enough for the modern samples to be used as a robust and useful model.

Normally, the process of validation is carried out by using an independent set of known composition, which is prepared in the same manner as the standards, but whose components are not included in the calibration set. However, in this case certain adjustments had to be made to the overall methodology, as it is extremely difficult to know the exact composition of any sample when using IR as the main analytical technique. Given these analytical constraints, it was considered feasible to use the subset and the calibration set as a method of validation, where the subset validated the calibration, which then set the standard for the data as a whole.

8.4 Results of SCAN in Species Identification

As the model is at its optimum once the goodness-of-fit and prediction have been validated, it is possible to demonstrate the positive aspects of the results with a high level of confidence. These can be demonstrated by observing the actual levels of identification achieved, from coarse to extremely fine.

1. **Coarse Clustering**: It proved possible to cluster taxonomic groups such as sub- families and tribes.
2. **Fine Groupings**: It also proved possible to clearly distinguish groups at the level of genera.
3. **Finer Separations**: The finer the analysis, the more it became possible to define different species within individual genera. By occupying their unique place within the numerical matrix, it offered the prospects for identifying 'unknowns' at species level.

Although all of the regression techniques may provide a template, designed and constructed by way of the criteria contained within the matrix, not all are adequate to be utilized on their own.

When dealing with natural materials, anomalies are likely to occur; and, while some of these may be inexplicable, they can often be explained by referring to the history of the sample. For example, group 12 contains five members, two of which are totally incorrect; however, when compared to the botanical information, it was possible to explain the cause of these aberrant members: samples 131-135 are immature. As a consequence, the lipid fraction will be different in terms of the amounts of specific triacylglycerols which are present during and at the end of maturation, for the same species.

8.4.1.1 Triticum

As the model is now considered to have reached its optimum level of performance, the groups that it creates should be comparable to those selected by the morphological information. These groupings are contained within the matrix, which initially shows the genus level, but on closer inspection may also show the species, or even subspecies levels. This level of identification is achieved by analysing one genera at a time, to ensure the maximum degree of accuracy allowed by the constraints of the model, The first genus selected was *Triticum*, which was subsequently split into its individual species, primarily by recognizing the differences between each, numerically and spectrally, and then comparing these individual groups with the botanical identification. This technique enabled the separation of each distinct species, a subspecies, and one that remained unidentifiable due to the non-diagnostic character of the initial spectrum. This was sample 106, *Triticum uratu*, which was consequently not numerically valid, and would have been recorded on the matrix as an outlier and not a true component of the model. As FTIR is not a separation technique, it cannot identify the actual separate components of the samples. Identification at the matrix level is, therefore, carried out by observing the calculated differences and similarities between the peaks and their values. **Table 8.20** indicates how many species of *triticum* are represented, the quantity of each, and their identifying samples numbers. **Fig. 8.29** shows examples of spectra, which visually and numerically illustrate the differences and similarities between the species, thereby demonstrating where the different groupings arise.

Table 8.20. Number of *Triticum* species in modern sample

SPECIES	NO.	SAMPLE NOS.
T. araraticum	**2**	**111, 112**
T. boeoticum	**2**	**105, 107**
T. uratu	**1**	**106**
T. dicoccoides	**5**	**108, 110, 113, 114, 115**
T. dicoccum	**4**	**109, 117, 118, 123**
T. palaeocolchicum	**1**	**119**

Both *T. araraticum* and *T. dicoccoides* are wild emmers: however, a significant dissimilarity was apparent on comparison and was attributed to variation at species level. These variants are represented genetically as AAGG for *T. araraticum* and AABB for *T. dicoccoides*. Quite often, unexpected differences appeared which stemmed from the closeness of the species in terms of breeding. This was particularly apparent when *T. dicoccoides* and *T. dicoccum* were compared. As mentioned above, *T. dicoccoides* is a wild emmer, being the wild precursor of the domestic emmer: *T. dicoccum.* In view of this close link, distinction was thought highly improbable. However, there were clear differences between the two, specifically around the C=C region, where it was apparent that *Triticum dicoccoides* (Sp.1) does not contain this bond, possibly implying a lack of unsaturated esters of fatty acids (see Fig. 8.29).

Two further species showing clear similarities are *T.*

palaeocolchicum and *T. dicoccum*. This is due to them both being types of domestic emmer, and it appeared that this was a highly significant factor in morphological terms, as *T. palaeocolchicum* showed little similarity to *T. dicoccoides*, as would be expected of the wild precursor.

Each species was compared to the other in order to look for both similarities and differences; when similarities were indicated between species, it also implied that there were none between those remaining. The situation was somewhat different for *T. bocoticum*, as it bore no resemblance to any of the other species, its distinction or uniqueness being totally reliant on its differences.

The final species analysed from the genus *Triticum* was *T. uratu*. Due to its nondiagnostic spectral information, it was impossible to compare it with the others. The data was unsound, and as a consequence, any group assignment would be unreliable.

8.4.1.2 Hordeum

The second major genus to be split into species groupings was *Hordeum*. The following table indicates how many species there are, their individual sample numbers, and how many there are of each species.

Table 8.21 Hordeum species which illustrate differences and similarities between and within species.

SPECIES	NO.	SAMPLE NOS.
H. sativum	**1**	**116**
H. marinum	**3**	**104, 71, 74**
H. bulbosum	**6**	**103, 102, 101, 68, 78, 79**
H. distichum	**3**	**100, 90, 91**
H. tetrashichum vulgare	**1**	**99**
H. hexastichum	**1**	**98**
H. hexastichum* var *nudum	**3**	**97, 96, 95**
H. vulgare	**3**	**94, 93, 92**
H. genticulatum	**3**	**69, 72, 73**
H. murinum	**7**	**75, 76, 77, 85, 86, 87, 88**
H. murinum* var *glaucum	**1**	**70**
H. murinum* var *leporinum	**1**	**66**
H. spontaneum	**6**	**67, 80, 81, 82, 83, 84**

Classifications within the *hordeum* group represent an important factor in the identification of grasses within this project, as this is the genus to which many of the archaeological specimens are likely to belong. Unlike the *triticum*, classification is not straightforward, as there are variations related to the conventions of naming, for both varieties and species. While this obviously does not affect the chemistry, it can create confusion regarding species comparisons, when attempting to establish where similarities and differences lie. For example, the name *H. sativum* was previously a blanket name for the overall domestic species, while the name *H. vulgare* is now more commonly used. Still further confusion may arise from the fact that the name *H. vulgare* was commonly used to refer to a specific four rowed variety, *H. tetrastichum*. As a consequence of these complications, it was necessary to verify the sample names once again, to ensure that the correct format was being used. (Perscom Zohary and Hillman 1997). Once these were established, the overall results were ready to be defined via the recognized botanical data. Where this research is concerned, the initial identifications were made utilizing the biochemical and spectral information only. The independent results were then compared to established data, in order to qualify the findings.

The primary comparisons were made between *Hordeum spontaneum* and *H. distichum*, the two showing close similarities. *H. spontaneum* is the wild progenitor of the domestic barleys, and *H. distichum* is its closest domestic variety. The second set of comparisons was made between *H. distichum* and *H. vulgare*, which showed clear distinctive similarities, as they are effectively from the same biological species. There are subtle genetic differences, which are expressed phenotypically by the number of rows of grains, these being two rowed for *H. distichum* and four rowed for *H. vulgare*. The next comparisons were made between *H. vulgare* and *H. hexastichum*: these too showed the expected similarities both belonging to the same biological species. The differences can be seen genetically, and are expressed phenotypically as four rowed grains and six rowed grains, respectively. However, due to the fine distinctions between these genetic differences, they are not visible by the chemometric analytical technique. Although the spectra are not identical, it is extremely difficult to ascertain which part of the spectra is responsible for these genetic differences.

Other comparisons which show close similarities are *H. vulgare* and *H. tetrastichum var vulgare,* both of these are four rowed domestic barleys, (in fact *H. vulgare* is considered to be a synonym for *H. tetrastichum*). Therefore, their spectra would be expected to resemble one another, though not to be identical, as the samples were not monozygotic. There would be differences but these would be attributed to their 'individuality'. *H. hexstichum* and *H. distichum* also showed similarities: in fact it would be expected that *H. distichum* (2 rowed), *H. tetrastichum* (4 rowed), and *H. vulgare* (4 rowed) would show similarities, as they belong to the same biological family. All told, they would also all be likely to share similarities with *H. spontaneum*, which is, as previously mentioned, the wild progenitor of domestic barleys in general.

Finally there were comparisons between groups which, although evidently belonging to distinct species, did show certain tenable similarities. In the case of this group, it was easier to make clearer distinctions via their differences, as their similarities, if they occur at all, are less readily described. These are listed in sequence, the adjacent pairs being most closely related. This level of information is not evident via spectroscopic analysis, as the similarities between the more distant species are subtle, and usually only visible by way of genetic determination. These are *H.geniculatum, H. marinum, H. murinum, H. bulbosum* and *H. spontaneum*, and the group also includes the different varieties such as *H. murinum glaucum*, which is obviously close to *H. murinum*. The relationships between these groups are complex, but there are some which did show differences, such as *H. murinum*, which was readily distinguishable from *H. marinum*.

Each species was compared to the others in order to determine similarity; where none is stated, the sample was considered to be either a different variety or a different species. In such a case, the sample was labelled according to the finding. In spite of all the information available, it proved quite difficult to determine where the species boundaries could be drawn, especially for the *hordeum* genus. However, by combining the chemical, morphological and botanical information, it was possible to make a number of species differentiations.

8.5 Relationships between Taxa

The following group are all forms of domestic barley, and are all one biological species. Overall, the morphological differences distinguishing these varieties (e.g. the number of rows of grains, and whether the husks are fused to grain) involve gene differences at only 3-4 chromosomal loci. However, this is not to say that the particular specimens utilized in this project did not differ in respect of a string of genes controlling various aspects of their physiology.

Hordeum sativum (= domestic barley in general), *H. distichum* (= *H. distichon* in an older spelling), (= two-rowed hulled domestic barley), *H. tetrashichum* (= 4-rowed hulled domestic barley), *H. hexastichum* (= 6-rowed hulled domestic barley), *H. hexastichum* var. *nudum* (6-rowed naked domestic barley), and *H. vulgare* (= synonym for *H. tetrastichum* or, sometimes *H. sativum*.)

The name "*H. sativum*" is the blanket name for the overall domestic species, although many people now use the name "*H. vulgare*" as the blanket name. (This latter label is confusing, as the same name traditionally refers to a specific variety – 4-rowed barley, which is how it was used on the specimens of this name in the present work.) "*H. sativum*" is only ever used as a blanket name; however, when the specimens were used in this project, data was taken from each individual label as written on the storage tubes, in order to keep to the exact information as given from the original collection notes.

Hordeum spontaneum is the immediate progenitor of the domestic barleys (see above). The key features distinguishing them involve only 2-5 genes, and they are mostly regarded as no more than subspecies of the same species (*H. vulgare* subsp. *spontaneum* and *H. vulgare* subsp. *vulgare*). Biologically, however, they are no more than VARIETIES of the same species. The domestic variety closest to *H. spontaneum* is the one which is the most primitive, namely *H. distichum* (= *H. distichon*).

H. marinum, H. bulbosum, H. geniculatum, H. murinum, H. murinum var. *glaucum, H. murinum* var. *leporinum.* The two varieties of *H. murinum* are, of course, mere varieties of the same species: *H. murinum.* Otherwise, this group are all different species. They can be ordered in a sequence in which adjacent species are the most closely related. This would be as follows: *H. geniclatum, H. marinum, H. murinum, H. bulbosum, H. spontaneum, H. sativum/vulgare* (i.e. the domestic barleys, in following order: var. *distichum, H. tetrastichum, H. hexasticum, H. texastichum* var. *nudum*).

8.6 Identification of Archaeological Samples

Once the model content is verified, it is then possible to use this information as a means of identifying the unknown samples, by way of a comparative template. Each of the archaeological samples was tentatively assigned to a group, by utilizing the minute morphological detail (Hillman 1995). The spectral and biochemical criteria were used to either support this group or species allocation, or to propose another. Ideally, by combining all of the information gathered, both the new and that from prior research, it should be possible to identify the unknown specimens with a high degree of confidence.

The table below indicates how many possible species there are among the archaeological samples, their individual sample numbers, and how many there are of each species.

Table 8.22. Archaeological samples and their allocated genera.

GENUS	NO.	SAMPLE NOS.
Hordeum	**30**	**7, 8, 9, 10, 11, 12, 13, 14, 15, 16, 17, 18, 19, 20, 21, 22, 23, 24, 25, 26, 27, 28, 30, 31, 32, 33, 35, 39, 56, 58,**
Bromus	5	**1, 2, 3, 43, 48,**
Stipa	8	**5, 6, 29, 34, 36, 37, 38, 40,**
Crypsis	2	**4, 51,**
Aegilops	5	**41, 53, 54, 55, 57,**
Elymus	2	**42, 46,**
Lolium	3	**44, 45, 47,**
Tragus	1	**49,**
Festucoid	1	**50,**
Vulpia	1	**52,**

8.6.1 Recognition of Genera and Species from Archaeological Specimens

The collation of all the data enabled the recognition of both genus and in some cases species, from samples previously demonstrating ambiguities, derived from the effects of the preservation processes on the seed morphology. As can be seen by the end results, there was almost total concurrence on the tentatively assigned genera, which was enhanced by the allocation of the respective species, for almost all of the samples. This was particularly evident for the *Hordeum*, where there are a number of species which appear almost indistinguishable to all except the well-informed botanist. Nevertheless, when the same technique was applied to those samples tentatively assigned to groups other than *Hordeum*, both *Stipa* and *Bromus* followed suit, with little or no deviation.

As it was not only the largest of all the groups, but also the most likely source of all the archaeological samples, the first group of samples to have their initial identifications verified (for both genus and species) were the *Hordeum*. A small number of anomalies did appear, two samples produced spectra which were totally non diagnostic, therefore rendering identification impossible; one could not be placed within the current criteria, and three produced data totally different to that of the other *hordeum* samples. The data from these three samples, was so strong and significant, that it was evident that they belonged to an entirely different genus. The most likely candidate was considered to be *Lepturus*, as its seed samples appeared almost indistinguishable morphologically from those of the small-grained *hordeum*, but quite different chemically. This genus allocation was later verified morphologically (Hillman 97), in order to ensure that the information collected regarding these samples covered all aspects of the botanical data.

Once the main genera had been covered, it was then possible to look at the more ambiguous, and less well documented samples. Utilizing the same criteria applied to the *hordeum, stipa* and *bromus*, it was possible to verify some of the tentative groupings; however, it was evident that this was not possible for a small number of genera. Where verification was proving to be difficult, the information on that sample was compared to other, closely related genera. This investigative step demonstrated the closeness of many species, in biochemical terms, as in a number of cases the sample was correctly identified as far as the genus, but not at species level. This clearly indicated that a number of the seed samples could not be separated morphologically, but required a species specific compound or compounds which provided a chemical pattern, and by means of which each could be individually identified.

All told, once all of the archaeological samples had been investigated, 95% or more were identified down to either genus and species or just genus; this result demonstrated conclusively that the technique was producing positive results. Thus, it satisfied another core objective, which was integral to the project: the identification of archaeological samples via a chemical method, utilizing prior botanical information as a tool of verification.

8.7 The Appropriateness of the Statistical Analysis

When analysing chemical data once all of the laboratory work is complete, it is necessary to ensure the suitability of the numerical techniques selected. For this project, the initial design indicated which of these would be the most relevant for the questions being put forward. In addition, the large computer capacity allowed time for a test series to be run in order to try different numerical techniques. As modern systems are able to study large data sets in detail, in relatively short periods of time, it is possible to carry out significant experimentation with these techniques.

8.7.1 Variability

Expected results are part of the overall experimental design; they effectively suggest the initial direction the experiment should take, given the type of data being used. There may, however, be divergence from this suggested direction, anomalies occurring which cannot be predicted at the outset, as their effects are invisible until analysis has taken place. The most important aspect of this divergence is the magnitude: results appearing at too great a distance from what was anticipated may imply incorrect procedure in both the chemical and numerical techniques. On the other hand, the slight variation occurring in this project implies the predicted level of minor error, to be anticipated in experiments with natural extracts and a semi-quantitative analytical system The core questions are answered, and the model is a reliable template for the selection of genus from the unknown samples, therefore making it possible to group the previously unidentifiable archaeological specimens.

8.8 Effects of Seed Morphology

One of the core questions arising from this research is related to seed morphology: specifically whether differences or similarities between species are reflected by the spectral data. Where the modern identified species are concerned, it is possible to see that morphological differences are defined by the spectral results, albeit minute in some cases. Nevertheless, it is precisely these differences that allow for the utilization of this material as a template for the selection of the ancient specimens. Where the archaeological samples are concerned, morphological identification entails certain inherent problematic issues, chief among these is the fact that the remains analysed in this project were charred, which tends to render all of the seed material similar in external features. The extensive charring of virtually the entire range of archaeological materials meant that accurate specification was impossible if external characteristics were exclusively relied upon; this was one factor which suggested the employment of the type of analysis chosen. Once the chemical extracts were made, the use of spectral analysis made it possible to see that, despite their apparent physical similarities, some of the seeds were in actual fact from different genera, species and subspecies. This justified the use of the selected analytical methods as a means of resolving the identification problems, where the use of a single technique alone shown to be inadequate.

8.8.1 The Effects of Seed Size and Type on Extract Quantity and Quality

Although seed size and type are properly morphological characteristics, they can nevertheless have significant effects on the chemical extracts in terms of quantity and concentration of the required compounds. When dealing with natural materials such as seeds, size irregularity is not necessarily an indication of different genera, as variations both physical and biochemical in nature may occur within species. These variations may be environmentally initiated, and could comprise numerous factors (such as water and nutrient availability during maturation).

Seed size was included in the experimental considerations, because of the way the modern and archaeological extractions were made. As previously mentioned, a ratio of five modern seeds to one ancient sample was used per extraction; if seed size was a major factor, then this procedure could pose problems. In the event, as the results demonstrated that no problems arose, there was, therefore, no need for any modifications of the overall procedure.

This was visually evident when comparing spectra, where five smaller seeds produced a spectrum of greater proportions than five larger, implying that the former contained higher levels of the required extract than the latter. As different species are being compared, it is necessary to point out that in this instance the spectral interest is purely on dimensions, not bond and peak location. (Fig.8.28) This is a comparison of two modern species, the small seeds being represented by *Cenchrus ciliaris* and the larger by *Hordeum hexastichum* var *nudum*, two distinctively different species varying quite markedly in size. A further comparison was made of two modern types, *Triticum dicoccoides* and *Triticum dicoccum*, which are from the same genus but produce different spectra. One produces a spectrum with high levels, while the other produces a spectrum which, although still diagnostic, is of quite low proportions. (Fig. 8.29)

As previously mentioned, the modern seeds were extracted in sets of five, which was considered an appropriate number: while not excessive in terms of comparison with archaeological samples, it would produce enough extract to be diagnostic.

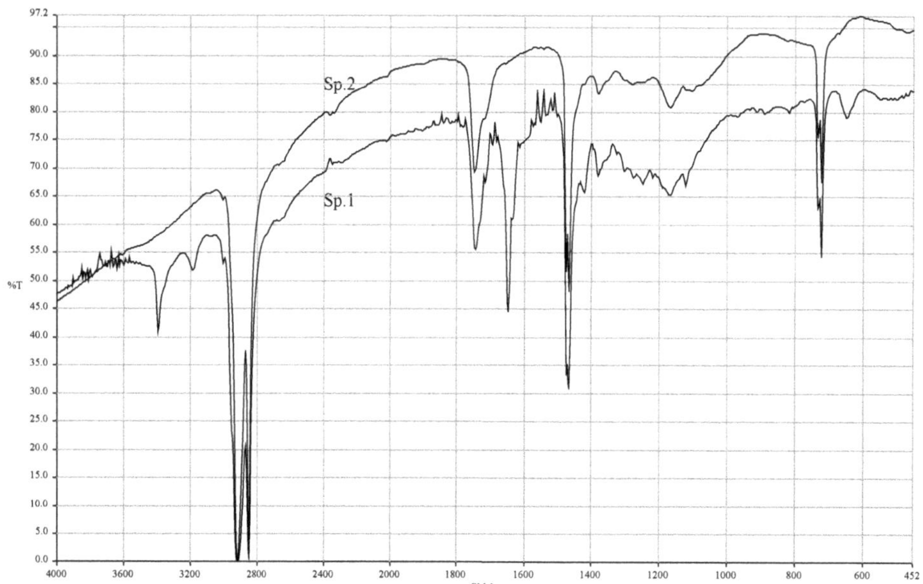

Comparison of spectra from two modern samples Sp.1 Cenchrus ciliaris (small) Sp.2 Hordeum hexastichum (large)

Fig. 8.28

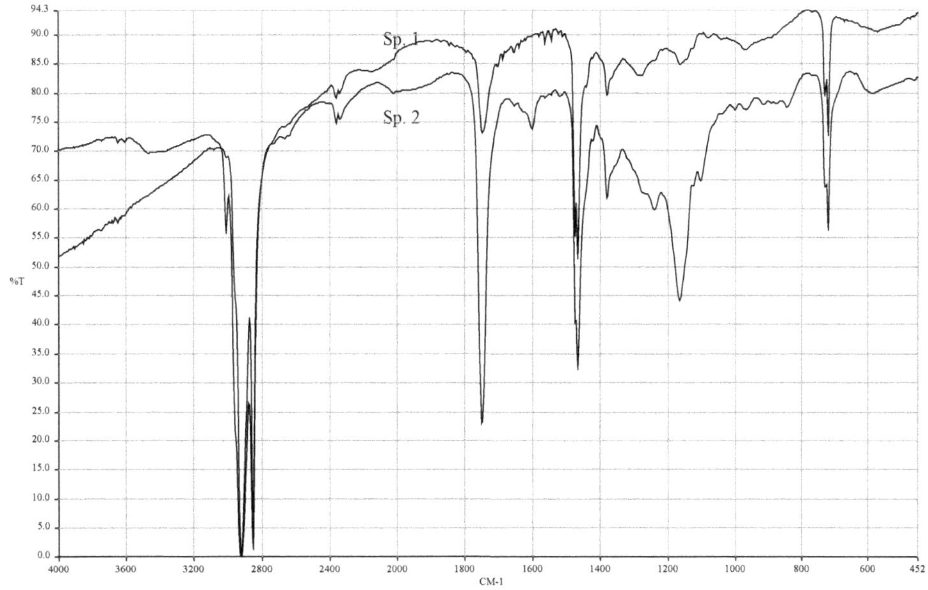

Comparison of spectra from two modern samples Sp.1 Triticum dicoccoides (Large) Sp. 2 Triticum dicoccum (large)

Fig. 8.29

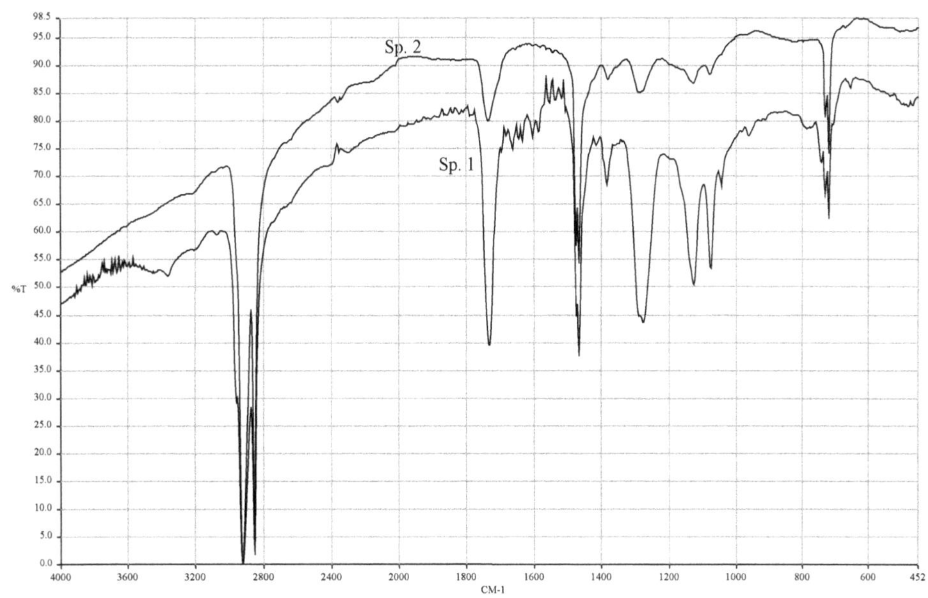

Comparison of Sp.1 Archaeological sample Hordeum? Sp. 2 Modern sample Hordeum bulbosum

Fig. 8.30

However, as demonstrated by the spectral comparisons above, size must be used with considerable caution, as larger dimensions do not necessarily yield larger extracts. A still more vivid illustration of this fact was provided when spectral comparisons were made between archaeological samples where, in order to prevent cross species contamination only one seed was available for each extraction. Here, a small seeded sample, tentatively identified as *Hordeum* (Hillman 1995), produced a spectrum of extremely high proportions, with peaks indicating extract content higher than its modern equivalent. (Fig. 8.30)

These observations answered yet another key question posed by project, concerning the need for a biochemical approach: this concerns the construction of an internal picture of a retrievable chemical compound, a picture which functions as a type of map to aid in species identification. As can be seen from the illustrations above, they clearly support the requirement for this type of analysis to function as an important adjunct to the morphological studies. Often, the external features of seeds may not accurately represent the specimen as a whole.

8.8.2 The Effects Of Genotype Change in Seed Morphology and Biochemistry

Genotype changes may occur over time as responses to environmental influences, as climatic variations alter rainfall patterns, nutrient availability and nutrient type. They may also however, occur as a result of human intervention. For example, hunter-gatherers collected seeds from wild grasses, their actions may have initiated a selection process that would eventually show up in the ancestral genotype. This may possibly have been a consequence of the selection of plants with specific characteristics, such as those which yielded the seeds more readily, or those with seeds held tighter in the car. Whatever the causality, pressure exerted on these plants will eventually be phenotypically manifested as changes in the morphology.

Where such changes in morphology are evident, it is also likely that the biochemistry of the seeds may have altered through time, a factor of particular relevance in those instances where seeds that were once wild grasses are now important cultigens. Fortunately, however, there remain isolated sources of the wild precursors containing biochemical information which is still relatively close to that of the original seeds gathered from the wild. Regardless of the specific origins of this biochemical data, these modern wild and domesticated grains are the sole source of information upon which to base the identification of the archaeological specimens; there is no other comparative source. As a consequence, it is important for this research to demonstrate these biochemical differences, primarily those which obtain between the modern genera and, where possible, species; and then utilize this as an aid in the classification and identification of the archaeological specimens. Fig.8.31 indicates, by spectral patterns, the differences between samples of the same genus *Triticum araraticum* and *Triticum dicoccoides*. This is an instance in which variation has evolved over time, separating the two, by both morphology and biochemistry, even though they are both wild emmers.

Whatever these differences or similarities are, it is important to be able to recognize them, particularly when looking at wild and modern species of the same genera, as these are visible variables which are incorporated into the model, and will effect the level of identification of the archaeological specimens. The fact that the identification has been positive to the level of species for these specimens indicates that all of the relevant variables have been considered and incorporated into the identification schema.

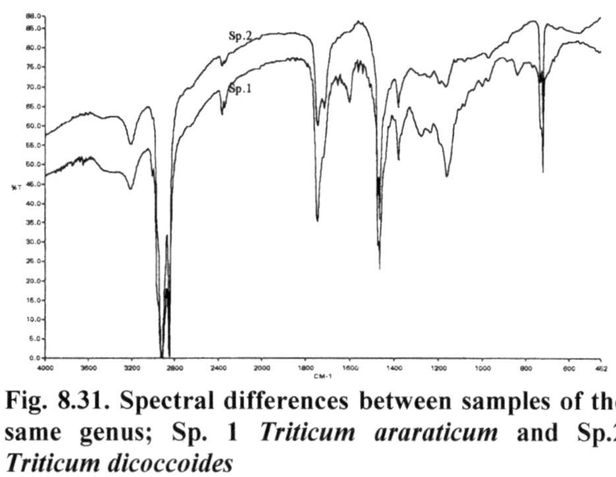

Fig. 8.31. Spectral differences between samples of the same genus; Sp. 1 *Triticum araraticum* and Sp.2 *Triticum dicoccoides*

8.9 CONCLUSION

Given the extent and rigour of the experimental analyses to which they were subjected- both chemical and numerical - it is clearly evident that lipids were present in the ancient samples. As stated above, the experimental design took into account a broad range of theoretical and methodological considerations gathered from various sources, including materials which are as yet unpublished. The work of Evans and McLaren, in addition to previous experimentation by Cave, had, by making use of the simple expedient of spectral comparisons between ancient seeds and their contemporary counterparts, demonstrated the continued presence of lipids in the interior of the former. Furthermore, extensive practical work carried out in Evans' laboratory had provided conclusive evidence that the charring process had caused no deterioration in lipid content; seeds charred under a wide variety of conditions had continued to yield their lipid content when put through the extraction process. Had the lipid components failed to persist through the charring processes, and consequently remained unavailable in terms of structural information, there would have been no basis for the spectral comparisons included in the current research; the project would therefore, of necessity, been confined to an investigation of the manner in which such material is lost in the passage of time. However, this was not the case.

The work of Eglinton has also demonstrated the durability of lipids, showing that they are very likely to persist over extremely lengthy periods of time once the initial phase of the decay process has closed off This brings us once more to Hillman's hypothesis regarding the prophylactic capabilities of the outer shell, which are, he suggests, reinforced by the action of charring, and combine with the desiccating burial environment to produce a protective sheath across which nothing passes in either direction.

CHAPTER 9

DISCUSSION OF RESULTS

9.0 Introduction

The results reveal the key objectives to have been answered, and in some cases with far greater success than originally anticipated. To answer even part of the inquiry is a substantial step forward, particularly when dealing with archaeological materials that can be extremely ambiguous (e.g. charred seeds with an altered physical appearance due to heating or abrasion, which renders them unidentifiable from their morphology). Indeed, because the techniques themselves are on trial as much as the materials being tested, it is not unfair to say that any level of results that can be deemed positive is a success. Overall the process of analysing spectra numerically as applied here has proved to be reliable to the level of the statistically predicted goodness-of-fit which in this case was 82% at its optimum.

To be able to reliably identify charred grass remains at the genera level and also the species level was extremely encouraging. However it must be pointed out that the 82% goodness-of-fit left at least 18% unaccounted for. This will inevitably have produced anomalies within the results that cannot be ignored. Indeed, this 18% may help explain not only where the deviations from the norm lie and how they affect the results, but also what their origins are. If it is possible to identify the source of these errors, then it may be possible to eliminate them from the outset should a similar or follow-up programme ever be undertaken.

9.1 Project Analysis

To discuss the results in any detail regarding the way in which they may or may not have answered the key objectives, it is essential to first consider the project as a whole and how its basic design affected the outcome.

The initial experimental design divided the project into four main components, each part being totally reliant on the success of the preceding component. It was, therefore, essential to ensure that each component was producing the required positive outcome before embarking on the next step of the experiment. As indicated in section 8.1, the four main components are as follows;

1. Processing of botanical samples: that is the seed preparation for both the archaeological and modern specimens.
2., Wet chemistry: the extraction procedures for the two solvents applied to both the modern and archaeological seed samples.
3. Instrumental analysis: the utilization of Fourier Transform Infrared as the principal system for producing spectroscopic data.
4. Numerical analysis: chemometrics as the preferred method of data handling for the accumulated spectra.

Each of these areas produced positive results albeit at different levels. The botanical preparation was totally successful in the sense that all of the modern seeds were cleaned of husks before being utilized in the reflux procedure. The main question here was whether there was actually any need to remove these outer coatings, i.e. would leaving them attached have altered the seeds chemical components to the point where comparisons with the ancient materials was invalidated on account of the charred remains lacking these coverings. This question was already partially answered by Evans 1990 who had demonstrated in previous work that these husks do contain their own lipids and should therefore be eliminated when modern and ancient samples are to be compared.

Where the wet chemistry is concerned the high level of success was easily predicted especially where the modern materials are concerned, as it is a standard technique for the extraction of compounds from an organic matrix such as plant material. The same was true of the targeting of specific compounds by using particular solvents for which these substances have an affinity albeit only for the modern material, as it is not possible to know the level of deterioration these compounds have reached in the ancient specimens. Nevertheless, the targeting process will also bring out the products of this deterioration, which is information of potential value in its own right.

That is because deterioration products such as aldehydes or ketones can create problems with the final data analysis if they are not recognized at the outset. As they are when the spectra are finally converted to numbers. Therefore if these compounds are seen to be present in the initial extraction, it would be wise to run an alternative sample, if available, in order to prevent these problems at the very beginning.

Previous research (see chapter 1) has already demonstrated that infrared data can provide a valuable method for comparing modern and ancient samples. However, the testability and repeatability of these comparisons depends on the method used to affect the comparisons. For example, most of the samples both modern and archaeological produced spectra of potential use in further analysis. However, the final numerical analysis demonstrates that spectra from samples extracted in hexane were unusable for use in chemometric analysis. Nevertheless, because all samples were extracted through two solvents, hexane and chloroform, it was possible to use the latter for the subsequent identification analyses, since they contained all the essential information required for diagnostic peak recognition and numerical conversion to matrices.

One of the major considerations of the experimental design was how to handle the bulk data that was to accumulate from the FTIR analysis. As previously mentioned, visual interpretation, as the only means of spectral comparison, was considered inappropriate because the huge size of the data set would render this unworkable and would inevitably create errors effecting the whole of the results section. As a consequence, the chemometric numerical analysis was considered as the best option, as this allowed for not only greater accuracy but also the possibility of repeating either all or any part of the project to test the reliability of the numerical outcome. This introduced objectivity to the spectral interpretation, as it did not rely totally on the opinion of the analyst.

The overall success of the numerical analysis can be measured to a certain extent by the goodness-of-fit of 82%. It is therefore possible to say that as the fourth major component of the project, the chemometric analysis proved to be successful but not without problems, however these can be understood and dealt with once the technique becomes more widely used and accepted in spectral interpretation.

9.2 The Chemical Regime

In a project that is looking for answers to an essentially archaeological problem, is it necessary to adhere to a regime normally designed for chemistry?

When the project was first designed, aspects such as the botanical identification of the modern seed samples were well established, while others such as the utilization of chemometrics for the identification of compounds from archaeological materials was previously untried. In consequence, it was essential to apply a regime to the chemical and instrumentation programme that would allow each step to be recorded, such that any errors that occurred would be quickly recognized and resolved and allow the development of a system of standardization in which each part is subject to the same treatment and allow the results to be vigorously quantified.

This systematisation, which primarily consists of calibration and validation, proved successful in helping to produce a good reliable data set which, in turn, enabled the construction of a robust model to be used as a template for the identification of the archaeological specimens. As cited in section 8.3.6 of chapter 8, the validation was somewhat unorthodox. It was carried out by exposing both the calibration set, and the random subset, to the same sort of analyses as the modern identified data set. (Normally the validation would be performed using an independent set of known composition that is not included in the actual calibration set.) Since some form of validation was essential, this seemed to be the most logical one available and turned out to be the most workable of the possibilities.

9.3 The Numerical Analysis

The utilization of the chemometrics system involved a series of numerical analyses in order to see if their format would fit the type of data generated by the instrumental analysis of archaeological data. If this project had been contemplated a decade ago, the application of so many numerical computations would have been almost impossible. However with the technology now available, it was possible to run a whole series of analyses to assess the value of each according to the results from their goodness-of-prediction and goodness-of-fit computations. The outcome of these analyses are presented in their entirety in both the results section of the text (chapter 8) and part two of the appendix, as it was considered important to show just how many types of statistics were available and how they compared in terms of results obtained from the same matrix analysis. As can be seen some performed well, such as the principal components regression (PCR). By contrast, the ordinary least squares (OLS) proved to be poorly suited for these data. All told it seems that the selection of a numerical analysis to find a positive result in terms of model production and the identification of the archaeological specimens was a success, as these tasks produced a relatively high order of goodness-of-fit.

9.4 SCAN: Species Identification

As previously mentioned there are several main components of the project as a whole each requiring a positive outcome if the next part is to work at its optimum. One of the most important aims was the identification of the archaeological seed remains, and the selection of the most effective way to carry this out. SCAN chemometrics, the computerized statistical package for spectral data, offered a particularly effective system for achieving this.

Once the model construction was complete and the calibration and validation had been carried out to ensure it was functioning at its optimum, it was then possible to assess how the SCAN analysis would work on the different groupings from the most coarse clusters to the more refined regression analysis for genera confirmation or species selection. As pointed out in section 8.4 these precise levels of identification were achieved, and can be cited as a source of reference with a high degree of confidence.

9.5 Analysis of Archaeological Remains

The identifying archaeological remains of caryopses using the chemometrical system of analysis was undertaken "blind". Firstly, many specimens were received with no indication of their possible taxonomic affinity, either because they had an aberrant morphology or were very poorly preserved. Secondly, even with specimens where Hillman or other archaeobotanists had been able to use the surviving morphology to tentatively suggest some level of identification, a] when I recorded an identity based on my own analyses, I made no cross-reference to my record of the name originally suggested; b] some of the names used were outdated synonyms, and I was (and am) insufficiently familiar with nomenclatural niceties to know which were synonyms of which; c] I was (and am) insufficiently familiar with grass taxonomy to have known whether what turned out to be alternative identities (indicated by my chemical data) represented closely related species, different genera, different tribes, or even different sub-families.

It was therefore only when I worked through my results with my supervisor, Gordon Hillman, that we finally compared the two sets of identifications (one based, often very tentatively, on morphology), the other based on my chemical criteria, and he was able to tell me whether any of the identifications either accorded with prior expectations, overturned them, or represented a closely related taxon.

9.5.1 Results from the Analysis of Archaeological Remains

Overall there was considerable accordance, at least at the level of subfamily, tribe and genus (see table below), but there were also some interesting exceptions. For example, several specimens very tentatively assigned *Aegilops* (with at least one question mark in each case) turned out to be *Hordeum*, Similarly, *Vulpia* turned out to be the closely related *Loliolum*, and *Tragus* turned out to be *Cynodon* a closely related member of the same tribe which, while in the same tribe, is an entirely different genus. These anomalies together with the good fits are explained below.

1. Coarse-Clustering: It proved possible to cluster taxon-omic groups such as sub- families and tribes.
2. Fine Groupings: It also proved possible to clearly dist-inguish groups at the level of genera.
3. Finer Separations: The finer the analysis, the more pos-sible it became to distinguish different species within individual genera. By occupying their unique place within the numerical matrix, it offered the prospects for identifying 'unknowns' at species level.

The results mostly confirmed Hillman's identifications at the generic level based on morphology. The chemistry was just able to take the identifications that crucial step further, and identify them to species. This is clearly demonstrated with both the *Stipa* and *Hordeum*, where it proved possible to identify the archaeological specimens down to genus level using morphology alone, but required chemical criteria to separate them into their different species.

9.5.1.1 Hordeum

Wild *Hordeum* species, the small-grained species of this genus are extremely difficult to distinguish using morphological criteria (cf. Nesbitt 1997). However, using chemical criteria, it has generally proved possible to distinguish between each of the species of this genus, explored so far, barring the different forms (hitherto referred to as separate species) within the *Hordeum murinum* aggregate, e.g. "*Hordeum glaucum*", "*Hordeum leporinum*" and "*Hordeum murinum*" in the strict sense.
Virtually all the 30 grains tentatively referred on the basis of morphology to *Hordeum* were shown by chemical criteria to be of that genus. However all 7 specimens originally labelled as "*Hordeum* ?? *marinum* type" turned out to belong to the adjacent *Hordeum murinum* aggregate. Similarly, 17 specimens turned out to be *Hordeum bulbosum*, although only 3 had originally been assigned to this species. An unexpected result involved the 3 specimens which did not fit any of the *Hordeum* aggregate; these were tentatively assigned to the *Lepturus* group, although further work is required in order to be absolutely positive regarding this designation. Finally there were 3 specimens that were non-diagnostic: their spectral results were such that any sort of comparison would be inconclusive, and any numerical interpretation would produce an outlier.

9.5.1.2 Stipa sp., the feather grasses

Similarly to *Hordeum*, from morphological and micro-morphological criteria have not allowed archaeobotanists to distinguish the remains of grains of the many species of this critical genus. They could only group them together into crude size categories approximately in two major sections of the genus, and even then with only considerable uncertainty. Once again, however chemometrical analysis of FTIR spectra has allowed me to distinguish between each of the 7 species investigated so far, and to identify archaeological specimens (in respect of these 7 species) with complete confidence.

It was possible to confirm that all the archaeological specimens referred to as *Stipa* by Hillman were, indeed, of this genus. Most (but not all) of the specimens described as "large-grained *Stipa*" proved to be *Stipa holosericea* (syn. *S. lagascae*), the dominant large-grained *Stipa* species of present-day Syrian moist steppe. Similarly, most (but not all) specimens labelled as "small-grained *Stipa*" or "*Stipa parviflora* type?" proved to be *Stipa capensis*.

In a few cases (1) however, small-grained specimens proved to be *Stipa holosericea* and 1 large-grained specimen proved to be *Stipa capensis* .

Two of the ancient grains were exceptionally large, and Hillman tentatively referred to them as *Stipa gigantea*. My analyses revealed both of them to be no more than over- sized grains of *Stipa holosericea.*

9.5.1.3 Aegilops

Similarly to the procedure for *Hordeum* a number of specimens were very tentatively assigned to "?*Aegilops* type" on the basis of their morphology. However, the spectra from these archaeological samples appeared not to fit any modern specimen genus either spectrally or numerically. As a consequence it was necessary to analyse all potential genera and species, in order to find the taxonomic group they best fitted. This was carried out entirely by matrix numerical comparisons without the added bias of possible related species, as my unfamiliarity with grass taxonomy means I am not able to find results to suit a preferred outcome. Finally all five had the closest possible fit with the genus *Hordeum*, with three fitting *Hordeum bulbosum* and the remaining two

Hordeum murinum, a result that complemented the actual specimens originally already assigned to the *Hordeum* aggregate.

9.5.1.4 Bromus

The identification of the specimens tentatively assigned to *Bromus* was quite straightforward, with the chemical results confirming the initial morphological identification at the level of genus. However to assign any of these specimens to particular *Bromus* species proved to be quite difficult, as a) the differences between them were very slight, such that no particular assignment could be definite. b) The modern species included in the project represented less than $^1/_5$ of the very large number of species that grow in South West Asia. (The decision to include relatively few *Bromus* species reflects the fact that the grass remains from Abu Hureyra include few specimens recognizable as belonging to this genus. In order to prevent errors, therefore, none of them were given a species identification.

9.5.1.5 Tragus

My results indicated that two grains identified as *Tragus* by the morphology, were actually from the closely related genus *Cynodon*.

9.5.1.6 Elymus

In the case of *Elymus* there were only two archaeological specimens. One of these clearly fitted the numerical dimensions of *Hordeum bulbosum*, while the second one was not so clearly definable. As previously stated, where this lack of clarity occurs, the sample cannot be competently identified to species until further analysis can improve the information available, and/or until missing taxa are added to the modern model.

9.5.1.7 Crypsis

Where Crypsis was concerned, I again examined only two specimens. Only one clearly fitted in the genus *Eragrostis*, but the confident assignment of an actual species proved to be difficult and so none was given. The second specimen also lacked clarity; consequently it was not possible to assign either an alternative genus or a feasible species.

General: Overall it is possible to see from the examples discussed that the numerical system worked successfully at the relatively high level of precision accorded by both the goodness-of-fit and the goodness-of-prediction. As both were working at this level of confidence the confirmed genera and assigned species can be considered as accurate, and it is perhaps not so surprising that the results fit in with other quite independent data, such as Hillman's ecological models.

9.6 Archaeological Significance of the New Levels in Precision of Identification The new levels of precision in identification of grass grains allows us to address entirely new questions concerning archaeology and palaeoecology. This is true even at sites such as Abu Hureyra where there has already been 25 years research on plant remains involving a battery of many of the latest developments in archaeobotanical methodologies. A few examples must suffice.

9.6.1. Was the start of cereal cultivation really triggered by advancing aridity?

Hillman's (in press) evidence front Abu Hureyra suggests that, once the cold, dry conditions of the period called the Younger Dryas began to 'bite', a series of plants that provided staple food that the local hunter-gatherers had hitherto gathered from the wild began to decline in availability: first the most drought-sensitive in the form of acorns from the local oaks; then the slightly less drought-sensitive wild lentils and vetches together with some of the *Cruciferae*; then the wild wheats and ryes; then the more drought resistant feather-grasses (*Stipa* species)-, and finally the very drought-resistant shrubby Chenopods. It was during the middle stages of this series of reductions in availability of wild foods that Hillman has found that some of the people of Abu Hureyra started to cultivate certain of their cereals as crops. However, the data are still very coarse, and to identify more precisely the stages in advancing desiccation and to pin- point the stage at which they started to cultivate, requires that we distinguish between, for example, the different species of feather-grass (*Stipa* species). Each of the *Stipa* species exhibit different degrees of drought tolerance, but hitherto it has been impossible to distinguish them from morphological criteria. (Hillman *et al.* 1989; Nesbitt 1997)

However, the results of this project indicate that the grain remains of the key *Stipa* spp. can now be identified to species. So, once we have identified sufficient numbers of the charred remains, we can plot the pattern of their changing abundance through time, and follow the decline in each of the species present, and work out precisely how desiccation was affecting the local supply of wild foods. This would then provide the means of testing Hillman's (1996, in press) model for the adoption of cultivation during the early 9th Millennium bc (uncal.) being triggered by the loss of specific, key wild food grasses, and for this, in turn, being due to advancing desiccation associated with the Younger Dryas episode.

9.6.2. Where did they travel to collect their grain of wild wall-barley grasses? Did hunter-gatherers travel deep into the drier zones of steppe for such relatively low-grade food?

As with the *Stipa*, the charred caryopses of different species of *Hordeum* wall-barley grasses (*Hordeum* species - excluding the cultivated species) are almost impossible to identify from their morphology. This is particularly frustrating in the first phase of the Epipalaeolithic levels of Abu Hureyra, as these grasses not only appear to have been a gathered food at this stage, but are also characteristic of different habitats and could therefore provide a useful tool for helping to plot the seasonal movements of hunter-gatherers. Again, however, the chemical criteria demonstrated that it was possible to distinguish between each of the key species. Provisional results suggest that the principal species gathered in this earliest phase was *Hordeum bulbosum* rather than *H. murinum* or a member of the *murinum* aggregate as initially assumed by *Hillman et al.* (1989). The ecological significance of this identification is that, while *Hordeum bulbosum* forms its dense stands in rocky terrain, screes and other disturbed areas with moderate rainfall that could potentially support park-woodlands and the damper sub-zone of woodland steppe (Hillman 1996). By contrast, *H. marinum* and near eastern members of the *Hordeum murinum* aggregate form really dense extensive stands (from which it would be worth hunter-gatherers gathering grain in bulk) primarily in salty or damp hollows in dry steppe.

Hillman's 1990 model for the environment of Abu Hureyra during phase 1 of the Epipalaeolithic suggests that, although it was rapidly becoming drier, it still supported rich woodland-steppe, and that near dry steppe probably existed only at some distance to the south cast. At this point, therefore, gathering grain from *H. marinum* and/or members of the *Hordeum murinum* aggregate would have involved the people of Abu Hureyra in long treks into the steppe. My new identifications suggests that most of the grain came from *Hordeum bulbosum* which could have grown on cliffs at the edge of the Euphrates Valley close to Abu Hureyra. This conclusion, therefore fits Hillman's ecological models, and suggests that the local hunter-gatherers did not needlessly expend energy on gathering grain from more distant stands.

9.6.3. How was it that the people of Abu Hureyra were gathering grain of a desert species (*Tragus*) at the start of the Epipalaeolithic when conditions were quite moist?

By using my chemical-numerical criteria to reassess the identification of grain suggested by morphology to be *Tragus*, it was possible to resolve another ecological enigma. *Tragus* is a grain that was often gathered as food by pastoralist peoples in areas such as the Sahara and Rajasthan deserts. Ecologically, however, it did not fit with the other species recovered from Abu Hureyra. Since other evidence regarding the climatic conditions of that time indicated that the environment was relatively moist, and that the nearest desert was at least a hundred miles away, it seemed very unlikely that the specimens assigned to *Tragus*, were actually of that particular genus. As noted above, my results indicated that the "Tragus" grains actually derived from the closely related (but slimmer grained) *Cynodon dactylon*, Bermuda grass. Although archaeobotanist always make allowance for seeds inflating during the charring process, it transpires that *Cynodon* inflates far more than most. Clearly, therefore, identifications of *Cynodon* based on morphology need to make greater allowance for the grains puffing up during charring even more than most other grass grains.

However, the major issue is that *Cynodon* is ecologically quite different in its requirements from *Tragus*. More specifically, *Cynodon dactylon* or 'Bermuda Grass' would have grown around the salty edges of backswamps and wadi-mouths of the nearby Euphrates Valley, not only putting them within easy reach of the hunter-gatherers of Abu Hureyra but also fitting closely with Hillman's ecological models. This is therefore a very coherent result, demonstrating how the chemistry was able to resolve a misidentification involving confusion between two members of the tribe *Cynodonteae* and consequently explain what appeared to be an ecological contradiction.

Reconstructing the nature of the ancient resource base relies upon knowing which plants are characteristic of a particular environment, and resolving ecological anomalies of this sort allows a much more coherent approach to the entire process.

9.6.4. How did the giant feather-grass, *Stipa gigantea*, a plant of the Central Asian Steppes, come to be available to the Epipalaeolithic hunter-gatherers of Abu Hureyra?

A further ecological anomaly arose when Hillman tentatively identified two grains as belonging to the "*Stipa gigantea* type" (Hillman *et al.* 1989). Today this particular species grows only in Central Asia and North East Iran and requires forms of steppe that Hillman's ecological models suggest never existed in the Abu Hureyra area.

Hillman's identification of these two grains had been based on their exceptionally large size, which fell well outside the upper end of the frequency distribution of sizes of the other grains for the site, the largest of which he had tentatively referred to the "*Stipa holosericea* type". Chemical analysis finally revealed the grains to be *Stipa holosericea*, after all, albeit exceptionally large ones, and this fitted perfectly with Hillman's ecological models. This clearly demonstrates the value of precise identification down to species level of the sort that could not be achieved using morphology.

9.6.5.Why did the Epipalaeolithic population of Abu Hureyra bother to gather grain from the prostrate tussocks of *Crypsis* when there were much more readily harvestable small-grained grasses available?

Using grain morphology, Hillman identified two batches of the tiny grains of "*Crypsis* type" (Hillman *et al.*). However, this posed him a problem of accounting for how they arrived on site and became charred. He only assigned plants to the category of "gathered foods" in cases where there were ethnographic parallels to their use as grain foods among recent hunter-gatherers occupying analogous resource environments (Hillman 1998). With *Crypsis* and/or equivalent small-grained prostrate mat-grasses, he was unable to trace such parallels, which is not surprising in view of its prostrate habit and the corresponding difficulty of gathering its grains.

Again chemometrical analysis came to the rescue. The "*Crypsis*-type" grains turned out to be from another grass with equally tiny grains, and relatively long embryos, namely *Eragrostis* (one of the "love grasses") which grows sufficiently tall to have been readily harvested by the usual methods employed by hunter-gatherers, and which has been widely used as a food grain by hunter-gatherers in recent times (Hillman 1998). Indeed, one species of *Eragrostis* (*E. teff*) has even been taken into cultivation and domesticated (Simmonds, 1976).

This new identification therefore makes sense of another apparent anomaly. Ecological implications are unaffected in this case, as the local species of *Eragrostis* and *Crypsis* occupy closely similar habitats (Hillman, l998).

9.7 How has this research aided the identification of archaeological materials?

As noted above, the modern samples analysed were predominantly wild, small seeded grasses, which were largely identifiable from their morphology, robust and easy to handle. By contrast, the archaeological samples were extremely fragile because of the initial charring process and subsequent treatment, somewhat less easy to handle, and identified only tentatively or to the level of genus. Indeed, for many of them tentative identification had been suggested at only the more general level. However, many of them are archaeologically important, as they appear to have served as staple foods for early hunter-gatherers of South Western Asia. A sound identification of this material therefore aids not only the understanding of what grain foods were gathered, but also in the cultural development of these people, for example-relative to the origins of farming.

CHAPTER TEN

CONCLUSION AND FURTHER WORK

10.0 Introduction

Once a project has been completed, it is possible to consider its design, methodology and results in retrospect. Scientific problems do not remain static, and consequently there is always scope for improvement; it is possible to analyse the strengths and weaknesses of the finished product and amend where necessary in order to improve performance should the experimental procedure be repeated.

10.1 Summary

This project utilized a system of modelling and prediction. A series of known modern species was employed to generate a model capable of identifying unknown archaeological samples. A number of problems arose in relation to this objective, and certain central questions were posed: which modern species would be most likely to represent the ancient ones? How many species would adequately cover the whole range of ancient possibilities? What would be the most practicable way of carrying out the analytical processes in the allocated time to produce the best results?

Looking back through the project as a whole it is possible to see how each of these core objectives was answered; by closely examining the specific components of design, methodology and results, it is possible to see what these components were able to achieve when they worked both independently and in combination. The first major test was for the overall project design itself: was it indeed possible to identify archaeological specimens by means of a combined use of chemical and numerical techniques? Although this type of dual approach has been deployed with some success in other areas of the field, it is quite new to archaeological chemistry. As a consequence, the technique itself was as much on trial as were the individual components and hypotheses.

One of the major aspects of the experimental design was *optimisation*; that is, its ability to utilize each facet of the project (comprising chemical extraction, and both instrumental and numerical analysis) to its maximum potential in order to ensure the identification of the unknown samples. A retrospective appraisal demonstrated the achievement of optimisation across the various discrete methodological strands. For example, the chemical extraction of the triacylglycerols from the seed matrix proved to be successful particularly for the chloroform extracts, for both the modern and the archaeological samples. By using solvents which have an affinity for the requisite compounds, it was possible to target the matrices with the greatest diagnostic potential, thereby avoiding arduous processes of trial and error which can be both time consuming and expensive.

Once the sample extracts from both hexane and chloroform had been produced, they were analysed using FTIR in order to obtain the best possible spectra for both the spectral library and the ensuing numerical analysis. Although a great many of the hexane extracts contained few (if any) diagnostic regions, this fact was demonstrated before the numerical analysis was carried out. As a result, time was not wasted constructing matrices from hexane spectral numbers, and they had no influence on the final results.

In this way, the overall project design was optimised by eliminating certain processes that were eventually shown to be non-essential with respect to the final outcome, thereby allowing more time and resources for the positive aspects of the analysis. In particular, the outcome of each successive step was assessed in terms of its role in the broader methodology, thus ensuring that the desired result had been obtained before proceeding to the next. As a consequence, errors were minimized, and their incorporation into the experimental procedure as a whole was prevented.

The project has successfully integrated the results of experimental wet chemistry, instrumentation and numerical analysis, although each of these areas was independent to the extent that they could be refined separately before contributing to the overall outcome. As a consequence, each section produced the best results possible at this stage in their development. Indeed, when combined they answered the main objectives of the project at the 82% level of confidence, which implied that the modern model was accurate and the identifications of the ancient remains were surprisingly reliable. However, it must be pointed out again that there was also an 18% error calculation, which may account for certain anomalies with regards to a few of the identifications. While these anomalous data present no serious problems for the work, they should not be entirely discounted.

10.2 Further Work

This research project has, by utilizing the procedures laid out in the experimental design, clearly answered its core objectives. Nonetheless, time has moved on, and new ideas have developed that would still further improve the outcome. The constraints of time, however, do not permit them to be applied as part of this thesis.

As a consequence of these developments, the following suggestions are made for additional research to be carried out, should it eventually prove possible. (I follow the basic format of the original project.)

10.2.1 Incorporation of additional modern taxa in the model

The taxa selected for the original project had to include the most important tribes and genera likely to have been available at Abu Hureyra during the Epipalaeolithic and early Neolithic era: these were taxa such as *Stipa* and wild *Hordeum* spp.

In order to achieve the main objectives of the project, it was necessary to include as many taxa as possible; however, it was not feasible to cover all potential candidates due to time constraints. As a result, some major genera such as *Bromus, Eragrostis* and *Aegilops* were represented by only a few species. Indeed, certain minor genera that could theoretically have been present at Abu Hureyra had to be omitted altogether, these included *Puccinellia, Catapodium, Koeleria, Trisetum, Polypogon, Boissiera, Aristida,* and *Dactyloctenium*. Ideally, all taxa that may have been available at Abu Hureyra should have been incorporated into the model; but-as ever practicalities intervened, and demanded some form selection.

If, furthermore, the model based on modern grasses were to serve for the identification of assemblages of grass remains for excavations in other parts of S.W. Asia, yet more taxa would need to be incorporated. For example, identification of remains from hunter-gatherer sites in the Jordan Valley such as Ohalo II would need to include a range of *Puccinellia* species and many additional members of the millet tribe *Paniceae*.

10.2.2 How many of each species?

Another methodologically important question to consider was: how many of each separate infra-specific component species should be used to ensure all of the possible components of variation within the species itself are covered (for example, the variation between populations within species [especially populations from profoundly different habitats]; variation between plants within populations, such as heads and ears, and grains within spikelets)? This was a major problem with the current project: time constraints made it impossible to cover the optimum numbers for modelling and predicting; and it was only possible to use a relatively small number of samples of any one target species.

Overall, then, a new project would increase the numbers of species used in the project, making sure that at least five seeds for each of the additional population samples were available, in order to standardize all the numbers used in the extraction, calibration, validation and the final statistical analysis. Such a measure would enhance significantly the reliability of the statistical results, as it gives a greater scope for improving still further the goodness-of-fit and hence the prediction of unknowns (Increasing information correspondingly increases stability.)

10.3 Numerical Analysis

The numerical analysis included in the current project assumes a proficiency in both statistics and chemistry. The major problem here is that archaeologists, for whom this system was devised, are not normally trained in both of these disciplines as well as in their own areas of archaeology. Ideally, time is needed to refine this system, in order to find ways of reducing the required level of proficiency in statistics and chemistry for the operator, while at the same time retaining the necessary sophistication of the computer package as a whole. If the system presented in this thesis is to be used as a tool for the routine identification of archaeological plant remains, it is essential to explore all the possibilities related to its simplification in terms of operation, that is rendering it more *'user friendly'*. The basic idea is therefore to retain the data from the modern samples as a model, while allowing the introduction of unknown archaeological samples to be identified by the system through a series of optimum statistical analyses, where the programme can select the most useful algorithm according to the information being introduced. The outcome could then be printed out to give the operator the desired information, including the rationale for the choice of analytic system, and what it means.

The intention would therefore be to reduce the amount of technical operations, so that the novice can be tutored in its use over a relatively short period; generally, the user will be interested primarily in the results, and not in the totality of the application. There are, however, a number of chemical steps that cannot be avoided; the initial input must be introduced in the form of spectral data, as the chemometrics is designed specifically *for* these types of data. Nevertheless, most laboratories will have technicians who can, if necessary, undertake the necessary preparations.

10.4 Overall Requirements

The proposed development of this system of identification would justify the funding of a further project to cover both those aspects that were included in the original research and the further extension of its applications. Thus far, the results have shown the system to be successful, and further work would render the system more user friendly. Once subjected to the proposed enhancements, one may envisage the use of the system being extended to not only the archaeobotany of further archaeological sites, but also to adjacent areas of botanical research, including the re-evaluation of the taxonomy of grasses, or, indeed, any other plant group.

The subsequent research could also extend beyond the chemical extractions and spectral interpretation and go on to explore new technical applications related to the numerical analysis. However, as the work has not yet been completed for archaeological applications, the proposed series of refinements is likely to require another project of equal length. As with the present project, the ensuing one should also be treated as, essentially, a chemical research programme, methodically recording and checking each step to ensure that the results from the extraction, standardization, calibration, validation and instrumentation are accurate, reliable and fully repeatable. Finally, with the inclusion of all these new facets, it is likely that this system will produce results of high caliber. In addition to the accurate identification of unknown archaeological specimens, the project should find application in a number of other areas of archaeology and biology.

REFERENCES

Bowden B.N. 1963. Sorting out Grasses. *New Scientist* 360: 94-96.

Brereton R.G. 1990. *Chemometrics: Applications of Mathematics and Statistics to Laboratoy Systems.* Chichester, England. Ellis Horwood Limited.

Brereton R.G. ed. 1992. *Multivariate Pattern Recognition in Chemometrics.* Amsterdam. Elsevier.

Cable M. & F. French. 1942. An Australian Aboriginal seed grinding and its archaeological record: a case study from the Western Desert. In: D.R. Harris & G.C. Hillman (eds), *Foraging and farming: the evolution of plant exploitation,* 99-119. London: Unwin and Hyman.

Champman D. 1965. *The Structure of Lipids by Spectroscopic and X-ray techniques.* London. Methuen.

Christie W.W. 1973. *Lipid Analysis.* Oxford. Pergamon Press.

Clayton W.D. and Renvoize S.A. 1986. *Genera Graminum. Grasses of the World.* London. HMSO.

Clayton W.D and Renvoize S.A. 1992. System of Classification for the Grasses in *Grass Evolution and Domestication.* Ed. Chapman GP. Cambridge University Press. Pp.338-353.

Colledge S.M. 1988. Scanning Electron microscope studies of the pericarp layers of some wild wheats and ryes. In: S. Olsen (ed.), *Scanning electron microscopy in Archaeology,* 225-36 Oxford: British Archaeological Reports (International Series) 452.

Colledge S.M. 1994. Ph.D. Thesis, Institute of Archaeology, University College London.

Davis R. and Frearson M. 1992. *Mass Spectroscopy.* Chichester. John Wiley & Sons Inc.

Dickerson R.E., Gray H.B., Darensbourg M.Y. & Darensbourg D.J. 1984. *Chemical Principals.* California. The Benjamin/Cummings Publishing Company Inc.

Downing D. 1983. *Algebra. The Easy Way.* New York. Barron's Educational Series Inc.

Eglinton G. and Curry G.B. 1991. Preface. In: *Molecules Through Time: Fossil Molecules and Biochemical Systematics.* London. Proceedings of the Royal Society 20-21 March.

Eglinton G. and Logan G.A. 1991. Molecular Preservation. In: *Molecules Through Time: Fossil Molecules and Biochemical Systematics.* 315-328. London. Proceedings of the Royal Society 20-21 March.

Ferraro J.R. and Krishnan K. 1990. *Practical Fourier Transform Infrared Spectroscopy.* Industrial and Laboatory Chemical Analysis. New York. Academic Press Inc.

Galliard T. and Mercer E.I. 1975. *Recent Advances in Chemistry and Biochemistry of Plant Lipids.* Vol. XVI. Arranged by the Phytochemical Society. University of East Anglia. London Academic Press.

George W.O. and Willis H.A. ed. 1990. *Computer Methods in UV, Visible, and IR Spectroscopy.* Cambridge. Royal Society of Chemistry.

Griffiths P.R. 1975. *Chemical Infrared Fourier Transform Spectroscopy.* New York. John Wiley & Sons Inc.

Gunstone F.D. 1967. *An Introduction to the Chemistry and Biochemistry of Fatty Acids and their Glycerides.* London.Chapman Hall.

Guyot D. 1995. *Introduction to Experimental Design.* Camo AS Trondheim, Norway.

Guyot D. 1995. *Calibration: Multivariate Calibration in Practice.* Camo AS Trondheim, Norway.

Harlan J.R. 1989. Wild-grass seed harvesting in the Sahara and Sub-Sahara of Africa. In: D.R. Harris & G.C. Hillman (eds.), *Foraging and Farming: the evolution of plant exploitation,* 79-98. London: Unwin and Hyman.

Harwood J.L. and Russell N.J. 1984. *Lipids in Plants and Microbes.* London. Allen & Unwin.

Harwood L.M. & Moody C.J. 1989. *Experimental Organic Chemistryy. Principles and practice.* Oxford. Blackwell scientific publications.

Helbeck H. 1970. The plant husbandry at Hacilar. A study of domestication and cultivation. In: J. Mellaart, *Excavations at Hacilar.* Edinburgh. University Press.

Hillman G.C. 1996. Late Pleistocene changes in wild plant-foods available to hunter-gathers of the northern Fertile Crescent: possible preludes to cereal cultivation in: *The Origins and Spread of Agriculture and Pastoralism in Eurasia* eds. Harris DR. UCL Press. Pp.159-203.

Hillman G.C., Colledge S.M. & Harris DR. 1989. Plant-food economy during the Epipalaeolithic period at Tell Abu Hureyra, Syria: dietry diversity, seasonality, and modes of exploitation in: *Foraging and Farming* eds. Harris DR. & Hillman GC. Unwin Hyman. Pp.240-268.

Hillman G.C. and Davies MS. 1990. Measured domestication rates in wild wheats and barley under primative cultivation, and their archaeological implications in: *Journal of World Prehistory.* 4(2): 157-222.

Hillman G.C., Jones C.E.R. and Robins G.V. 1982. The potential of examining archaeological materials with pyrolysis-mass spectrometry: an exploratory exercise. Paper presented at the 5th *International Symposium on Analytical Pyrolysis*, Vail, Colorado, April 1982.

Hillman G.C., Mason S., de Moulins D., and Nesbitt M. 1996. Identification of archaeological remains of wheats: the 1992 London workshop in: *Circea.* 12(2): 195-209.

Hillman G.C., Wales S., McLaren F., Evans J. & Butler A. 1993. Identifying problematic remains of animal plant foods: a comparison of the role of chemical, histological and morphological criteria in: *World Archaeology.* Vol. 25: 94-121.

Hitchcock C. & Nichols B. 1941. *Plant Lipid Biochemistry.* London. Academic press.

Jones C.E. 1982. *Pyrolysis Mass Spectroscopy in Acrhaeology: An Exploratory Exercise.* London. Institute of Archaeology.

Jones S.B. and Luchsinger A.E. 1987. *Plant Systematics.* Singapore. McGraw-Hill Book Company.

Kenner C.T. 1971. *Analytical Determinations and Separations: A Text Book in Quantitative Analysis.* New York. Macmillan.

Kislev M., Melamed Y., Simchoni O., and Marmorstein M. 1997. Computerized Key of Grass Grains of the Mediterranean Basin. In: *Lagascalia.* 19 (1-2): 289-294.

Korber-Grohne U. 1981. Distingishing prehistoric cereal grains of *Triticum* and *Secale* on the basis of their surface patterns using SEM. *Journal of Archaeological Science* 8: 197-204.

Korber-Grohne U. and Piening U. 1980. Microstructure of the surfaces of carbonised and non-carbonised grains of cereals as observed in scanning electron and light microscopes as an additional aid in determining prehistoric findings. *Flora (Jena.).* 170: 189-228.

Lindsay S. 1987. *High Performance Liquid Chromatography.* Chichester. John Wiley & Sons Inc. ACOL.

Macko S.A. and Engel M.H. 1991. Assessment of indigeneity in fossil organic matter: amino acids and stable isotopes. In: *Molecule Through Time: Fossil Molecules and Biochemical Systematics.* 367-374. London. Proceedings of the Royal Society 20-21 March.

Malinowski E.R. 1991. *Factor Analysis in Chemistry.* New York. John Wiley & Sons Inc.

Mark H. 1991. *Principles and Practice of Spectroscopic Calibration.* New York. John Wiley & Sons Inc.

Martens H. and Naes T. 1991. *Multivariate Calibration.* Guilford. John Wiley & Sons Inc.

Massart D.L., Vandeginste B.G.M., Deming S.N. and Michotte Y and Kaufman L. 1988. *Chemometrics: a textbook.* Amsterdam. Elsevier.

Masterton W.L. and Slowinski E.J. 1978. *Chemical Principals with Qualitative Analysis.* WB Saunders Company. Philadelphia.

Miller J.C. and Miller J.N. 1993. *Statistics for Analytical Chemistry.* Third Ed. London. Ellis Horwood PTR Prentice Hall.

Morgan E. 1991. *Chemometrics: Experimental Design.* London. ACOL. John Wiley & Sons.

Pavia D.L., Lampan G.M and Kriz G.S. 1979. *Introduction to Spectroscopy.* Washington. Saunders Golden Sunburst Series.

Press F. and Siever R. 1986. *Earth.* 4th Ed. W.H. Freeman & Co. New York.

Schonkopf S., Guyot D. 1995. *Multivariate Calibration in Practice.* Camo AS. Trondheim, Norway.

Simmonds N.W. (Ed.) 1976. *Evolution of Crop Plants.* London. Longman Group Limited.

Skibo J.M. 1992. *Use Alteration: Absorbed Residues.* New York. Plenum Press.

Smith L.A. and Thompson G.A. 1987. Analysis of Lipids by High Performance Liquid Chromatography in: *High Performance Liquid Chromatography in Plant Sciences.* Ed. H.F. Linskens and J.F. Jackson Springer-Verlag. Berlin.

Solomons T.W.G. 1990. *Fundamentals of Organic Chemistry.* 3rd. Edition Wiley and Sons.

Stewart J.H. 1932-34. Ethnography of the Owen's Valley Paiute in: *Publications in American Archaeology and Ethnography.* University of California. 33: 233-350.

Stumpf P.K. & Hammond J.L. 1975. Articles on Fatty Acid Biosynthesis in Plants in *Recent Advances in the Chemistry and Biochemistry of Plant Lipids.* Ed. T. Galliard and E.I. Mercer. London. Academic press.

van Zeist W. and Buitenhuis H. 1983. A botanical study of Neolithic Erbaba, Turkey. *Anatolia 10*: 47-81, and *Palaeohistoria* 24.

Whittow J.B. 1984. *Dictionary of Physical Geography.* London. Penguin Books Ltd.

Willet J. 1991. *Gas Chromatography.* Chichester. John Wiley & Sons Inc.

Wright I.K. 1994. Ground-stone tools and hunter-gatherer subsistence in Southwest Asia: implications for the transition to farming. In: *American Antiquity.* 59(2), 238-263.

KEY

The key provides access to the understanding of the text as a whole; it provides answers to problems and words for interpreting and extracting the meaning of the data.

Part II of the appendix consists of an extended version of all of the numerical techniques deployed in the analysis of the data. In order to adequately comprehend the results and the manner of their derivation from the raw data, it is advisable to read the two volumes in tandem; this is especially the case with regard to Chapter 8, 'Final Results' and Chapter 9, 'Discussion of results'. Each paragraph heading within the main text will include a reference to a specific page and paragraph in the appendices; this reference should be looked up in order to supply both a comprehensive description of the technique used and an explanation of the, tables and graphs in which the results are displayed.

It will be noted that descriptive captions have been omitted from the graphs and tables; this is due to the fact that the SCAN programme, does not permit the making of any alterations, additions or omissions whatsoever to the tabular or graphical displays of results. This feature of SCAN is a defensive mechanism deliberately written into the programme to guarantee the integrity of the data; it protects the final results from manual tampering or from any form of creative enhancement into which temptation might lead the weary researcher anxious for a positive experimental outcome.

The modern seed listing has been included twice; firstly by order of the taxonomic groupings, and secondly in the numerical order, which is the number each sample was allocated when it was first received for analysis.

Numerical techniques utilized in the chemometric analysis; including a brief description and where located in the appendix.

Covariance Matrix- page 112.
Calculates pair-wise values and prints them lower triangle matrix- covariance matrix is the first product moment of the two variables x and y above their means.

Matrix Condition- page 112.
Quantifies the collinearity between variables and gives warnings about constant and highly collinear columns. Collinearity can make some classifications and regression results arbitrary or even impossible.

PCA- page 112.
1: By **SVD**- will use covariance matrix to either calculate as many compounds as there are input variables, or as many components as specified.
2: By **NIPALS**- uses covariance matrix to calculate just 2 confounds, or the number of components as specified. Calculates eigenvalues, eigenvectors and p.c. scores of a data matrix: = eigen analysis.

Factor Analysis- page 118.
Calculates factors from covariance matrix, by using either as many factors as there are input variables or the number specified. F.A. extracts the underlying factors and performs orthogonal rotation to make the factors more interpretable; factors are calculated by PCA.

HCA- page 121.
Perform agglomerative clustering of objects. Also starts with each object in its own cluster. Step 1: the two objects in closest proximity are joined. Step 2: either third object joins first two or two further objects join different cluster. Each step results in fewer clusters than previously, until finally all objects are resolved into one cluster.

OLS- page 126.
Calculates a parametric linear model for a single response, and calculates unbiased least squares coefficients. OLS is used if the data set is well determined, i.e. where there are more observations than predictors and the predictions are not highly correlated. This model assumes that the response is a linear function of the predictors, and that the errors are identically and independently distributed.

PCR- page 130.
Calculates parametric linear model for a single response, and calculates biased non-least squares coefficients. It is used where the data set is under-determined, i.e. where there are fewer observations than predictors and/or the predictors are highly correlated. Also assumes that the response is a linear function of predictors, and that most of the information about the response is in a high variance subspace of predictors, i.e. the first few PCs.

PLS- page 135.
Calculates a parametric, biased linear model which may contain one or several responses. If there are several responses, they are modelled together in a multivariate way. It is used when the data set is under-determined; i.e. where there are fewer observations than predictors and/or the predictors are highly correlated. PLS attempts to calculate a more predictive model than OLS.

ACE- page 140.
Calculates a non-linear parametric regression model, where the response is estimated as the sum of functions of the predictors. The only restriction on the functions is that they must be smooth. It is used when the poor performance of the OLS model stems from non-linearity, and is only applicable where there are sufficient degrees of freedom, i.e. Where the data set is well determined, containing more observations than predictors.

NPLS- page 141.
Calculates non-linear PLS model for one or more responses, where the inner relationships are non-linearized by smoothers. It is a biased method, used to model non-linear predictor- response relationship when

the data set is under-determined; i.e. where there are fewer observations than predictors, and/or the predictors are highly correlated.

Numerical seed Listing

1. *Trachynia distachya*
(robust form)
GCH 8967
(L.) Link

2. *Crithopsis delileana*
GCH 8676 (-)
(Schultes) Roshev.

3. *Taeniatherum cf. caput- medusa*
GCH ?
(L.) Nevski.

4. *Heteranthelium*
GH/SC7186 (-)
Hochst.

5. *Eremopyram*
GCH 2861 (-)
(Ledeb.) Jaub. & Spach

6. *Henradia ?* or *Parapholis incurva*
GCH 8360
C.E. Hubbard

7. *Bromus tectorum*
GCH 8151 (H)
(L.) Nevski.

8. *Bromus madritensis*
GCH 8354 (-)
(L.)

9. *Bromus nr. squarosus*
GCH/SMC 7559
(L.)

10. *Lolium temulentum*
B-152
(L.)

11. *Bromus cf. intermedius*
(extremely dense infl)
GCH 8113 (H)
Guss.

12. *Vulpia cf myuros*
GCH 8232 (-)
(L.) C.C. Gmelin

13. *Loliolum (nardurus) sublatus sublatum*
GCH 8072 (-)
(Banks & Sol.)

14. *Panicum millanceum* var. *luteum*
B165
(L.)

15.*Setaria italica* var.*flavesceus*
B181
(L.)

16. *Oryzopsis miliacea*
B159
(L.) Benthan & Hooker ex Aschers.& Schweinf.

17. *Cynodon dactyla*
GCH $5086^{1}/_{2}$ (-)
(L.) Pers.

18. *Beckmannia eracaeformis*
HB-UL
(L.) Host

19. *Aeluropus cf. littoralis*
GCH 5011
(Gouan) Parl.

20. *Laphochloa phleoides*
GCH 8789 (-)
(Vill.) Rchb.

21. *Stipa holosericea (lagascae)*
GCH
(L.) Roemer & Schultes

22. *Oryzopsis viresceus*
B1233 (-)
(Trin.) Beck

23. *Digitaria*
GCH 2128 Heister ex Haller

24. *Festuca pratensis* ®
Hudson

25. *Digitaria sanguinalis*
B97
(L.) Scop.

26. *Dactylis glomerata ssp. Hispanica*
B1209
(Roth) Nyman.

27. *Cynosurus nr echinatus*
GCH/SMC 7120 (H)
(L.)

28. ?*Sphenopus divaricatus*
GCH 8090 (-)
(Gouan) Reichb.

29. *Sclerochloa sp.*
GCH/SMC 7203 (-)
P. Beauv.

30. *Phalaris nodosa*
B1239 (-)
Murray

31. *Echinochloa cf. -crus-galli*
GCH 2400 (H)
(L.) Beauv.

32.*Phalaris canariensis*
seed merchant
(L.)

33. *Oryzopsis coerulescens*
B1232 (-)
(Desf.) Hackel

34. *Stipa sp gigantea*
grown at woodlands B185
Link.

35. *Sorghum halepense* B184
(L.) Pers.

36. *Cutandia sp.* GCH
Willk.

37. *Eragrostis cf. minor*
GCH 3938 (-)
(L.) Host

38. *Setaria pumila = (S. glauca)*
B1248
(Poiret) Roemer & Schultes

39. *Stipa capensis*
K 293/2
Thunb.

40.*Setaria verticillata*
B142
(L.) P. Beauv.

41. *Stipa parviflora*
GCH 8739 (H)
Desf.

42. *Stipa hohenackeriana*
GCH 8749 (H)
Trin. & Rupr.

43. *Bromus tectorum*
GCH/SMC 7184
(L.) Nevski.

44. *Parapholis sp.* GCH
C.E. Hubbard
45. *Rhizocephalus*
GH/SC 7595
Boiss.

46. *Bromus nr. lanceolatus*
GCH 8333 (H)
Roth.

47. *Eremopoa persica*
GCH 8122 (H)
(Trin.) Roshev.

48. *Psilurus incurvus*
GCH 9019 (-)
(Gouan) Schinz & Thell.

49. *Cenchrus ciliaris*
GCH 3320 (-)
(L.)

50. *Echinaria*
GH/SC 7170 (H)
Desf.

51. *Poa cf. reuteriana*
GCH 8155 (H)
(L.)

52. *Paspalidium geminatum*
GCH 3940 (-)
(Forsk.) Stapf.

53. *Panicum milleacum* var. *nigrum*
B166
(L.)

54. *Panicum milleacum gordia*
(L.)

55. *Taeinatherum caput-medusa ssp asper*
GCH 50001 G
(L.) Nevski.

56. *Bromus sp.*
GCH 2504 (H)
(L.)

57. *Stipa holosericea (lagascae)*
GCH 9015 (H)
(L.) Roemer & Schultes

58. *Stipa capensis tortilis* big
GCH 8665 (H)
Thunb. Desf.

59. *Parapholis sp.*
GCH 8217 (H)
C.E. Hubbard.

60. *Stipa parviflora*
GCH 8739 (H)
Desf.

61. *Stipa ehrenbergiana (barbata)*
GCH 8718 (H)
Trin. & Rupr. Desf.

62. *Stipa holosericea*
GCH 4166 (H/P)
(L.) Roemer & Schultes

63. *Bromus cf. intermedius*
GCH 8113 (H)
Guss.

64. *Eregrostis minor*
GCH 3938 (-)
(L.) Host.

65. *Stipa joannis (pennata)*
HB Hung 1697
Celak

66. *Hordeum murinum* var. *leporinum*
GCH 8230 (-)
(L.)

67. *Hordeum spontaneum*
RCH/SMC
C. Koch emend. Bacht.

68. *Hordeum bulbosum* (med)
GCH 8964(-)
(L.)

69. *Hordeum genticulatum*
GCH 2515(H)
All.

70. *Hordeum murinum glaucum*
GCH 1830
(Steudel) Tzvelev.

71. *Hordeum marinum*
Syria, El Korum
GCH...(-)
Hudson

72. *Hordeum genticulatum*
W. Turkey
RMN 4480
All.

73. *Hordeum genticulatum*
Central Turkey
NM 1204 All.

74. *Hordeum marinum*
W. Turkey
RMN 4487
Hudson

75. *Hordeum murinum*
E. Turkey
RMN 1258
(L.) Sensu Boiss.

76. *Hordeum murinum*
DJS 405
Syria
(L.) Sensu Boiss.

77. *Hordeum murinum*
RMN 4301
Israel
(L.) Sensu Boiss.

78. *Hordeum bulbosum*
RMN 4256
Israel
(L.)

79. *Hordeum bulbosum*
RMN 4483
W. Turkey
(L.)

80. *Hordeum spontaneum*
RMN 4223
Israel
C. Koch emend. Bacht.

81. *Hordeum spontaneum*
RMN 4481
W. Turkey
C. Koch emend. Bacht

82. *Hodeum spontaneum*
RMN 218
SE. Turkey
C. Koch emend. Bacht

83. *Hordeum spontaneum*
Nevo T-12-4
Iran C. Koch emend. Bacht

84. *Hordeum spontaneum*
Nevo T-19-34
Turkey
C. Koch emend. Bacht

85. *Hordeum murinum*
GCH 1803
East Turkey
(L.) Sensu Boiss.

86. *Hordeum murinum*
GCH 2505
East Turkey
(L.) Sensu Boiss.

87. *Hordeum murinum*
RMN 4460
West Turkey
(L.) Sensu Boiss.

88. *Hordeum murinum*
RMN 4631
Central Turkey
(L.) Sensu Boiss.

89. *Hordeum bulbosum*
DHF 419
Turkey
(L.)

90. *Hordeum distichon*
RMN 1088
Central Turkey
(L.)

91. *Hordeum distichon*
1819
Nth Turkey
(L.)

92. *Hordeum vulgare*
RMN 1634
West Turkey
(L.)

93. *Hordeum vulgare*
RMN 1815
N. Turkey (L.)

94. *H. vulgare*
RMN 2246
S. Turkey
(L.)

95. *Hordeum hexastichon nudum*
PBI 3855
1972
(L.)

96. *Hordeum hexastichon nudum*
Czechoslovakia
Kuhn 1970
(L.)

97. *Hordeum hexastichon nudum*
(black)
PBI 4408
Ethiopia 1975
(L.)

98. *Hordeum hexashichum*
GCH 5054 a)
Nth Turkey
(L.)

99. *Hordeum tetrashichum vulgare*
(black)
Reading 1969
(L.)

100. *Hordeum distichon*
GCH 3492 (G)
Central Turkey
(L.)

101. *Hordeum bulbosum*
(+/- black)
GCH 8964 (-)
E. Central Turkey
(L.)

102. *Hordeum bulbosum*
White (1 seed)
GCH 8964 (-)
E. Central Turkey
(L.)

103. *Hordeum bulbosum*
GCH 1802
white
E. Central Turkey
(L.)

104. *Hordeum marinum*
GCH/SMC 7006
Hudson

105. *Triticum baeoticum*
(Cemisgezek)
GCH 3776
L. (Boiss.)

106. *Triticum urartu*
Icarda
sy-20047
Tuman

107. *Triticum baeoticum*
(Karadag)
GCH 3341
L. (Boiss.)

108. *Triticum dicoccoides*
narrow, white
+/- mature
PBI (Israel)
(Koern.) Koern. Ex Schweint.

109. *T. dicoccum*
GCH 7501
(Schuebler) Heckel

110. *T. dicoccoides*
wide, spiked
black
Poyarkova 1.
(Koern.) Koern. Ex Schweint.

111. *T. araraticum*
PBI. D1985
Jakubz

112. *T. araraticum*
PBI. 115001
Jakubz

113. *T. dicoccoides*
wide, spiked
white (immature)
Poyarkova 2.
(Koern.) Koern. Ex Schweint.

114. *T. dicoccoides*
PBI Israel
129 B
(Koern.) Koern. Ex Schweint.

115. *T. dicoccoides*
Jordan
black, narrow
GCH 8966
(Koern.) Koern. Ex Schweint.

116. *H. sativum*
6.row
GCH 3673
Lam.

117.*T.dicoccum*
Turkey, white
GCH 5031
(Scheubler) Hackel.

118.*T. dicoccum*
(Spain)
LPCH 27
(Scheubler) Hackel.

119. *T.turgidum ssp. palaeocolchium*
PBI 1993
A.Love & D. Love

120. *Secale vavilovii ararat*
pop 5
GH/FM/DZ 12,002
Grossh.

121. *Secale vavilovii ararat*
pop1
GH/DZ/FM 12,016
Grossh.

122. *Secale montanum*
Uzunyayli
segetal robust
GCH
Guss.

123. *T. dicoccom* LPCH 73
(Schuebler) Hackel

124. *Secale montanum*
GCH 3095
Guss.

125. *Secale vavilovii ararat*
pop 4
Dz/FM/GH 12,028
Grossh.

126. *Tragus racemous*
H.B. Bordeaux
(L.) All.

127. *Crypsis schoenoides*
B+B 261
(L.) Lam

128. *Crypsis alopecuriodes*
GCH. 2399
(Piller & Mitterp.) Schrader.

129.*Oryza sativa*
(cult) immature
RMN 2907a
(L.)

130. *Avena barbata*
RMN 2204
Pott. Ex Link

131. *Koeleria cristata*
(immature)
GCH 7194
(L.) Pers.

132. *Trisetum flavescens*
(immature)
RMN 1425
(L.) P. Beauv.

133. *Aegilops speltoides* var *ligustica*
GCH 3806
(savignone) Bornm.

134. *Eremoprum compactum*
GCH 7191
(Ledeb.) Jaub. & Spach

135. *Agrostis stolonifera*
145/45/9
(L.)

APPENDIX

CONTENTS

PART I (a)

Modern Seed Listing- Poeacea List of modern seeds with identities grouped according to - Subfamily, Tribe, Sub-tribe Genus, Species, and sometimes variety.

List of the Modern Specimens used in this Project

Subfamily Bambusoideae

Tribe Oryzeae
Oryza sativa 129. (cult) immature RMN 2907a

Subfamily Pooideae

Tribe Stipeae
Stipa holosericea (S. lagascae) 21. GCH
Stipa holosericea (S. lagascae) 57. GCH 9015(H)
Stipa holosericea 62. GCH 4166(H/P)

Stipa gigantea 34. Grown at woodlands. B185

Stipa capensis 39. K293/2

Stipa parviflora 41. GCH 8739(H)
Stipa parviflora 60. GCH 8739(H)

Stipa hohenackeriana 42. GCH 8749(H)

Stipa capensis (S. tortilis) 58. big GCH 8665(H)

Stipa ehrenbergiana (S. barbata) 61. GCH 8718(H)

Stipa joannis (S. pennata) 65. HB Hung 1697

Oryzopsis miliacea 16. B 159

Oryzopsis virescens 22. B 1233(-)

Oryzopsis coerulescens 33. B 1232(-)

Tribe Poeae
Festuca pratensis R24

Poa cf. *reuteriana* 51. GCH 8155(H)

Lolium temulentum 10. B-152

Vulpia cf. *myuros* 12. GCH 8232(-)

Loliolum (Nardurus) 13. *sublatus.* GCH 8072(-)

Psilurus incurvus 48. GCH 9019(-)

Cynosurus nr *echinatus* 27. GCH/SMC 7120(H)

Dactylis glomerata ssp. *hispanica* 26. B1209

Eremopoa persica 47. GCH 8122(H)

Sphenopus divaricatus 28. GCH 8090(-)

Sclerochloa sp. 29. GCH/SMC 7203(-)

Cutandia sp. 36. GCH

Echinaria capitata 50. GH/SC 7170(H)

Tribe Hainardieae 6.
Parapholis incurva GCH 8360

Parapholis sp. 44. GCH
Parapholis sp. 59.

Tribe Aveneae
Subtribe Aveninae
Avena barbata 130. RMN 2204

Rostraria (Lophochloa) phleoides 20. GCH 8789(-)

Subtribe Phalaridinae
Phalaris nodosa 30. B1239(-)

Phalaris canariensis 32. seed merchant

Subtribe Alopecurinae
Beckmannia erucaeformis 18. HB-UL

Rhizocephalus falcatus 45. GH/SC 7595

Tribe Bromeae
Bromus (Anisantha) cf. *tectorum* 7. GCH 8151 (H)
Bromus (Anisantha) nr. *tectorum* 43. GCH/SMC 7184

Bromus madritensis 8. GCH 8354(-)

Bromus nr. *squarosus* 9. GCH/SMC 7559

Bromus cf. *intennedius* 11.
(extremely dense infl.) GCH 8113(H)
Bromus cf. *intermedius* 63. GCH 8113(H)

Bromus nr. *lanceolatus* 46. GCH 8333(H)

Bromus sp. 56. GCH 2504(H)

Trachynia distachya 1(robust). GCH 8967

Tribe Triticeae
Taeniatherum cf. *caput medusae* 3. GCH
Taeniatherum caput-medusae ssp. *asper* 55. GCH 5001 G

Crithopsis delileana 2. GCH 8676(-)

Heteranthelium piliferum 4. GH/SC 7186(-)

Hordeum bulbosum 68. GCH 8964(-) A
Hordeum bulbosum 78. RMN 4256 Israel A
Hordeum bulbosum 79. RMN 4483 W. Turkey A
Hordeum bulbosum 89. DHF 419 Turkey A
Hordeum bulbosum 101. (+/- black)
GCH 8964 (-)E. Central Turkey A
Hordeum bulbosum 102. White (1 seed)
GCH 8964 (-)E. Central Turkey A
Hordeum bulbosum 103. GCH 1802 white E. Central Turkey A

Hordeum geniculatum 69. GCH 2515 B
Hordeum geniculatum 72. W. Turkey RMN 4480 B
Hordeum geniculatum 73. Central Turkey NM 1204 B

Hordeum murinum var. *glaucum* 70. GCH 1830 C

Hordeum murinum var. *leporinum* 66. GCH 8230(-) D

Hordeum murinum 75. E. Turkey RMN 1258 D

Hordeum murinum 76. DJS 405 Syria D
Hordeum murinum 77. RMN 4301 Israel D
Hordeum murinum 85. GCH 1803 East Turkey D
Hordeum murinum 86. GCH 2505 East Turkey D
Hordeum murinum 87. RMN 4460 West Turkey D
Hordeum murinum 88. RMN 4631 Central Turkey D

Hordeum marinum 71. GCH. 11, xxx Syria, El Korum E
Hordeum marinum 74. W. Turkey RMN 4487 E
Hordeum marinum 104. GCH/SMC 7006 E

Hordeum spontaneum 67. GCH/SMC F
Hordeum spontaneum 80. RMN 4223 Israel F
Hordeum spontaneum 81. RMN 4481 W. Turkey F
Hordeum spontaneum 82. RMN 218 SE. Turkey F
Hordeum spontaneum 83. Nevo T-12-4 Iran F
Hordeum spontaneum 84. Nevo T-19-34

Hordeum distichum 90. RMN 1088 Central Turkey G
Hordeum distichum 91. 1819 Nth. Turkey G
Hordeum distichum 100. GCH 3492(G) Central Turkey G

Hordeum vulgare 92. RMN 1634 West Turkey H
Hordeum vulgare 93. RMN 1815 Nth. Turkey H
Hordeum vulgare 94. RMN 2246 Sth. Turkey H

Hordeum tetrastichum vulgare 99. (Black) Reading 1969 I

Hordeum sativum 116. 6 row, hulled GCH 3673

Hordeum hexastichum 98. GCH 5054 a Nth. Turkey J
Hordeum hexastichum var *nudum* 95. PBI 3855 1972 K
Hordeum hexastichum var *nudum* 96.
Czechoslovakia Kuhn 1970 K

Hordeum hexastichum var *nudum* 97.
(Black) PBI 4408 Ethiopia 1975 K

Agropyron sp. GCH 2861(-)
Agropyron cristatum 134 GCH 7191

Secale montanum 124. GCH 3095 a

Secale montanum var *anatoliam* 122.
Uzunyayli, segetal robust GCH 12,035 b

Secale vavilovii Ararat 120. pop 5 GCH/DZ/FM 12,002 c
Secale vavilovii Ararat 121. pop 1 GCH/DZ/FM 12,016 c
Secale vavilovii 125. Ararat pop 4 GCH/DZ/FM 12,021 c

Triticum beoticum 105. GCH 3776 **A** (Cemisgezek)
Triticum boeoticum 107. GCH 3341 **A** (Karadag)

Triticum uratu 106. sy-20047 **B** Icarda

Triticum dicoccoides 108. PBI (Israel) **C**
narrow, white +/- mature
Triticum dicoccoides 110. wide, spiked black Poyarkova 1. **C**
Triticum dicoccoides 113. (immature)
wide, spiked white Poyarkova, Israel **C**
Triticum dicoccoides 114. PBI 129 B Israel **C**
Triticum dicoccoides 115. black, narrow GCH 8966 Jordan **C**

Triticum dicoccum 109. GCH 7501 **D**
Triticum dicoccum 117. GCH 5031 **D** white Turkey
Triticum dicoccum 118. LPCH 27 **D** Spain
Triticum dicoccum 123. LPCH 73 **D**

Triticum palaeocolchicum 119. PBI 1993 **E**

Triticum araraticum 111. PBI D1985 **F**
Triticum araraticum 112. PBI 115001 **F**

Aegilops speltoides var *ligustica* 133. GCH 7191

Subfamily Chloridoideae

Tribe Eragrostideae
Subtribe Monanthiochloinae

Aeluropus cf. *littoralis* 19. GCH 5011

Subtribe Eleusininae
Eragrostis cf. *minor* 37. GCH 3938(-)
Eragrostis cf. *minor* 64. GCH 3938(-)

Subtribe Sporobolinae
Crypsis schoenoides 127. B+B 261
Crypsis alopecuroides 128. GCH 2399

Tribe Cynodonteae
Subtribe Chloridinae
Cynodon dactylon 17. GCH 50861/2(-)

Subtribe Zoysiinae
Tragus racemosus 126. H.B. Bordeaux 189

Subfamily Panicoideae

Tribe Paniceae
Subtribe Setariinae
Panicum milleacum var *luteum* 14. B165

Panicum milleacum var *nigrum* 53. B166

Panicum milleacum 54.

Echinochloa cf. *-crus-galli* 31. GCH 2400

Setaria italica var *flavescens* 15. B181

Setaria glauca var *lutescens* 38. B1248

Setaria verticillata 40. B124

Paspalidium geminatum 52. GCH 3940(-)

Subtribe Digitariiae
Digitaria sanguinalis 23. GCH 2128
Digitaria sanguinalis 25. B97

Subtribe Cenchrinae
Cenchrus ciliaris 49. GCH 3320(-)

Tribe Andropogoneae
Subtribe Sorghinae
Sorghum halepense 35. B184?

PART I (b)

Archaeological Seed Listing Provisional identifications based on morphology.

Pages 108-109 list the Archaeological Specimens used in this Project

All were preserved by charring, and the vast majority came from the site of Tell Abu Hureyra in Syria (here abbreviated as "Abu H."). All were clearly grains of grasses, and the provisional identification originally assigned on the basis of their morphology is listed in each case.

In most cases only a generic name was suggested, and, even then, there was considerable uncertainty, "cf" = "compares with" or "resembles". "?" indicates a greater level of doubt, and "??" even more doubt. In some cases, the morphology allowed only an even vaguer attribution, and the specimen is merely referred to as "type"; e.g. *"Bromus* type" indicates that the specimen resembles *Bromus,* but could well be something different. *"Aegilops* type" indicates that the specimen might possibly be regarded as superficially resembling *Aegilops,* but even this is contestable.

AR001
Bromus or *hordeum* or *Lolium*
Abu H. 72 E.29

AR002
Bromus (a) type *(?Trachynia)*
Abu H. 72 E.29

AR003
Bromus sp.
Abu H. 73 / E.230

AR004
Crypsis aculeata type
Abu H. 73 / E.311

AR005
Stipa sp. (large)
Abu H. 73 E.330

AR006
Stipa sp. (small)
Abu H. E.311

AR007
cf. *Hordeum* sp. (medium-sized classic shape)
Abu H. 73 E.281
AR008
?? *Hordeum* sp. (v. narrow, atten. apex)
Abu H. 72 / E.52

AR009
? *Hordeum* sp. (v. wide, v. angular, grooved)
Abu H. 73 E.324

AR010
cf. *Hordeum* sp. (large, classic)
Abu H. 73 E.275

AR011
?? *Hordeum* sp. (v. small)
Abu H. 73 / E.326#2

AR012
?? *Hordeum* sp. +/- *marinum* type
Abu H. 72 / E.52

AR013
?? *Hordeum* sp. *marinum* type
Abu H. 72 / E.52

AR014
cf. *Hordeum* sp. *marinum* type
Abu H. 73 E.275

AR015
cf. *Hordeum* sp. (large, narrowish, +/- classic)
Abu H. 73 E.275

AR016
cf. *Hordeum* sp. (small, classic)
Abu H. 73 E.275

AR017
cf. *Hordeum* sp. (med., flat, elong.)
Abu H. 73 E.275

AR018
? *Hordeum* sp. (small, narrow)
Abu H. 72 / E.52

AR019
?? *Hordeum* sp. (v. wide)
Abu H, 73 E.326#2

AR020
? *Hordeum* sp. (*marinum* type)
Abu H. 72 / E.52

AR021
?? *Hordeum* sp. (v. wide)
Abu H. 73 E.326#2

AR022
?? *Hordeum* sp. (med. - large)
Abu H. 72 / E.52

AR023
?? *Hordeum* sp. (v. wide)
Abu H. 73 / E.326#2

AR024
cf. *Hordeum* sp. (med., +/- classic)
Abu H. 73 E.324
AR025
MT-90-1A seed 1 Mark Nesbit

AR026
MT-90-1A seed 4 Mark Nesbit

AR027
MT-90-1A seed 2 Mark Nesbit

AR028
Hordeum sp. (smalled grained)
Abu H. 73 E.274

AR029
Stipa sp. smalled grained
Abu H. 73 E.286 (meso)

AR030
Hordeum sp. (smalled grained) tiny, rounded TS when charred
Abu H. 73 E.230 (Neo)

AR031
Hordeum sp. (smalled grained) (tiny, ? *festuca)*
Abu H. 73 E.230

AR032
Hordeum sp. (smalled grained) (tiny, semi-flat, *narrow festucoid)*
Abu H. 73 E.230

AR033
Hordeum sp. (smalled grained) (tiny, v. short +/- flat, broadish)
Abu H. 73 E.230

AR034
Stipa sp. ? *S. gigantica* (v. big)
Abu H. 73 E.325 (meso)

AR035
Hordeum sp. (smalled grained) (small & flat & almost winged)
Abu H. 73 E.274

AR036
Stipa cf. (largish)
Abu H. 73 E.325 (meso)

AR037
Stipa sp. (large)
Abu H. 73 E. 305 (meso)

AR038
Stipa sp. (small grained)
Abu H. 73 E.325 (meso)

AR039
MT-90-1A seed 3 M. Nesbit

AR040
Stipa sp. (smalled grained)
Abu H. 73 E.313 (Meso)

AR041
?? *Aegilops type*
Abu H. 73 E.327

AR042
?? *Elymus* type
Abu H. 73 E.286 (Meso)

AR043
Bromus type
Abu H. 73 E.286

AR044
? Lolium type (v. short, & flat)
Abu H. 73 E.230 (Neo)

AR045
? *Lolium* narrow, (embryo end only)
Abu H. 73 E.230 (Neo)

AR046
?? *Elymus* type
Abu H. 73 E.286 (Meso)

AR047
Lolium type
Abu H. 73 E.230 (Neo)

AR048
Narrow *Bromus* or wide *Vulpia*
Abu H. 73 E.294

AR049
?? miniature *Tragus* type
Abu H. 73 E.294

AR050
? *Festucoid* (small 2.5 long, v. narrow, tapered towards apex, with full length hilum.)
Abu H. 72 E. 329

AR051
Crypsis-Cynodon (like in size & gen. form)
Abu H. 73 E.327

AR052
cf. *Vulpia* type
Abu H. 72 E.29 (Neo)

AR053
Aegilops cf. *squarrosa* type. (flat-backed, with elliptic outline, small)
Abu H. 72 E.29

AR054
Aegilops cf. *umbellata* type, (short)
Abu H. 72 E.29

AR055
Aegilops cf. *squarrosa* type (flat-backed, with elliptic outline, v. small)
Abu H. 72 E.29

AR056
Hordeum cf. *murinum* MT-4

AR057
? *Aegilops umbellata* type (but maybe a fraction short)
Abu H. 72 E.29

AR058
Hordeum cf. *murinum* MT-4

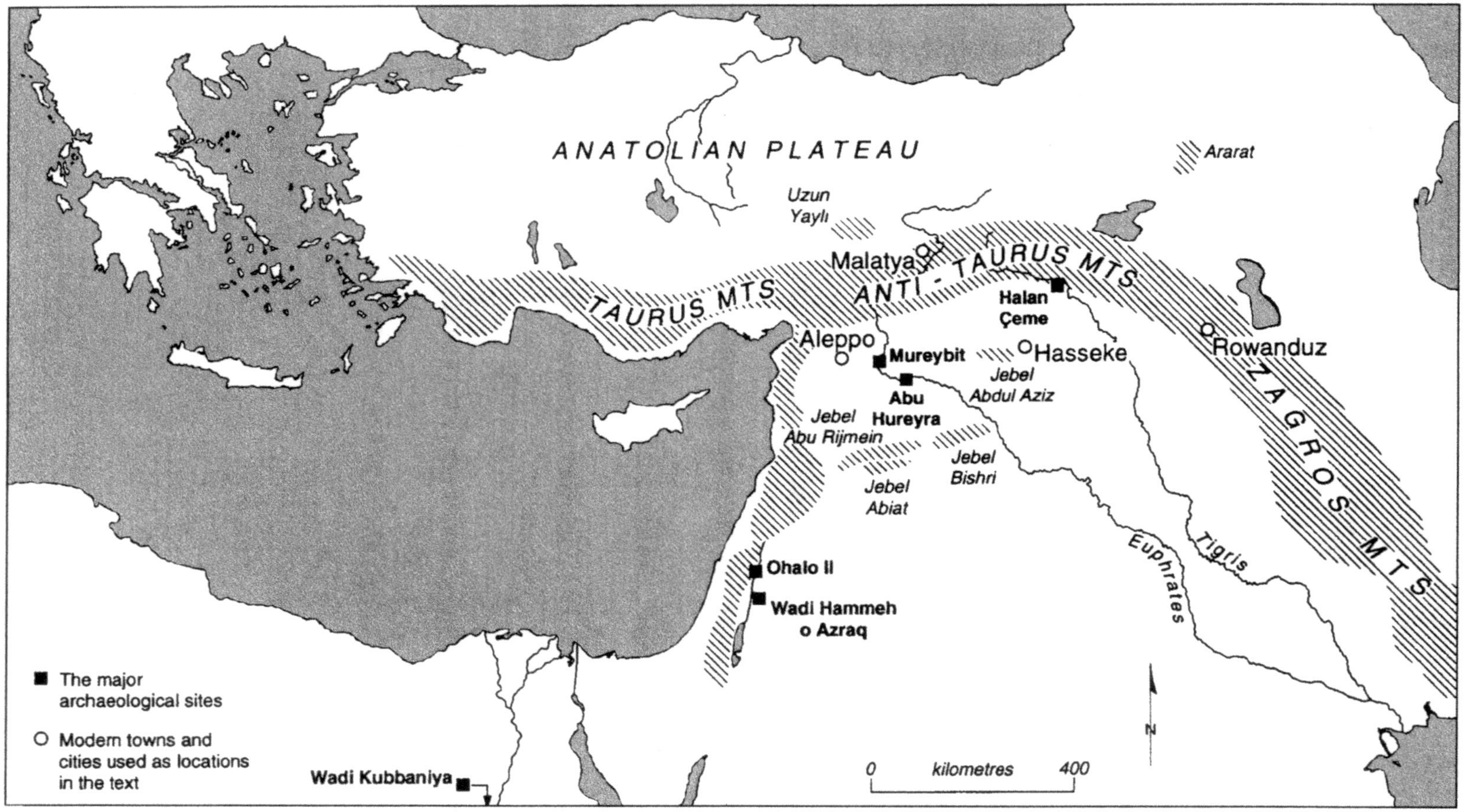

Location of archaeological sites. (Adapted by R. Prior from Hillman 1996: Figure 10.1)

PART II

The sequence followed below allowed a progressive level of refinement, the rationale and role in the project are discussed in the main part of the project.

The following entries correspond to headings for each section.

A. COVARIANCE MATRIX

The covariance command calculates pairwise values and prints them in a lower triangle matrix.

The covariance is the first product moment of two variables x and y about their means. In a sample of *n* objects, where variable *x* has mean $\bar{x}$ and variable *y* has mean $\bar{y}$, the covariance between x and y is calculated as:

$$s(x,y)^2 = \left[\sum_i (x_i - \bar{x})(y_i - \bar{x})\right]/(n-1)$$

The covariance is scale dependent, and unbounded. The pairwise covariance values are arranged in a symmetric covariance matrix. This matrix is a measure of the dispersion of multivariate data. The diagonal elements are the variances of corresponding variables. If the variables are first autoscaled, the covariance matrix (as in this case), the covariance matrix equals the correlation matrix.

The correlation is a scale independent measure of linear association between variables *x* and *y*. The most common measure of correlation is the Pearson correlation coefficient, also called product moment correlation coefficient. It is calculated as:

$$r(x,y) = \left[\sum(x_i - \bar{x})(y_i - \bar{y})\right] / \sqrt{\sum(x_i - \bar{x})^2 \sum(y_i - \bar{y})^2}$$

The correlation ranges between -1 and +1. A zero value indicates absence of correlation, while -1 or +1 indicate perfect negative or positive correlation. The correlation matrix is a symmetric matrix of the pairwise correlation coefficients of several variables.

B. MATRIX CONDITION

This is a must analysis before any multivariate analysis is attempted. This analysis quantifies the collinearity among variables and gives warnings about constant and highly collinear columns. Collinearity can make some classification and regression results arbitrary or even make the calculation impossible. In the latter case there will be an error message to this effect.

The output prints the condition number. This is the ratio of the largest and smallest eigenvalues calculated from the correlation matrix. A number larger than 1000 indicates collinearity. Should be careful with classification and regression performed on a predictor matrix with high collinearity. Biased methods, such as RDA, SIMCA and PCR, RIDGE, and PLS are specifically designed to handle collinearity.

This command also tells about columns that are essentially constant. Constant columns can make most modelling impossible, so they should be removed from the matrix before further analysis.

Matrix Condition

Condition Number = Largest Eigval / Smallest Eigval
3060.34082 = 8.52389 / 0.00279

Variables are highly collinear !

C. & D. PRINCIPAL COMPONENT ANALYSIS (PCA) BY SVD & NIPALS

PCA by Singular Value Decomposition (SVD) will use the covariance matrix; to either calculate as many components as there are input variables, or as many components as specified.

PCA by Non Linear Iterative Partial Least Squares (NIIPALS) will also use the covariance matrix to either calculate just two components or the number of components as specified.

It calculates eigenvalues, eigenvectors, and principal component scores of a data matrix. This analysis is also called eigenanalysis. There are two algorithms implemented in SCAN: SVD and NIPALS.

Principal component analysis, also referred to as eigenanalysis, calculates orthogonal linear combinations, **T**, of the autoscaled variables, **X**, based on the maximum variance criterion. Such linear combinations are called principal component scores. The coefficients, **L**, of the linear combinations are called loadings. These three quantities are related as follows:

T= XL or $^{t}im = \sum_j x_{ij} 1_{jm}$

Since $\mathbf{L}^{-1} = \mathbf{L}^{!}$, the original variables can be written as linear combinations of the components: $\mathbf{X}=\mathbf{TL}^{!}$ or $x_{ij} = \sum_m t_{im} 1_{mj}$

m = first principal component.

i = number of objects.

j = number of variables.

When there are many variables and only a few components are needed the NIPALS method is a more efficient alternative. It starts with the data matrix **X** (often autoscaled) and calculates the eigenvectors and principal components corresponding to the largest eigenvalue in an iterative algorithm. Once iteration converges, the component is subtracted from the matrix **X** and the calculation continues with the second component.

Scores: are the projection of data onto a new axis.

Eigenvalues: relative measures of length in descending order. PC1, PC2 and PC3.

Loadings: equation for line - how much of original variables used; i.e. linear combinations of original data.

The sum of the variances of all p components equals the sum of the variances of the original p variables. The principal components can be calculated according to the maximum variance criterion, i.e. each successive component is an orthogonal linear combination of the original variables such that it covers the maximum of the variance not accounted for by the previous components.

The SVD algorithm calculates all components together, so its space and time requirements increase with the number of variables. It is used when calculating all components from a moderate number of variables, therefore it is necessary to specify the variables for the analysis. It should be used when there are small or moderate numbers of variables, e.g. below 50.

Correlation Matrix

	v3400	v2960	v2872	v1736	v1712	v1640	v1452	v1376
v2960	0.562							
v2872	0.690	0.979						
v1736	0.326	0.866	0.800					
v1712	0.435	0.867	0.822	0.903				
v1640	0.688	0.687	0.740	0.414	0.550			
v1452	0.444	0.958	0.910	0.880	0.914	0.648		
v1376	0.451	0.934	0.882	0.910	0.919	0.621	0.962	
v1252	0.451	0.880	0.838	0.937	0.892	0.581	0.881	0.953
vl 196	0.395	0.907	0.853	0.941	0.922	0.546	0.944	0.960
vl 152	0.333	0.892	0.828	0.949	0.898	0.463	0.934	0.925
v676	0.480	0.830	0.808	0.724	0.843	0.749	0.892	0.855
sample	0.481	0.131	0.236	-0.120	-0.101	0.470	-0.035	0.033

	v1252	v1196	v1152	v676
v1196	0.955			
v1152	0.903	0.978		
v676	0.777	0.838	0.813	
sample	0.083	-0.023	-0.111	0.077

Covariance Matrix

	v3400	v2960	v2872	v1736	v1712	v1640
v3400	1.0000					
v2960	0.5619	1.0000				
v2872	0.6902	0.9787	1.0000			
v1736	0.3259	0.8657	0.8001	1.0000		
v1712	0.4348	0.8669	0.8221	0.9028	1.0000	
v1640	0.6879	0.6869	0.7396	0.4136	0.5498	1.0000
v1452	0.4436	0.9584	0.9105	0.8805	0.9136	0.6484
v1376	0.4507	0.9335	0.8818	0.9100	0.9186	0.6215
v1252	0.4511	0.8802	0.8380	0.9373	0.8922	0.5810
vl 196	0.3950	0.9071	0.8528	0,9408	0.9217	0.5457
vl 152	0.3329	0.8918	0.8284	0.9494	0.8976	0.4631
v676	0.4802	0.8295	0.8084	0,7240	0.8431	0.7493
sample	18.9317	5.1470	9.2817	-4.7119	-3.9884	18.5271

	v1452	v1376	v1252	v1196	v1152	v676	
v1452	1.0000						
v1376	0.9619	1.0000					
v1252	0.8809	0.9532	1.0000				
v1196	0.9436	0.9601	0.9549	1.0000			
v1152	0.9337	0.9247	0.9029	0.9785	1.0000		
v676	0.8915	0.8554	0.7772	0.8379	0.8133	1.0000	**sample**
sample	-1.3679	1.2817	3.2810	-0.9232	-4.3733	3.0340	1551.7598

The NIPALS algorithm calculates one component at a time, so its space and time requirements increase primarily with the number of components; it is used when calculating just a few components from a large number of variables. It should be used when there are more than 50 variables.

The output contains two tables. The first shows the eigenvalue for each component calculated, the proportion of the total variance covered by that component, and the cumulative proportion of variance covered up to that component. The second table shows the eigenvectors. The rows of this table correspond to the variables, the columns to the components. The results from these tables are represented graphically, there are five important graphs which are relevant to the project analysis.

Eigenvalue profile plot: This is a scatter or scree plot of the eigenvalues against the component ids. The points are connected forming a decreasing curve. This plot helps to determine the number of significant components.

Score plot: This is a scatter plot of objects (rows) projected onto the plane of the first two components, i.e., the first principal component scores are plotted against the second principal component scores. When the first two components cover most of the total variance, the configuration of the points on this plot closely reflects the original multidimensional configuration. Outliers and clusters can easily be spotted on this graph.

Loading plot: This is a scatter plot of the variables projected onto the plane of the first two components, i.e., the first component loadings are plotted against the second component loadings. The variable points are connected to the (0,0) point. The angles of the lines reflect the correlations among the variables.

Biplot: This is a scatter plot which contains both the objects and the variables projected onto the plane of the first two components. A biplot is an overlay of the score plot and loading plot. To bring the two plots onto a comparable scale, the loadings are multiplied by the corresponding eigenvalues. Biplot with row ids prints the row id of each score on the graph. A biplot is created only if the components were calculated from the correlation matrix.

Principal Component Analysis Calculated from Covariance Matrix by SVD

Eigenvalue	9.7232	1.2397	0.4172	0.1972
Proportion	0.810	0.103	0.035	0.016
Cumulative	0.810	0.914	0.948	0.965

Eigenvectors

Variable	PC1	PC2	PC3	PC4
v3400	-0.176	-0.656	-0.569	0.285
v2960	-0.310	-0.053	-0.078	-0.509
v2872	-0.302	-0.196	-0.179	-0.476
v1736	-0.293	0.286	-0.250	0.092
v1712	-0.300	0.128	-0.002	0.419
v1640	-0.222	-0.545	0.487	0.017
v1452	-0.312	0.061	0.157	-0.275
v1376	-0.313	0.093	0.031	0.022
v1252	-0.303	0.122	-0.153	0.300
vl 196	-0.310	0.184	-0.024	0.094
vl 152	-0.301	0.252	-0.048	-0.071
v676	-0.287	-0.097	0.535	0.261

Principal Component Analysis Calculated from Covariance Matrix by NIPALS

Eigenvalue	9.7232	1.2397	0.4172	0.1972
Proportion	0.810	0.103	0.035	0.016
Cumulative	0.810	0.914	0.948	0.965

Eigenvectors

Variable	PCI	PC2	PC3	PC4
v3400	0.176	-0.656	0,569	~0.285
v2960	0.310	-0.053	0,078	0.509
v2872	0.302	-0.196	0,179	0.476
v1736	0.293	0.286	0,250	-0.092
v1712	0.300	0.128	0,002	-0.419
v1640	0.222	-0.545	-0,487	-0.017
v1452	0.312	0.061	-0,157	0.275
v1376	0.313	0.093	-0,031	-0.022
v1252	0.303	0.122	0,153	-0.300
vl 196	0.310	0.184	0,024	-0.094
vl 152	0.301	0.252	0,048	0.071
v676	0.287	-0.097	-0.535	-0.261

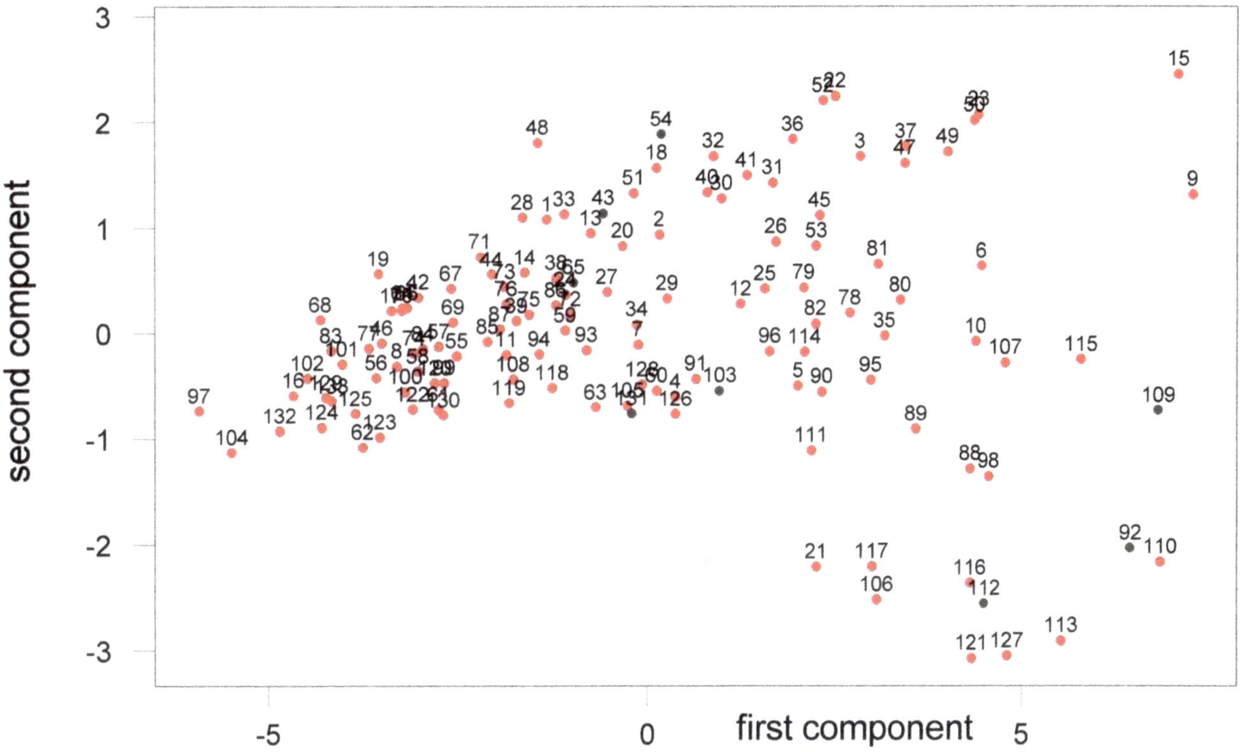

Principal Components Score Plot

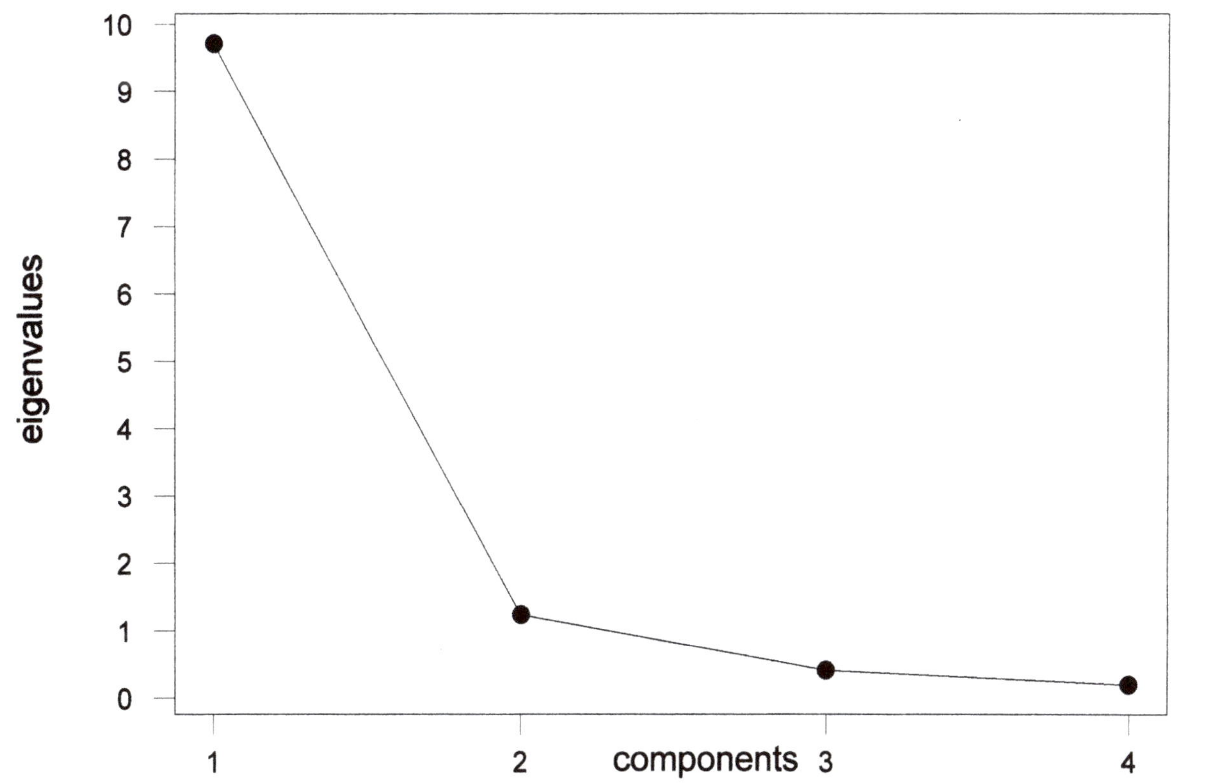

Principal Components Eigenvalue (Scree) Plot

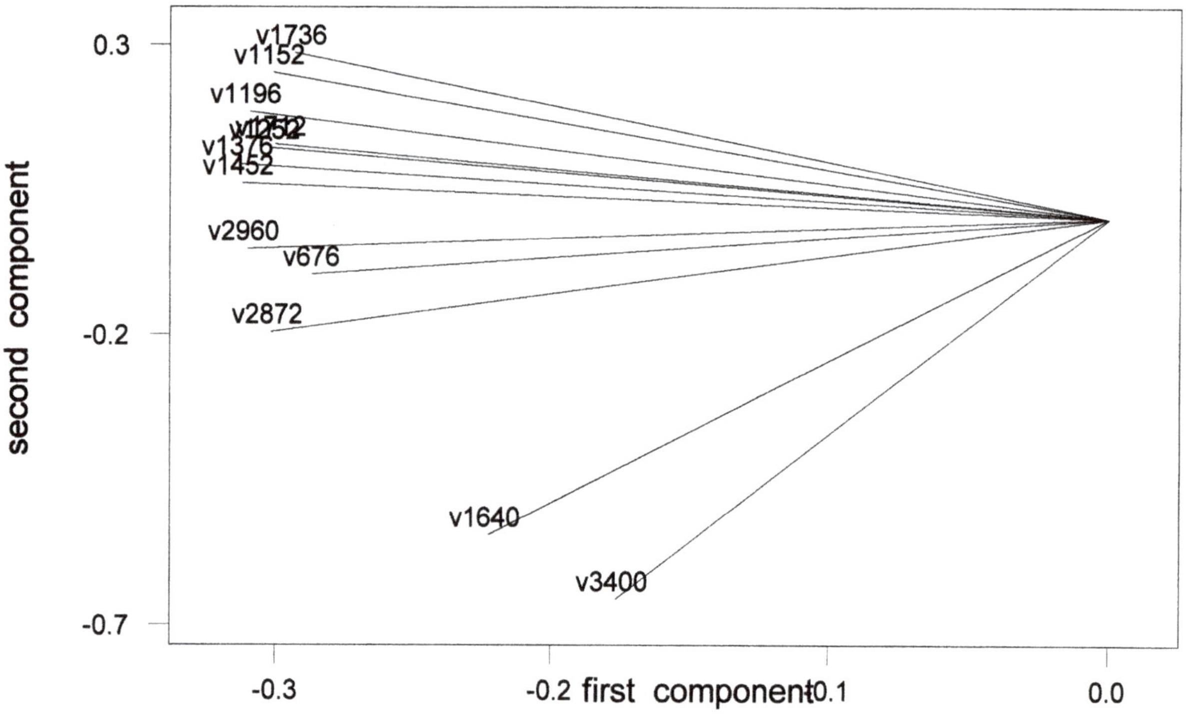

Principal Components Loading Plot

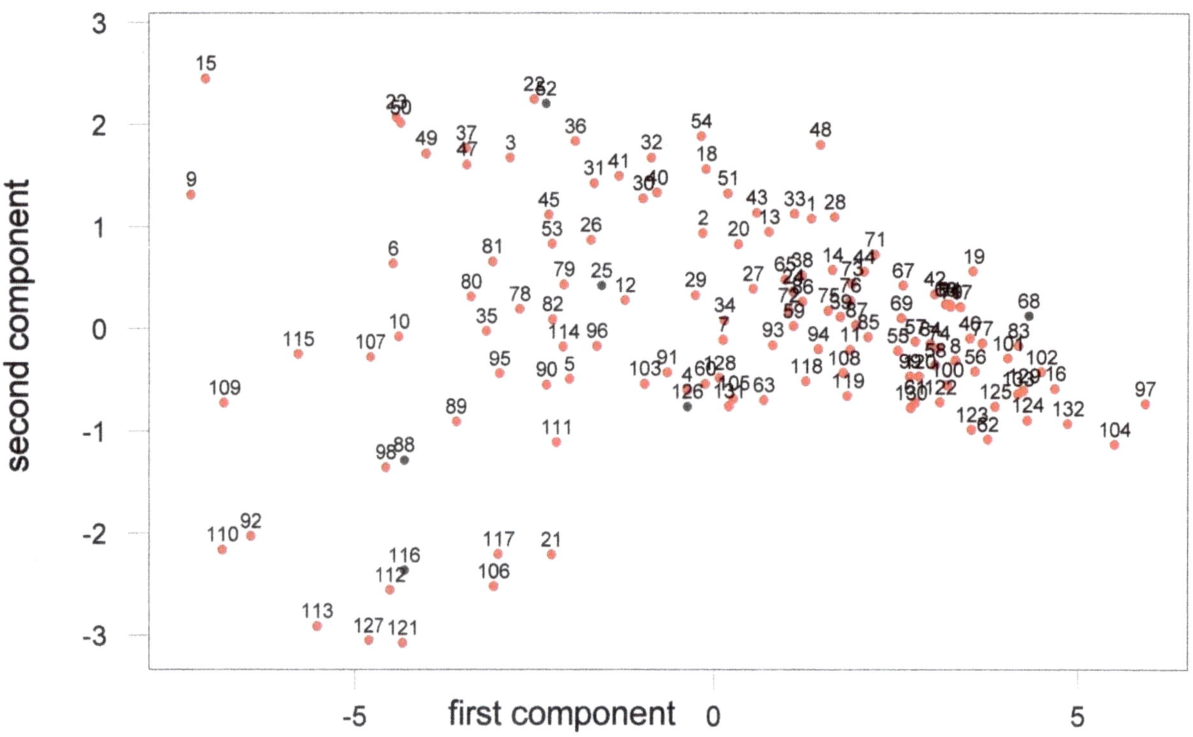

NIPALS Score Plot

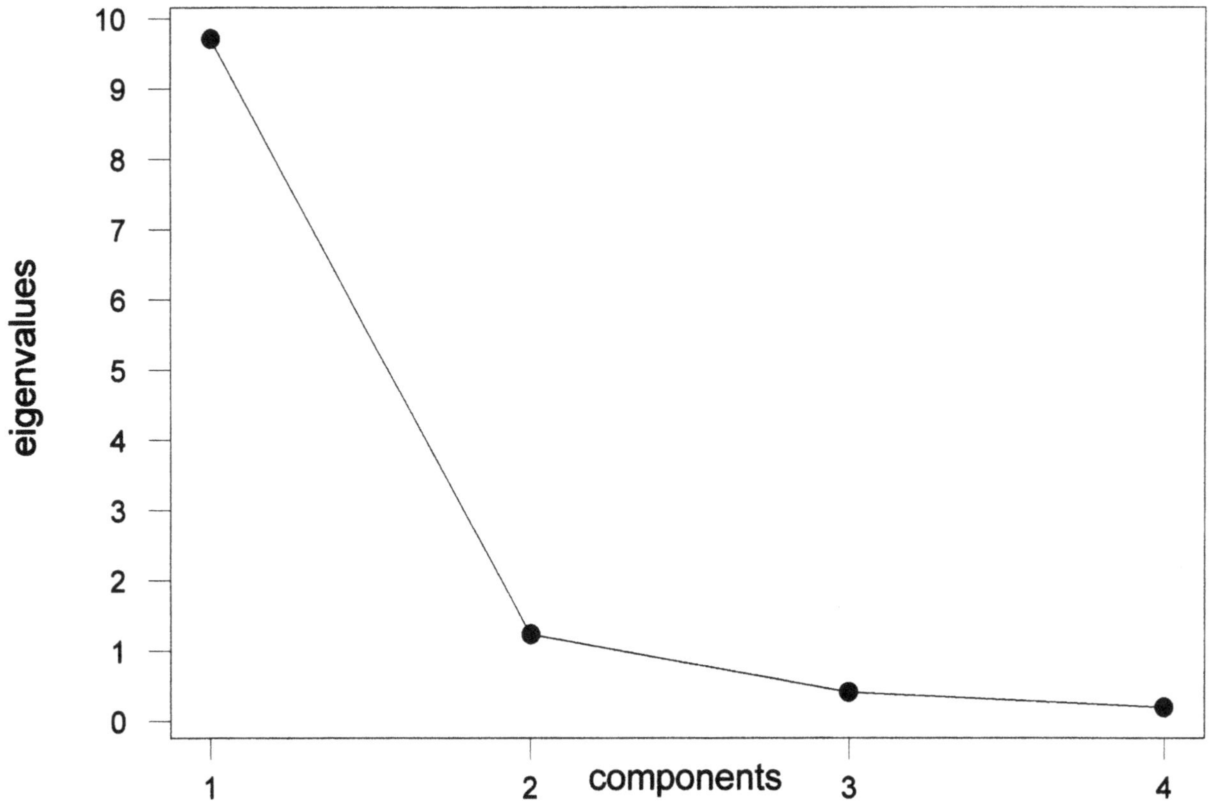

NIPALS Eigenvalue (Scree) Plot

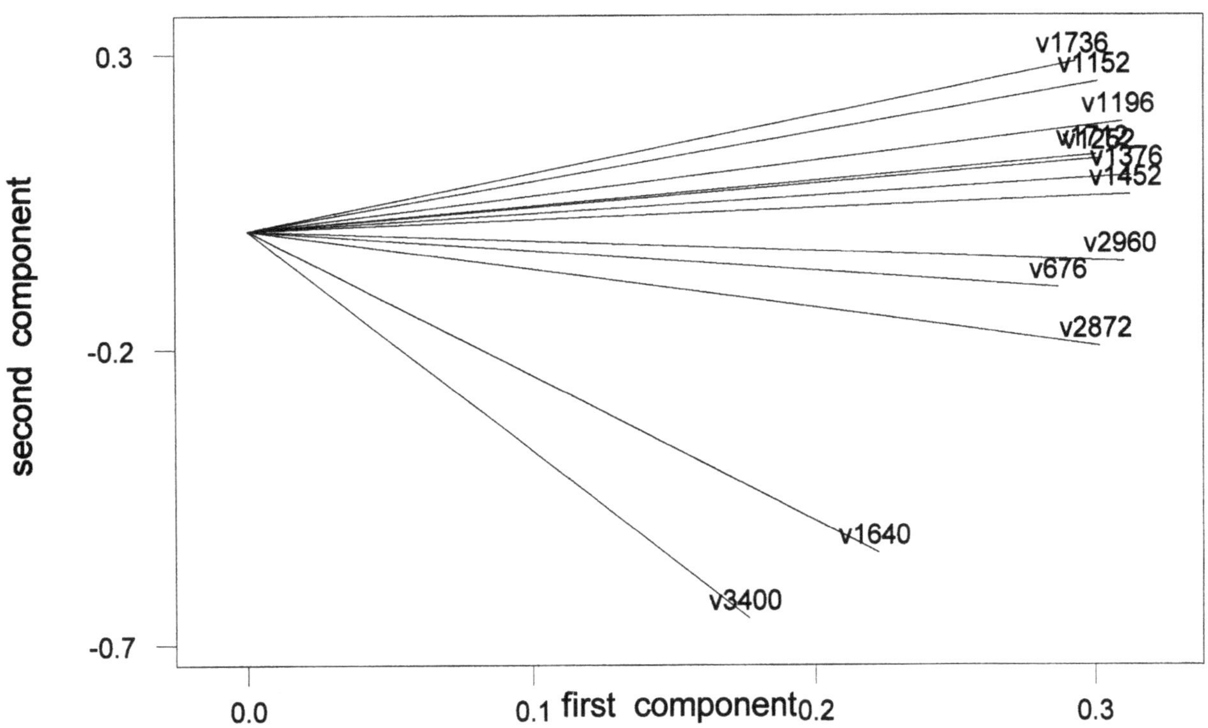

NIPALS Loading Plot

E. FACTOR ANALYSIS

Calculates the factors from a covariance matrix, by using either as many factors as there are input variables, or by a number specified. It can also perform orthogonal rotations, for this data the varimax rotation was utilized.

Factor Analysis extracts the underlying factors and performs orthogonal rotation to make the factors more interpretable; the factors are calculated by principal component analysis.

The output of FA produces are two tables. The first contains the factor loadings, their variances, and communalities. The second table contains the factor score coefficients. The rows of both tables correspond to the variables and the columns to the factors.

There are four important graphs representing the numerical output, these are:

Eigenvalue profile plot: This is a scatter plot of the eigenvalues against the factor ids. The points are connected forming a decreasing curve. This plot helps to determine the number of significant factors. This plot is also called a Scree plot. If a rotation was performed, both the rotated and unrotated eigenvalues are plotted.

Score plot: This is a scatter plot of objects (rows) projected onto the plane of the first two factors, i.e., the first factor scores are plotted against the second factor scores. When the first two factors cover most of the total variance, the configuration of the points on this plot closely reflects the original multidimensional configuration. Outliers and clusters can easily be spotted on this graph. If a rotation was performed, both the rotated and unrotated scores are plotted.

Loading plot: This is a scatter plot of the variables projected onto the plane of the first two factors, i.e., the first factor loadings are plotted against the second factor loadings. The variable points are connected to the (0,0) point. The angles of the lines reflect the correlations among the variables. If a rotation was performed, both the rotated and unrotated loadings are plotted.

Biplot: This is a scatter plot which contains both the objects and the variables projected onto the plane of the first two factors. A biplot is an overlay of the score and loading plot. If a rotation was performed, both the rotated and unrotated factors are plotted.

Factor analysis models a set of variables as a linear combination of a small number of common factors, obtaining a parsimonious description of the observed of measured quantities. These common factors are assumed to represent underlying phenomena that are not directly measurable. The factor model is:

$$\mathbf{X} = \mathbf{FL}' + \mathbf{u} \quad \text{or} \quad x_{ij} = \sum_m f_{im} 1_{jm} + u_{ij}$$

where each of the p observed variables x_{ij} is represented as a linear combination of M common factors f_{im} with coefficients 1_{jm} and a unique factor u_{ij}. The coefficients, 1_{jm}, are called factor loadings. The common factors account for the correlation among the variables, while the unique factor covers the remaining variance, called the specific variance, or unique variance. The unique factor is called residual error. The number of factors M determines the complexity of the factor model.

There two steps in factor analysis: the factor extraction and the factor rotation. In the first step, the underlying, unobservable latent factors are identified. In the second step, the extracted factors are rotated to obtain more meaningful, interpretable factors. The factor extraction involves determining the number of factors and calculating their coefficients, the factor loadings.

The most popular extraction method, the one implemented here is principal component factor analysis, where the analysis is based on the maximum variance criterion.

The second step in factor analysis is factor rotation, where the factors calculated in the factor extraction step are rotated into more interpretable factors. The one implemented here is the varimax rotation which maximizes the variance of squared factor loadings of a factor, i.e., it tries to simplify the loadings of the factors. Specifically, it rotates the original factors to obtain loadings; the goal is to obtain common factors that are composed of only a few variables. This rotation further increases the large factor loadings and large eigenvalues and further decreases the small ones in each factor.

Factor Analysis Principal Components from Covariance Matrix

Unrotated Factor Loadings and Communalities

Variable	Factor l	Factor 2	Factor 3	Factor 4	Community
v3400	-0.550	-0.730	-0.367	0.127	0.986
v2960	-0.968	-0.059	-0.050	-0.226	0.993
v2872	-0.940	-0.218	-0.116	-0.212	0.990
v1736	-0.915	0.318	-0.161	0.041	0.966
v1712	-0.937	0.142	-0.001	0.186	0.933
v1640	-0.694	-0.607	0.315	0.007	0.949
v1452	-0.974	0.068	0.101	-0.122	0.978
v1376	-0.975	0.104	0.020	0.010	0.961
v1252	-0.944	0.136	-0.099	0.133	0.937
vl 196	-0.966	0.205	-0.016	0.042	0.977
vl 152	-0.939	0.280	-0.031	-0.031	0.962
v676	-0.895	-0.108	0.346	0.116	0.946
Variance	9.7232	1.2397	0.4172	0.1972	11.5773
% Var	0.810	0.103	0.035	0.016	0.965

Rotated Factor Loadings and Communalities Varimax Rotation

Variable	Factor l	Factor 2	Factor 3	Factor 4	Community
v3400	0.170	-0.251	0.945	0.041	0.986
v2960	0.767	-0.367	0.345	0.388	0.993
v2872	0.677	-0.374	0.494	0.385	0.990
v1736	0.961	-0.102	0.150	0.100	0.966
v1712	0.883	-0.322	0.218	-0.039	0.933
v1640	0.232	-0.798	0.492	0.124	0.949
v1452	0.825	-0.440	0.181	0.267	0.978
v1376	0.872	-0.364	0.220	0.142	0.961
v1252	0.895	-0.247	0.273	0.022	0.937
vl 196	0.924	-0.289	0.168	0.104	0.977
vl 152	0.933	-0.230	0.101	0.168	0.962
v676	0.660	-0.690	0.182	0.015	0.946
Variance	7.2359	2.0882	1.7949	0.4584	11.5773
% Var	0.603	0.174	0.150	0.038	0.965

Factor Score Coefficients

Variable	Factor l	Factor 2	Factor 3	Factor 4
v3400	-0.057	-0.589	-0.880	0.642
v2960	-0.100	-0.047	-0.120	-1.147
v2872	-0.097	-0.176	-0.278	-1.073
v1736	-0.094	0.257	-0.387	0.207
v1712	-0.096	0.115	-0.002	0.944
v1640	-0.071	-0.490	0.754	0.038
v1452	-0.100	0.055	0.243	-0.618
v1376	-0.100	0.084	0.047	0.050
v1252	-0.097	0.109	-0.237	0.675
vl 196	-0.099	0.165	-0.038	0.213
vl 152	-0.097	0.226	-0.074	-0.159
v676	-0.092	-0.087	0.829	0.588

Rotated Factor Score Coefficients

Variable	Factorl	Factor2	Factor3	Factor4
v3400	-0.034	0.402	1.052	-0.518
v2960	-0.057	0.079	-0.031	1.154
v2872	-0.090	0.148	0.164	1.100
v1736	0.293	0.377	0.090	-0.177
v1712	0.251	-0.023	0.083	-0.918
v1640	-0.299	-0.850	-0.038	-0.047
v1452	0.001	-0.188	-0.237	0.602
v1376	0.125	-0.044	-0.051	-0.041
v1252	0.253	0.174	0.183	-0.637
vl 196	0.197	0.054	-0.039	-0.201
vl 152	0.187	0.126	-0.118	0.163
v676	-0.026	-0.763	-0.294	-0.616

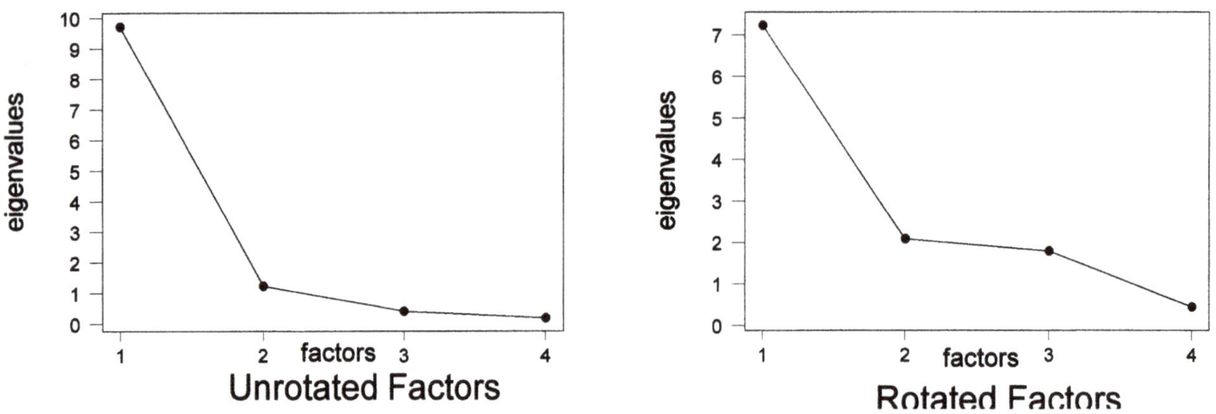

Factor Analysis Eigenvalue (Scree) Plot

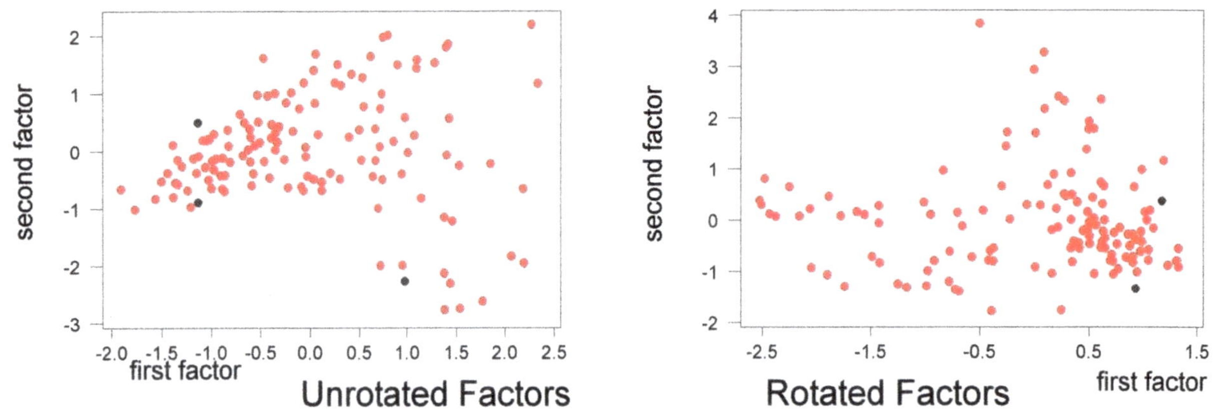

Factor Analysis Score Plot

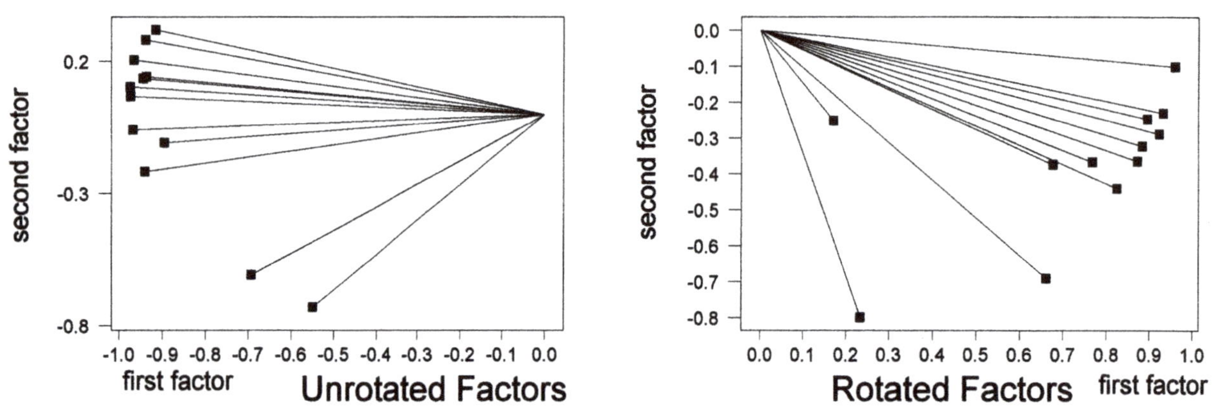

Factor Analysis Loading Plot

F. HIERARCHICAL CLUSTER ANALYSIS (HCA)

HCA performs agglomerative clustering of objects. The algorithm starts with each object in its own cluster. In the first step, the two objects closest together are joined. In the next step, either a third object Joins the first two, or two other objects Join together into a different cluster. Each step results in one fewer clusters than the step before, until, at the end, all objects are in one cluster.

The distance metric and linkage chosen determine how the distance between two clusters is calculated. The hierarchy of clusters can be represented by a binary tree, called a dendrogramme. A final partition, i.e. the cluster assignment of each object, is obtained by cutting the tree at a specified level. It is necessary to specify the variables for the analysis.

The output consists of two parts: what happened at each amalgamation step and a description of the final partition. At each amalgamation step, the two closest clusters are merged. The output contains the number of this step, the number of clusters after this step, the similarity level at which the two clusters were merged, the distance level at which the two clusters were merged, the id numbers of the two clusters just merged, the id number of the new cluster (this is the smaller id of the two merged clusters), and the number of objects in the new cluster.

The final partition is described by three tables. The first table gives the size of each cluster by number of member objects, within cluster sum of squared distances, average distance of member objects from the cluster centroid, and the maximum distance of member objects from the cluster centroid, also called the cluster radius. The second table contains the cluster centroids and the grand centroid. For autoscaled data the grand centroid column contains all zeros. The third table contains a matrix of the distances between cluster centroids. This symmetric matrix indicates how well the clusters are separated. Clusters that contain a single object (these clusters are called singletons) indicate outliers.

Hierarchical clustering helps to find clusters of objects (rows of a data matrix) in high dimensional space based on inter-object distances. There are a number of different ways to measure the distance between two objects. Clusters are defined by an agglomerative algorithm. We start with each object in its own cluster. In the first step, the two objects closest together are joined. In the next step, either a third object joins the first two, or two other objects join together into a different cluster. Each step results in one fewer clusters than the step before, until, at the end, all objects are in one cluster. Similarly with other numerical analysis, HCA calculations are graphically represented, the three most important are:

Dendrogramme: Displays the amalgamation of clusters in the form of a binary tree. The leaves, i.e. the terminal nodes at the bottom of the tree, are clusters containing a single object. The root, i.e. the top node of the tree, corresponds to a single cluster containing all objects. All other nonterminal nodes represent clusters formed during the amalgamation procedure. The vertical axis indicates the similarity level at which the clusters were formed. The tree can be cut at any similarity level. The cut determines a certain set of clusters.

Cluster centroid profile plot: Displays a scatter plot of cluster centroid coordinates versus the variable ids. The coordinate points of a cluster are connected to form a profile. Each profile is in a different colour. There are as many profiles as there are clusters in the final partition. This plot reveals how different the cluster centroids are i.e., how well separated the clusters are. The plot is available only if there are no more than ten clusters.

Object-cluster distance matrix plot: Displays a matrix plot that shows the distance of each object from each cluster centroid. The object points are colour coded according to their cluster membership. In an ideal situation, each object has a small distance from its own cluster and large distances from all other clusters. This plot helps to investigate the separation of clusters. The plot is available only if a final partition was obtained (either by specifying the number of clusters, or a similarity level at which to cut the tree), and if there are no more than ten.

Hierarchical cluster analysis produces a hierarchy of partitions of objects such that any cluster of a partition is fully included in one of the clusters of the later partitions. Such partitions are best represented by a dendrogramme. This strategy is different from nonhierarchical clustering, which results in one single partition. There are two main hierarchical strategies: agglomerative clustering and divisive clustering. Agglomerative clustering, used here, via SCAN, starts with n objects in n separate clusters (these are the leaves of the dendrogramme) and, after n-1 agglomeration steps, ends with all n objects in one single cluster (this is the root of the dendrogramme). In each step, the number of clusters is decreased by one, merging the two closest clusters. The first step is to calculate an n x n distance matrix based on of the following distance metric:

Euclidean Distance $$d_{ik} = \sqrt{\sum_j \left[x_{ij} - x_{kj}\right]^2}$$

The linkage you choose determines how the distance between two clusters is defined. At each step of the algorithm, there is a distance matrix. The entry d_{ik} of this matrix is the distance from cluster i to cluster k. At the beginning, when each cluster contains a single observation, d_{ik} is the distance from observation i to observation k. At each step, this matrix is reduced by one row and one column. The two rows (and two columns) s and t corresponding to the two closest clusters, are merged and are replaced by a new row (and column) i, corresponding to the resulting new cluster. An updating distance formula defines the elements of this new row (column) d_{ik} from the elements of the two old rows (columns) d_{sk} and d_{tk}.

Complete linkage was utilized for this data:

Complete linkage: (furthest neighbour linkage, maximum linkage), the distance between two clusters is the largest distance between all pairs of objects, one object in one cluster and the other object in the other cluster. This method tends to produce well separated, small, compact spherical clusters.

e.g. Let n_s, n_t, n_i be the number of observations in clusters s, t, and i respectively,

and $n_{sti} = n_s + n_t + n_i$.

Most of the agglomerative methods can be unified by the single formula:

$$d_{ik} = \alpha d_{sk} + \beta d_{tk} + \gamma d_{st} + \delta \left| d_{sk} - d_{tk} \right|$$

where the parameters $\alpha, \beta, \gamma, \delta$ are the following:

$\alpha = 0.5$, $\beta = 0.5$, $\gamma = 0.0$, $\delta = 0.5$

Hierarchical Cluster Analysis

Complete Linkage
Euclidean Distance
Amalgamation Steps

Step	Number of clusters	Similarity level	Distance level	Clusters joined		New cluster	Number of obs. in new cluster
1	132	98.55	0.196	64	66	64	2
2	131	97.74	0.306	64	70	64	3
3	130	97.36	0.357	58	74	58	2
4	129	97.12	0.390	129	133	129	2
5	128	96.82	0.431	23	50	23	2
6	127	96.73	0.444	55	57	55	2
7	126	96.71	0.445	85	87	85	2
8	125	96.63	0.456	72	86	72	2
9	124	96.21	0.514	101	102	101	2
10	123	96.12	0.525	56	77	56	2
11	122	96.03	0.538	30	40	30	2
12	121	96.01	0.540	124	129	124	3
13	120	95.93	0.552	105	131	105	2
14	119	95.75	0.575	120	122	120	2
15	118	95.73	0.579	67	71	67	2
16	117	95.57	0.600	39	76	39	2
17	116	95.56	0.601	58	84	58	3
18	115	95.51	0.608	75	94	75	2
19	114	95.50	0.610	31	41	31	2
20	113	95.30	0.637	17	46	17	2
21	112	95.29	0.639	59	72	59	3
22	111	95.22	0.648	78	81	78	2
23	110	95.22	0.648	1	28	1	2
24	109	95.18	0.653	83	101	83	3
25	108	95.09	0.665	45	53	45	2
26	107	95.03	0.673	124	125	124	4
27	106	94.90	0.691	79	82	79	2
28	105	94.82	0.701	106	117	106	2
29	104	94.50	0.745	31	36	31	3
30	103	94.37	0.763	118	119	118	2
31	102	94.31	0.771	51	54	51	2
32	101	94.31	0.771	19	42	19	2
33	100	94.29	0.773	105	128	105	3
34	99	94.29	0.774	67	73	67	3
35	98	94.22	0.783	55	58	55	5
36	97	94.07	0.804	30	32	30	3
37	96	94.04	0.808	120	130	120	3
38	95	93.98	0.816	13	20	13	2
39	94	93.96	0.819	39	85	39	4
40	93	93.95	0.820	2	18	2	2
41	92	93.67	0.857	104	132	104	2
42	91	93.66	0.859	14	44	14	2
43	90	93.62	0.865	99	100	99	2
44	89	93.41	0.893	38	65	38	2
45	88	93.35	0.901	1	33	1	3
46	87	93.26	0.914	17	64	17	5
47	86	93.16	0.927	25	26	25	2
48	85	93.14	0.929	68	83	68	4
49	84	93.08	0.938	90	95	90	2
50	83	92.96	0.954	62	123	62	2
51	82	92.76	0.980	88	98	88	2
52	81	92.75	0.982	16	124	16	5
53	80	92.40	1.029	55	99	55	7
54	79	92.27	1.048	75	93	75	3
55	78	92.12	1.068	60	63	60	2

56	77	92.08	1.072	55	120	55	10
57	76	92.05	1.078	2	30	2	5
58	75	92.00	1.084	22	52	22	2
59	74	91.97	1.088	121	127	121	2
60	73	91.89	1.098	67	69	67	4
61	72	91.89	1.111	112	121	112	3
62	72	91.76	1.116	107	*115*	107	2
63	70	91.74	1.119	111	114	111	2
64	69	91.73	1.121	13	43	13	3
65	68	91.62	1.135	27	38	27	3
66	67	91.45	1.158	47	49	47	2
67	66	91.35	1.172	78	80	78	3
68	65	91.30	1.178	34	60	34	3
69	64	91.28	1.181	17	56	17	7
70	63	91.17	1.196	39	59	39	7
71	62	90.84	1.241	79	96	79	3
72	61	90.79	1.247	3	37	3	2
73	60	90.45	1.294	6	10	6	2
74	59	90.26	1.319	8	19	8	3
75	58	90.20	1.328	16	68	16	9
76	57	90.10	1.342	61	62	61	3
77	56	90.03	1.351	108	118	108	3
78	55	89.98	1.357	103	105	103	4
79	54	89.82	1.380	89	90	89	3
80	53	89.77	1.386	97	104	97	3
81	52	89.48	1.425	1	13	1	6
82	51	89.25	1.456	39	108	39	10
83	50	89.04	1.485	14	67	14	6
84	49	88.97	1.494	17	55	17	17
85	48	88.47	1.562	23	47	23	4
86	47	88.30	1.586	22	31	22	5
87	46	88.24	1.594	27	75	27	6
88	45	88.16	1.605	91	103	91	5
89	44	88.15	1.605	112	113	112	4
90	43	87.85	1.646	4	5	4	2
91	42	87.75	1.660	25	79	25	5
92	41	87.74	1.661	92	110	92	2
93	40	87.35	1.714	24	27	24	7
94	39	87.33	1.717	21	106	21	3
95	38	87.06	1.752	2	51	2	7
96	37	86.31	1.854	112	116	112	5
97	36	86.14	1.877	91	126	91	6
98	35	86.03	1.892	35	45	35	3
99	34	85.60	1.950	7	29	7	2
100	33	85.31	1.991	8	17	8	20
101	32	85.07	2.023	78	89	78	6
102	31	84.96	2.037	11	14	11	7
103	30	84.95	2.039	1	24	1	13
104	29	84.82	2.057	92	109	92	3
105	28	84.79	2.060	3	23	3	6
106	27	84.33	2.123	88	107	88	4
107	26	82.97	2.307	8	16	8	29
108	25	82.61	2.356	4	12	4	3
109	24	82.50	2.370	25	111	25	7
110	23	82.49	2.373	11	39	11	17
111	22	82.09	2.426	7	34	7	5
112	21	81.15	2.554	1	48	1	14
113	20	81.12	2.557	22	35	22	8
114	19	78.66	2.892	25	78	25	13
115	18	78.64	2.894	8	61	8	32
116	17	76.98	3.118	1	2	1	21
117	16	76.94	3.124	4	91	4	9
118	15	76.23	3.220	6	88	6	6
119	14	75.97	3.255	9	15	9	2

120	13	75.83	3.274	3	22	3	14
121	12	74.36	3.474	21	112	21	8
122	11	72.91	3.671	8	97	8	35
123	10	72.13	3.776	7	11	7	22
124	9	70.93	3.939	1	7	1	43
125	8	69.97	4.068	4	25	4	22
126	7	65.06	4.733	6	21	6	14
127	6	63.81	4.904	9	92	9	5
128	5	57.86	5.709	3	4	3	36
129	4	52.99	6.369	3	6	3	50
130	3	46.62	7.232	1	8	1	78
131	2	39.15	8.244	3	9	3	55
132	1	0.00	13.548	1	3	1	133

Final Partition

Number of clusters: 5

	Number of observations	Within cluster sum of squares	Average distance from centroid	Maximum distance from centroid
Cluster 1	43	88.600	1.351	2.328
Cluster 2	36	139.874	1.886	3.202
Cluster 3	14	56.447	1.924	3.094
Cluster 4	35	43.857	1.025	2.359
Cluster 5	5	26.952	2.222	3.023

Cluster Centroids

Variable	Cluster 1	Cluster2	Cluster3	Cluster4	Cluster5
v3400	-0.1312	-0.7588	0.4183	0.9256	-1.0584
v2960	0.2178	-0.6803	-1.3037	1.2246	-1.8965
v2872	0.1509	-0.7380	-1.0278	1.2444	-1.8170
v1736	0.4664	-0.5679	-1,7582	0.9786	-1.8486
v1712	0.4033	-0.6635	-1.3724	1.0349	-2.0930
v1640	0.0116	-0.6651	-0.1485	0.9268	-1.3830
v1452	0.3454	-0.6287	-1.3962	1.1099	-2.3036
v1376	0.3902	-0.5931	-1.4833	1.0734	-2.4455
v1252	0.3836	-0.5614	-1.5632	1.0396	-2.1568
vl 196	0.3777	-0.5151	-1.6743	1.0473	-2.1834
vl 152	0.4124	-0.4784	-1.7518	0.9829	-2.0776
v676	0.2461	-0.6270	-0.8942	1.0218	-2.2514

Variable	Grand centrd
v3400	0.0000
v2960	0.0000
v2872	0.0000
v1736	0.0000
v1712	-0.0000
v1640	0.0000
v1452	-0.0000
v1376	-0.0000
v1252	0.0000
vl 196	-0.0000
vl 152	-0.0000
v676	0.0000

Distances Between Cluster Centroids

	Cluster 1	Cluster2	Cluster3	Cluster4	ClusterS
Cluster 1	0.0000	3.1334	5.7151	2.7716	7.9528
Cluster 2	3.1334	0.0000	3.0767	5.8192	4.8676
Cluster 3	5.7151	3.0767	0.0000	8.0251	3.0846
Cluster 4	2.7716	5.8192	8.0251	0.0000	10.5323
Cluster 5	7.9528	4.8676	3.0846	10.5323	0.0000

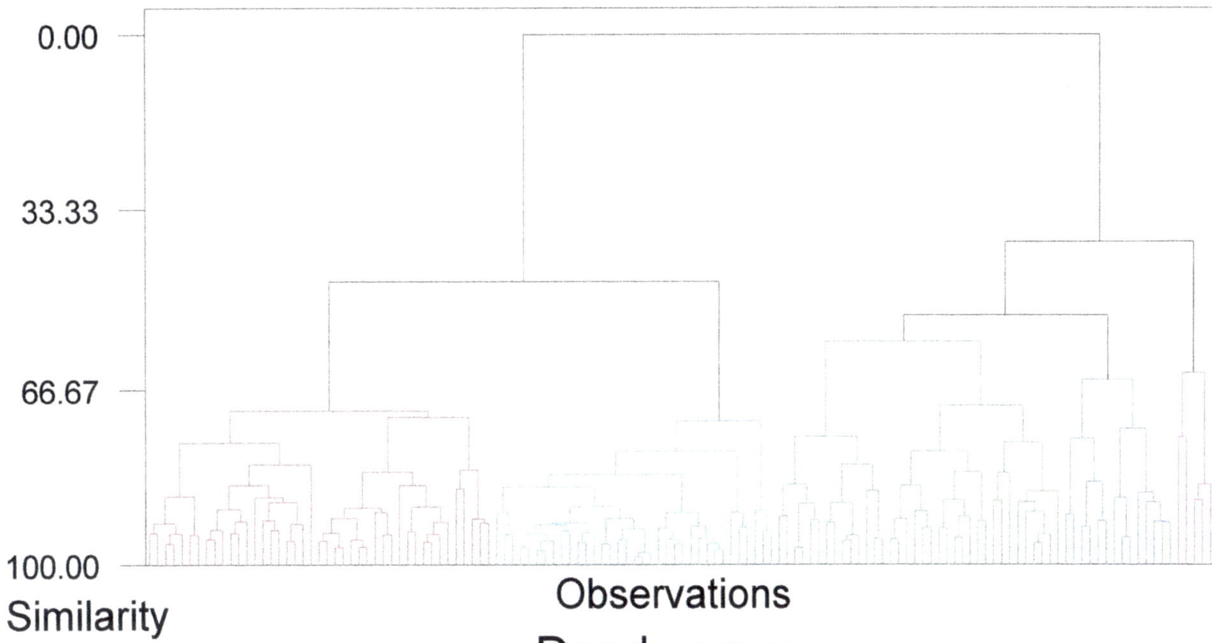

Dendrogram

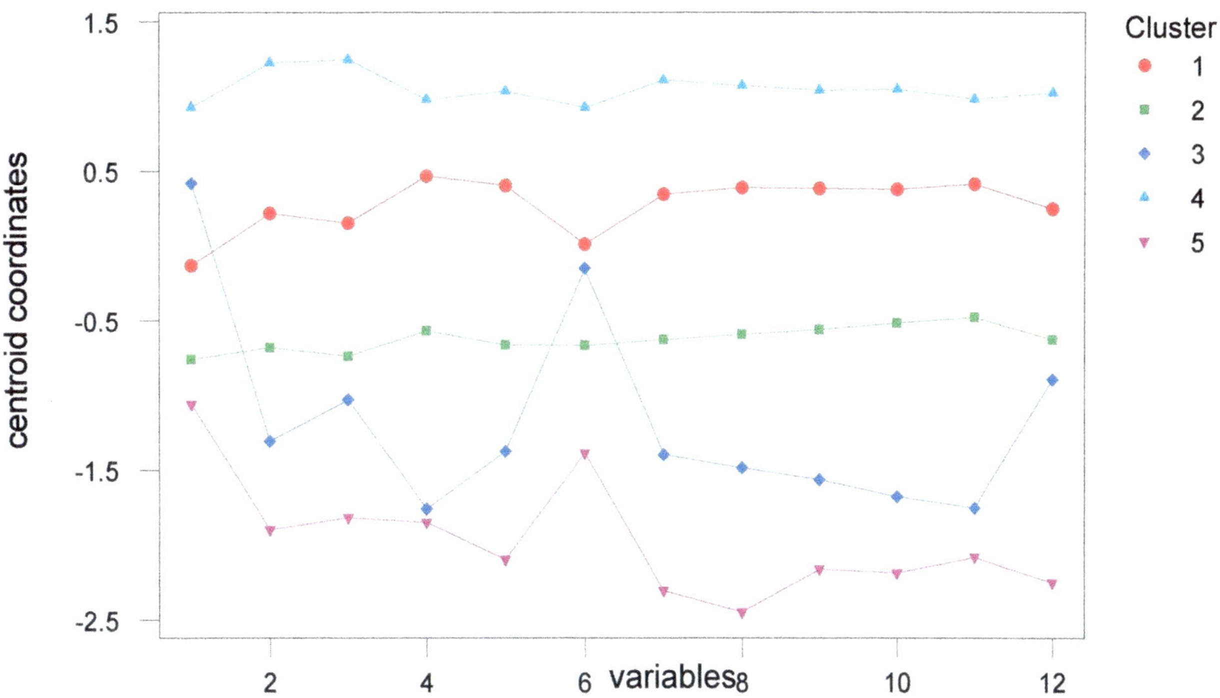

Hierarchical Cluster Centroid Profiles

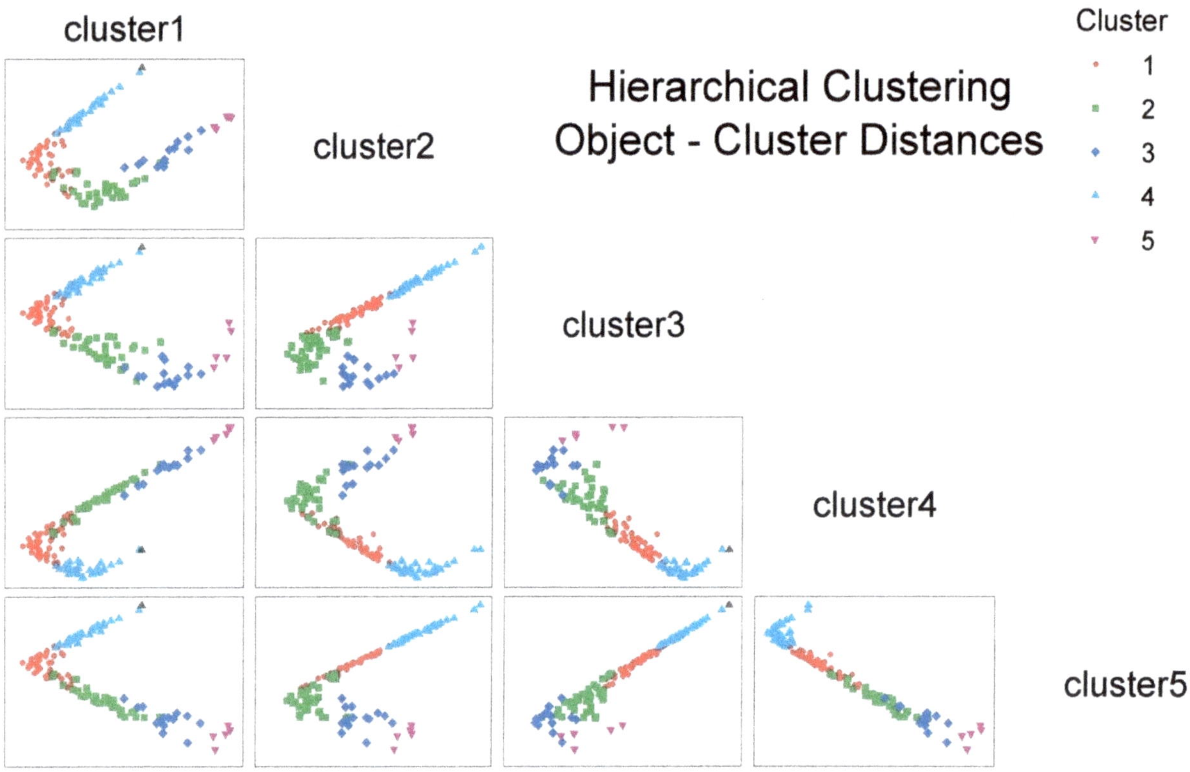

G. ORDINARY LEAST SQUARES (OLS)

OLS calculates a parametric linear model for a single response, and calculates unbiased, leastsquares coefficients. OLS should be used if the data set is well determined, i.e. if there are more observations than predictors and the predictors are not highly correlated. This model assumes that the response is a linear function of the predictors, and that the errors are identically and independently distributed.

There is a need to specify one or more predictor columns for the training set; and also a need to specify one response column for the training set.

The numerical output of OLS regression consists of three tables. The first, the Goodness-of-Fit and Goodness-of-Prediction table contains information about how well the calculated model fits the training set data and how well it would predict future data. The modelling capability (Goodness-of-fit) is measured by the residual sum of squares (Residual SS) and by the R-squared coefficient (R2). The predicting capability (Goodness-of-prediction) is measured by the residual predictive sum of squares (Resid PRESS) and by the cross-validated R-squared coefficient (R2cv).

The next table contains the regression coefficients and the importance of the predictors in the model. The importance of a predictor is measured by the standardized regression coefficient = (the regression coefficient) x (the standard deviation of the corresponding predictor). This table is presented graphically by the regression coefficient plot.

The Analysis of Variance table shows how much of the total variance is explained by the model. The total degrees of freedom (Df) and the total sum of squares (SS) have two components, model and residual. The mean squares (MS) are the sum of squares divided by the corresponding degrees of freedom. The standard error is the square root of the residual MS. The F value (F) and the corresponding probability calculated from the model and residual mean square indicate the significance of the model: a large F-value and small p value mean that the model is significant.

The above data is also graphically represented:

Regression coefficient plot: Displays two scatter plots: regression coefficients versus the predictor ids and predictor importances versus the predictor ids. These plots are especially informative when there are many predictors. The intercept is not shown in the plot because it often distorts the vertical scale. Can omit predictors with relatively small importances.

Response plot: Displays a scatter plot of the calculated response versus the true response. Two calculated response values are plotted: the ordinary fitted response (squares) and the cross-validated predicted response (diamonds). Observations where the cross-validated value differs significantly from the ordinary fit may be leverage points. Ideally this plot is close to the straight line, calculated = true. This line is also displayed on the plot.

Residual plots: Displays four scatter plots. Each plots one type of residuals (ordinary residuals, cross-validated residuals, standardized residuals, and jackknifed residuals) versus the fitted response values. These plots reveal nonlinearity, heteroscedasticity, and outliers. The horizontal line where residual = 0,0 is displayed on each plot. Ideally, residuals are distributed symmetrically around this line.

Normal residual plots: Displays four scatter plots. Each plots one type of residuals (ordinary residuals, cross-validated residuals, standardized residuals, and jackknifed residuals) versus the normal scores (quantiles from standard normal distribution). These plots are especially useful for checking the normality assumption of the residuals.

Leverage plots: Contains four scatter plots that reveal outliers and leverage points. The upper left plot shows leverages (diagonal elements of the hat matrix) versus the observation ids. The upper right plot shows Cook"s distances versus the observation ids. The lower left plot shows jackknifed residuals versus leverages (hat diagonals). The lower right plot shows leverages (hat diagonals) versus the squared standardized residuals. The vertical and horizontal straight lines indicate limits of normal values. The limits indicated by the lines on these plots are described in the theory.

OLS is the most popular regression method. In this method, the regression model is linear in both the coefficients and predictors.

$\mathbf{y} = \mathbf{Xb} + \mathbf{e}$

where **y** is the vector of response values, **X** is the matrix of the predictors, **b** is the vector of the regression coefficients, and **e** is the vector of errors.

The coefficients, **b**, are calculated as the value that minimizes the residual sum of squares or, equivalently, maximizes the correlation between a linear combination of the predictors and the response.

OLS calculates an unbiased model. However, it is suitable only for well-determined systems and is very sensitive to outliers. The estimated response and any statistic related to it are scale invariant. Since OLS is a linear estimator, the cross-validated residuals can be calculated from the ordinary residuals.

Regression diagnostics: using both numerical and graphical results, is a crucial step in regression analysis. Diagnostics help to check whether the assumption of the model are correct. Regression diagnostics in SCAN are based on the hat matrix, Cook"s distance, and various types of residuals.

The hat matrix, H, is an n x n symmetric, idempotent matrix of rank p, that projects the observed responses onto predicted responses, $\hat{y}$: $\hat{y} = \mathbf{H}\,\mathbf{y}$

H is often called the projection matrix. The trace of the hat matrix gives the degrees of freedom of the corresponding regression model. The diagonal element of H is the Mahalarlobis distance of the predictor vector. This is used as a test for the Euclidean distance. Mahalanobis is a unit distance measure which varies with the direction of measurement, it is highly directional and reflects both group shape and direction. The algorithm is based on simple and sound statistical foundations, and results can be given meaningful confidence limits or probabilities. In this case it is a technique used for confidently measuring spectral differences, as it can be calibrated to respond to sample-to-sample variations. This allows it to be used to identify materials which may have a broad specification and consequently quite variable spectra. Samples can be classified by their distance from individual group mean positions. The shorter the distance, the higher the probability that the unknown or test sample belongs to that group.

The hat matrix is an important quantity in influence analysis, as it the previously mentioned Cook"s distance; this measures the influence of observation on the estimated regression coefficient vector. The size of the regression coefficients reflect the relative importance of the corresponding predictors only if the predictors have equal variance, e.g., if they are autoscaled as the data here is. The standardized coefficients, i.e. coefficients multiplied by the standard deviation of the corresponding predictor, are a good measure of relative importance. In the ACE model, the predictor importance is calculated as the standard deviation of the corresponding nonlinear transformation.

An observation is considered influential if, compared to other observations, it has a relatively large impact on the estimated quantities such as the fitted response, the regression coefficients, and the standard error.

Ordinary Least Squares Regression (OLS)

Response Variable: sample
Goodness-of-Fit and Goodness-of-Prediction

Residual SS	R2	Resid PRESS	R2cv
63432	0.6903	78065	0,6189

Regression Coefficients, Predictor Importances

Intercept	v3400	v2960	v2872	v1736	v1712	v1640
68.14	-3.783	26.13	35.02	-13.04	-16.80	19.32

Importance	-3.783	26.13	35.02	-13.04	-16.80	19.32
v1452	v1376	v1252	v1196	v1152	v676	
-107.4	41.29	-11.76	41.70	-11.44	7.957	
-107.4	41.29	-11.76	41.70	-11.44	7.957	

Analysis of Variance

Source	Df	SS	MS	F	p
Model	12.00	141400	11783	22.29	0.0000
Residual	120.00	63432	528.6		
Total	132.00	204832			

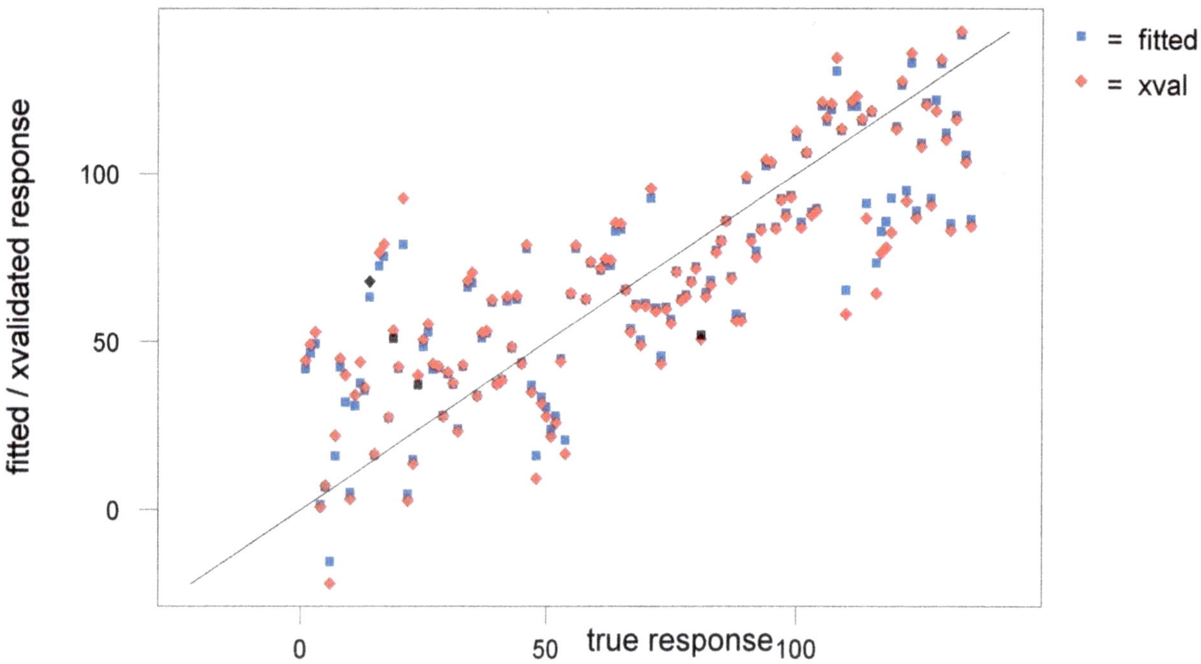

OLS Fit and Prediction

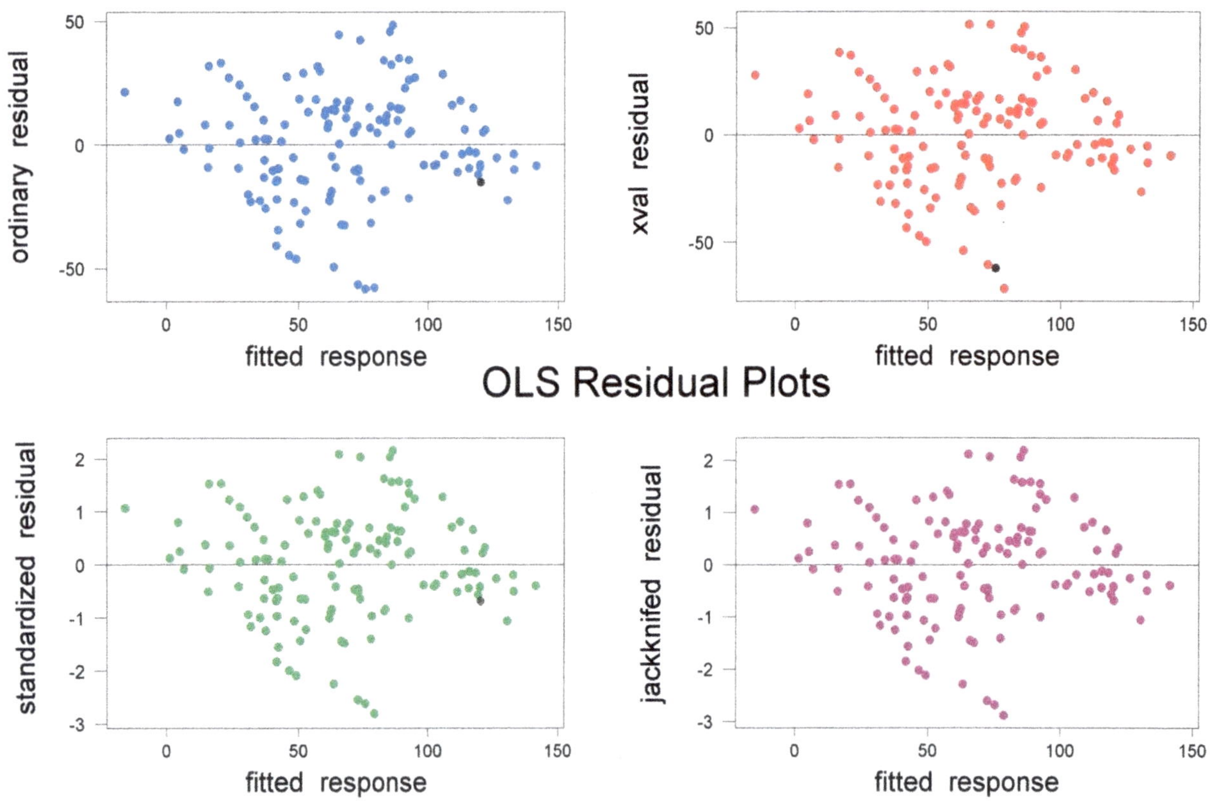

coefficient
0
-50
-100
2
4
6
8
10
12
OLS Regression Coefficients and Predictor Importances
importance
0
-50
-100
2
4
6
8
10
12

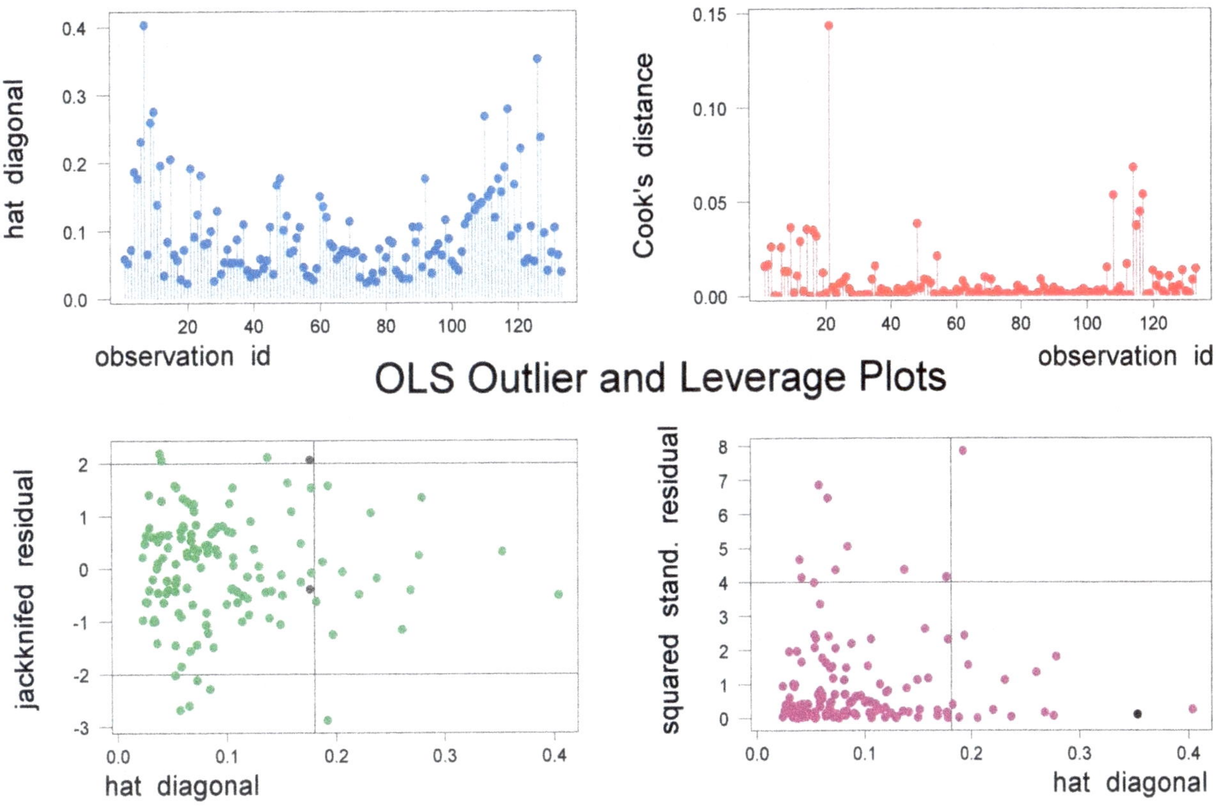
hat diagonal
0.4
0.3
0.2
0.1
0.0
20
40
60
80
100
120
observation id
Cook's distance
0.15
0.10
0.05
0.00
20
40
60
80
100
120
observation id
OLS Outlier and Leverage Plots
jackknifed residual
2
1
0
-1
-2
-3
0.0
0.1
0.2
0.3
0.4
hat diagonal
squared stand. residual
8
7
6
5
4
3
2
1
0
0.0
0.1
0.2
0.3
0.4
hat diagonal

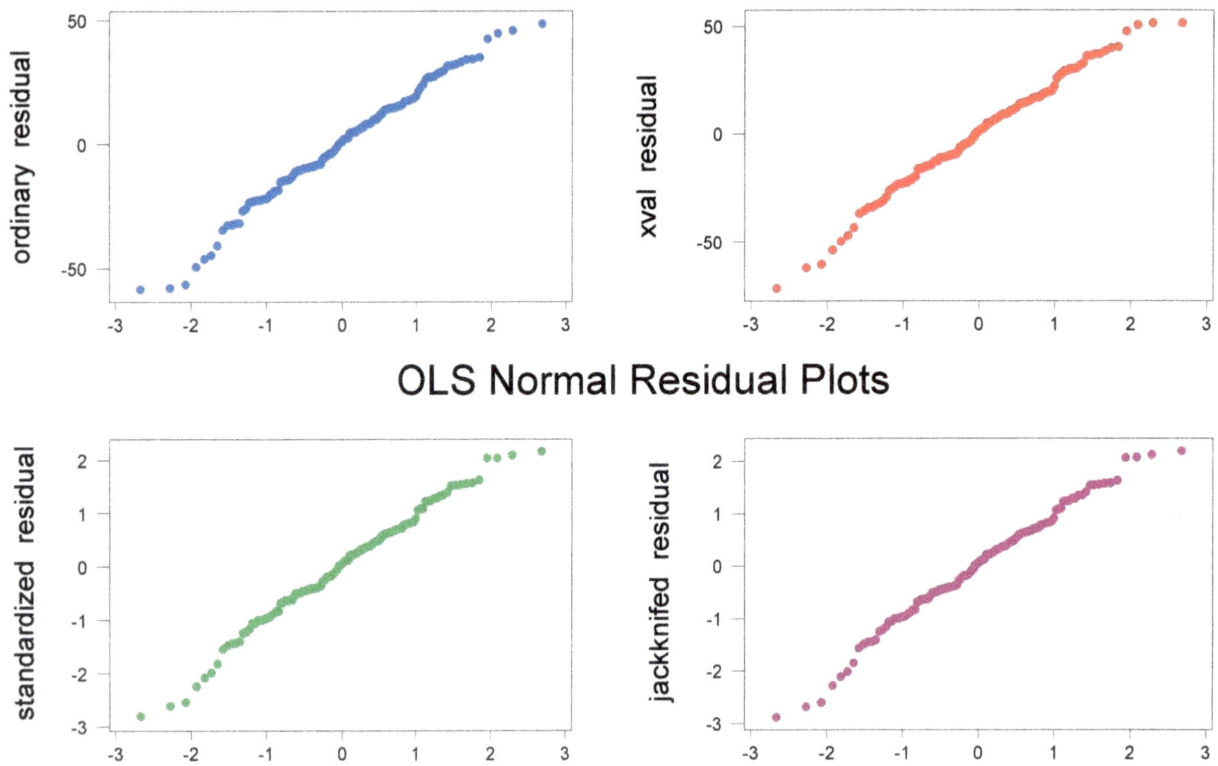

H. PRINCIPAL COMPONENTS REGRESSION (PCR)

PCR calculates a parametric linear model for a single response, and calculates biased, non least squares coefficients. PCR should be used when the data set is underdetermined, i.e. there are fewer observations than predictors and/or the predictors are highly correlated. PCR also assumes that the response is a linear function of the predictors, and that most of the information about the response is in a high variance subspace of the predictors, i.e. in the first few principal components.

There is a need to specify one or more predictor columns for the training set, and one response column for the training set.

The numerical output consists of three tables. The first, the goodness-of-fit and goodness-of-prediction table contains information about how well the calculated model fits the training set data and how well it would predict future data.

The table contains one row for each potential model that was calculated and crossvalidated. The modelling capability (goodness-of-fit) is measured by the residual sum of squares (Resid SS) and by the R-squared coefficient (R2). The predicting capability (goodness-of-prediction) is measured by the residual predictive sum of squares (Resid PRESS) and by the cross-validated R-squared coefficient (R2cv).

The next table contains the regression coefficients and the importance of the predictors in the model. The importance of a predictor is measured by the standardized regression coefficient = (the regression coefficient) x (the standard deviation of the corresponding predictor). This table is presented graphically by the regression coefficient plot.

The Analysis of Variance table shows how much of the total variance is explained by the model. The total degrees of freedom (Df) and the total sum of squares (SS) have two components, model and residual. The mean squares (MS) are the sum of squares divided by the corresponding degrees of freedom. The standard error is the square root of the residual MS. The F value (F) and the corresponding probability calculated from the model and residual mean square indicate the significance of the model: a large F-value and small p value mean that the model is significant.

Model selection: Cross-validate all components: where cross-validation is performed on all the components, it is also referred to as the leave one out cross-validation or leverage correction.

This tabulated information is also graphically represented by the following, as utilized in the results section:

Regression coefficient plot: Displays two scatter plots, regression coefficients versus the predictor ids and predictor importances versus the predictor ids. These plots are especially informative when there are many predictors. The intercept is not shown on the plot because it often distorts the vertical scale. Can omit predictors with relatively small importances.

Response plot: Displays a scatter plot of the calculated response versus the true response. Two calculated response values are plotted: the ordinary fitted response (squares) and the cross-validated predicted response (diamonds). Observations where the cross-validated value differs significantly from the ordinary fit may be leverage points. Ideally this plot is close to the straight line, calculated = true. This line is also displayed on the plot.

Residual plots: Displays four scatter plots. Each plots one type of residuals (ordinary residuals, cross-validated residuals, standardized residuals, and jackknifed residuals) versus the fitted response values. These plots reveal nonlinearity, heteroscedasticity, and outliers. The horizontal line where residual = 0.0 is displayed on each plot. Ideally, residuals are distributed symmetrically around this line.

Normal residual plots: Displays four scatter plots. Each plots one type of residuals (ordinary residuals, cross-validated residuals, standardized residuals, and jackknifed residuals) versus the normal scores (quantiles from standard normal distribution). These plots are especially useful for checking the normality assumption of the residuals.

Leverage plots: Contains four scatter plots that reveal outliers and leverage points. The upper left plot shows leverages (diagonal elements of the hat matrix) versus the observation ids. The upper right plot shows Cook"s distances versus the observation ids. The lower left plot shows jackknifed residuals versus leverages (hat diagonals). The lower right plot shows beverages (hat diagonals) versus the squared standardized residuals. The vertical and horizontal straight lines indicate limits of normal values. The limits indicated by the lines on these plots are described in the theory which follows.

Model selection profile plot: Displays a scatter plot of the ordinary fitted R-squared (squares) and the cross-validated R-squared (diamonds) versus the number of components. This plot shows the differences among potential models in terms of modelling and predictive capability. PCR is a non least squares (biased) regression method that models the response variable as a linear combination of the principal components with the higher variances (i.e., corresponding to the largest eigenvalues). First the predictors are autoscaled.

The PCR model can be expressed as $\mathbf{y} = \mathbf{T}(m)\,\mathbf{a} + \mathbf{e}$

where $\mathbf{T}(m)$ is the matrix for the first m principal component scores, calculated from the matrix $\mathbf{X}^! \mathbf{X}$, $\mathbf{a}$ is the vector of coefficients, and $\mathbf{e}$ is the vector of errors. Since the predictors are autoscaled, $\mathbf{X}^! \mathbf{X}$ is equal to the correlation matrix of predictors. The scores are calculated from $\mathbf{X}$ by $\mathbf{T}(m) = \mathbf{XL}(m)$ where $\mathbf{L}(m)$ is the matrix of the eigenvectors of $\mathbf{X}^! \mathbf{X}$, corresponding to the first (largest) m eigenvalues. The principal component scores are orthogonal vectors, and are linear combinations of the predictor variables. The coefficients, $\mathbf{a}$ are found using OLS, with $\mathbf{y}$ as the response and $\mathbf{T}(m)$ as the predictors. The regression equation can be expressed in terms of predictors, $\mathbf{X}$ as:

$$\mathbf{y} = \mathbf{T}(m)\,\mathbf{a} + \mathbf{e} = \mathbf{X}[\mathbf{L}(m)\,\mathbf{a}] + \mathbf{e} = \mathbf{Xb} + \mathbf{e}$$

therefore the coefficients for the principal components model with m components are given by: $\mathbf{b} = \mathbf{L}(m)\mathbf{a}$

The bias-variance trade off is controlled by the number of principal components m. The more components included, the larger the variance and the smaller the bias of the estimated regression coefficients. As the number of components increase, the PCR model converges to the OLS model. When m = p, PCR and OLS are the same. The optimal number of components is selected based on a goodness-of-prediction criterion, e.g., by cross-validation. PCR is a linear estimator. Therefore, the cross-validated residuals can be calculated from the ordinary residuals.

Ordinary residual: e_i is the residual from a model fitted to all observations.

Cross-validated residual: (also called deleted or predictive residual) $e_{(i)}$, for the ith observation is the residual from a model fitted to data with the ith observation excluded. For linear estimators such as OLS and PCR, the cross-validated residuals can be calculated from the ordinary residuals. For the nonlinear estimators such as PLS, NPLS, and ACE, the cross-validated residuals must be calculated by the leave-one-out technique, repeating the model calculation many times.

Standardized residual: r_i, is the residual of a model fitted to all observations. The residual is scale invariant. Each residual is scaled by its standard error, estimated from all observations.

Jackknifed residuals: t_i is the cross-validated residual standardized. Each residual is divided by its standard deviation, which is calculated without the ith observation. Jackknifed residuals have zero mean and unit variance, are scale invariant, and are more appropriate for detecting violations of assumptions in a regression model.

All these residuals are used as diagnostics, to check whether the assumptions of the model are correct.

Principal Components Regression (PCR)

Response Variable: sample

Optimal Number of Components: 10

Goodness-of-Fit and Goodness-of-Prediction

Comp	Df	Residual SS	R2	Resid PRESS	R2cv
1	131.00	203605	0.0060	210421	-0.0273
2	130.00	122577	0.4016	127487	0.3776
3	129.00	121921	0.4048	129067	0.3699
4	128.00	121193	0.4083	130535	0.3627
5	127.00	97097	0.5260	105774	0.4836
6	126.00	91406	0.5537	103082	0.4967
7	125.00	90985	0.5558	104230	0.4911
8	124.00	88105	0.5699	102592	0.4991
9	123.00	73463	0.6414	86368	0.5783
10	122.00	64952	0.6829	77082	0.6237
11	121.00	64942	0.6830	78451	0.6170
12	120.00	63432	0,6903	78065	0.6189

Regression Coefricients, Predictor Importances

Intercept	v3400	v2960	v2872	v1736	v1712	v1640
68.14	-4.032	24.29	34.28	-32.63	-11.82	18.00

Importance	-4.032	24.29	34.28	-32.63	-11.82	18.00
v1452	v1376	v1252	v1196	v1152	v676	
-95.31	31.10	17.08	1.749	18.92	5.446	
-95.31	31.10	17.08	1.749	18.92	5.446	

Analysis of Variance

Source	Df	55	MS	F	p
Model	10.00	139881	13988	26.27	0.0000
Residual	122.00	64952	532.4		
Total	132.00	204832			

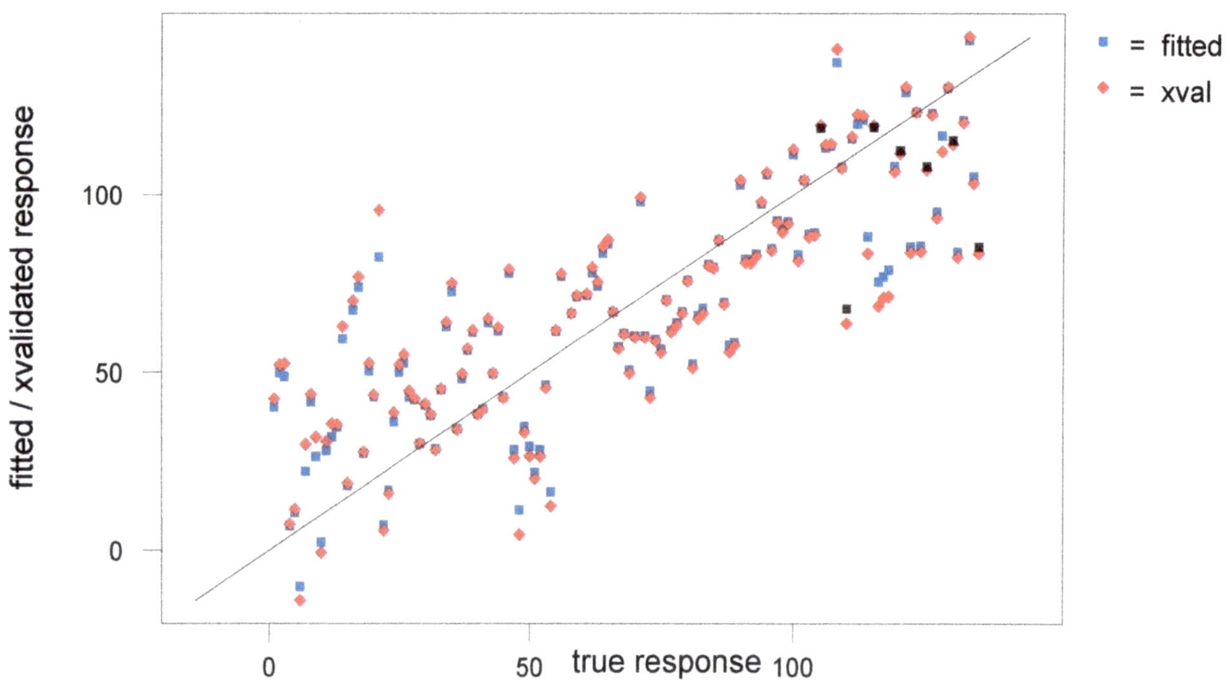

PCR Fit and Prediction

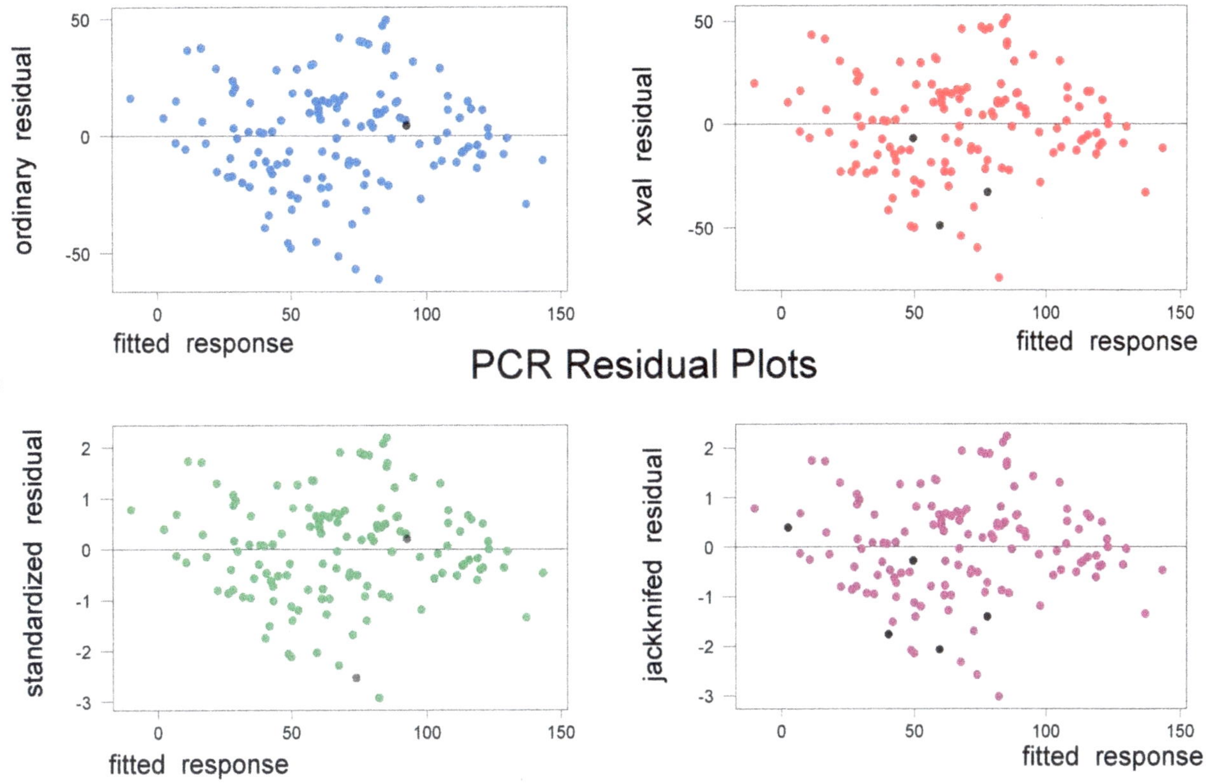
PCR Residual Plots
ordinary residual
xval residual
standardized residual
jackknifed residual
fitted response
fitted response
fitted response
fitted response

PCR Regression Coefficients and Predictor Importances
coefficient
importance

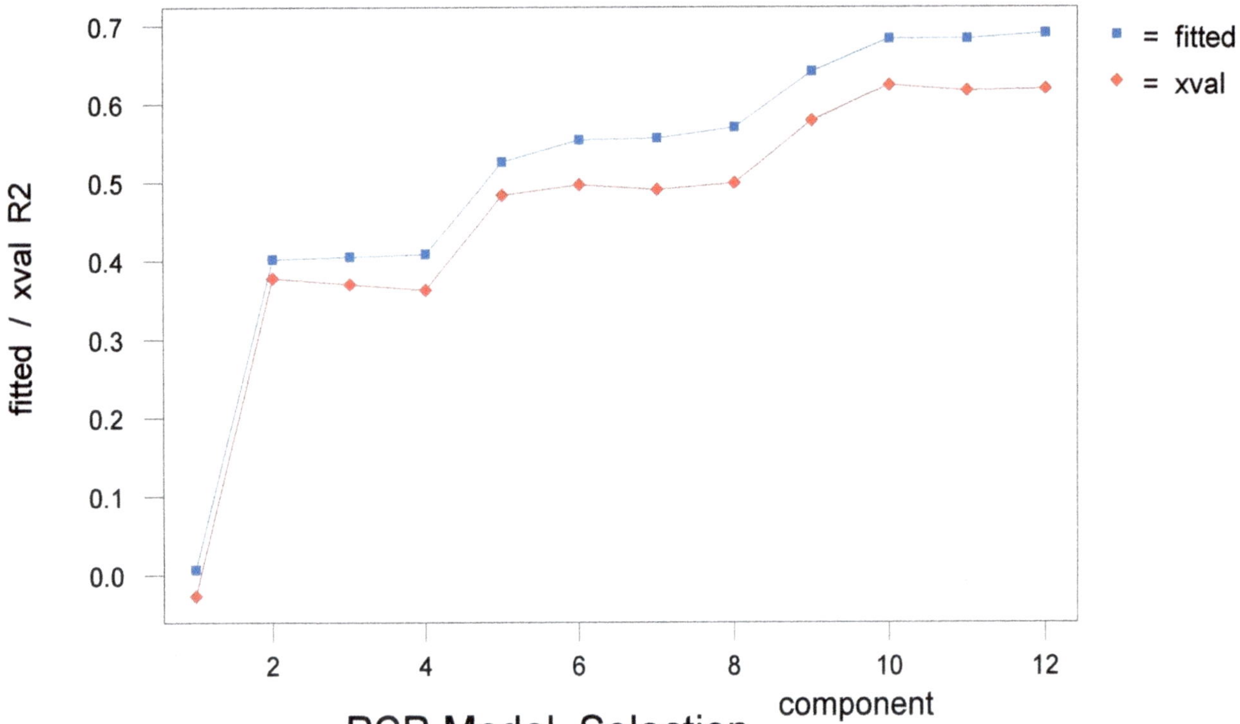

PCR Model Selection

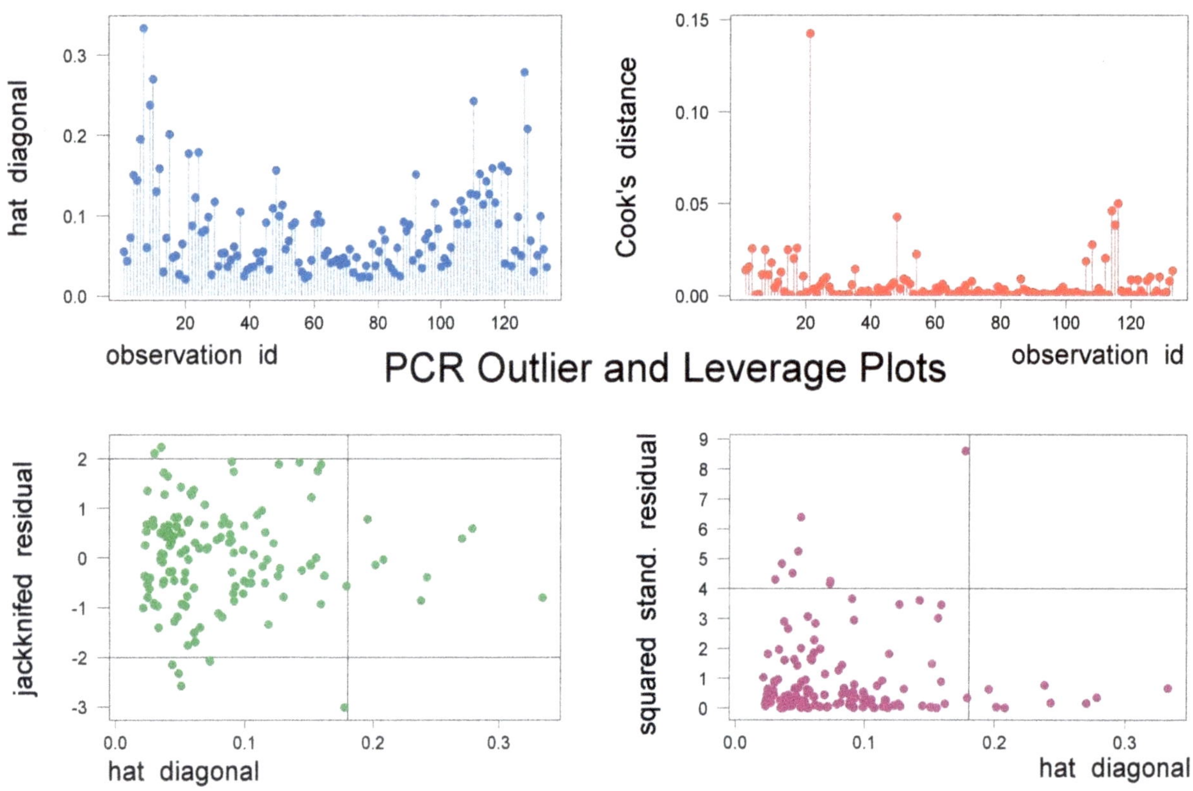

PCR Outlier and Leverage Plots

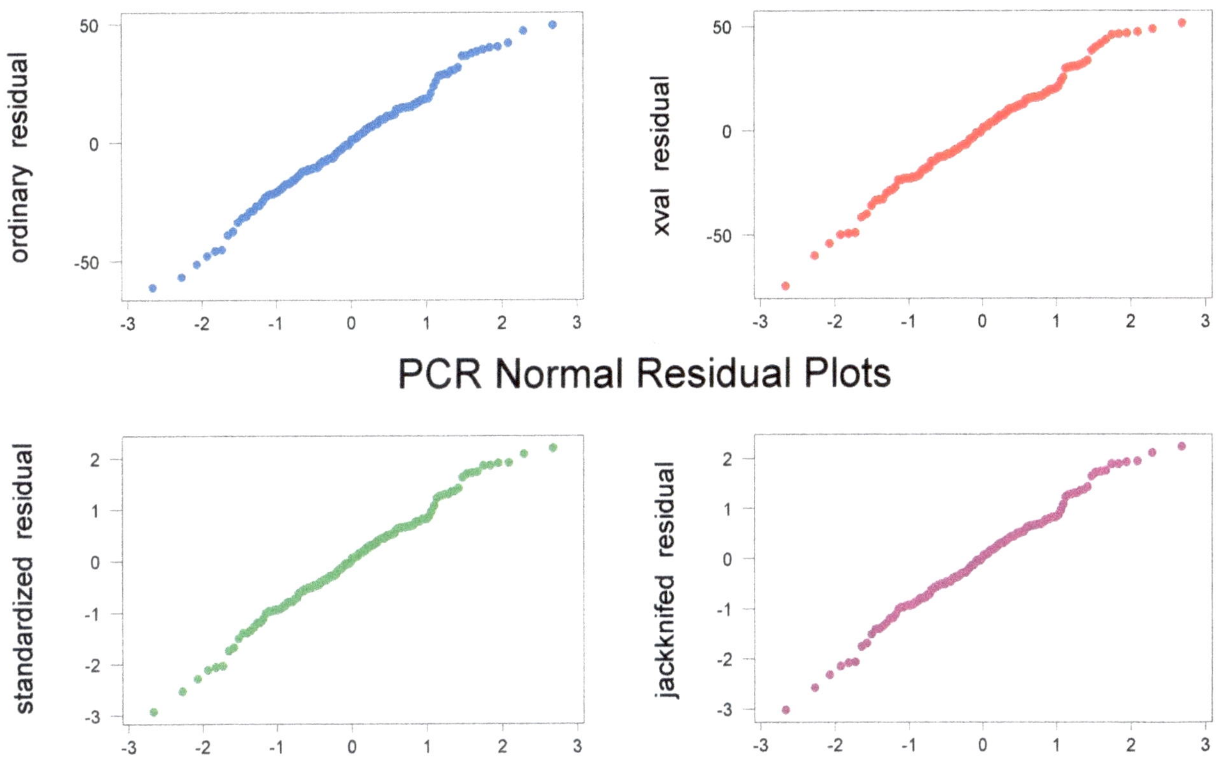

I. PARTIAL LEAST SQUARES (PLS)

PLS calculates a parametric, biased, linear model, which may contain one or several responses. If there are several responses, they are modelled together, in a multivariate way. Single response PLS is called one-block PLS or PLS 1. PLS with several responses, is called two-block PLS or PLS2.

PLS should be used when the data set is underdetermined, i.e. there are fewer observations than predictors and/or the predictors are highly correlated. PLS attempts to calculate a more predictive model than OLS, assuming that most of the information about the response (s) is in a subspace of high variance and of high correlation with the response (s), i.e. in the first few PLS components. When there is just one response, this model should be compared with PCR and RIDGE.

There are no exact formulae available in PLS to calculate cross-validated fit and residuals. These quantities must be estimated by actually leaving out observations and repeating the model calculation several times. Therefore if cross-validation is used, especially the leave-one-out, with a large number of observations, the calculation may take some time.

There is a need to specify one or more predictor columns and one or more response columns for the training set.

The numerical output consists of two tables for each response. The first, the Goodness-of-fit and goodness-of-prediction contains model validation parameters, i.e., information about how well each calculated model fits the training set data and how well it would predict future data.

This table contains one row for each potential model that was calculated and cross-validated. If the optimal number of components was specified then there is only one model. The first column of the table contains the number of components. The modelling capability (Goodness-of-fit) is measured by the residual sum of squares (Resid SS) and by the R-squared coefficient (R2). The predicting capability (Goodness-of-prediction) is measured by the predictive residual sum of squares (Resid PRESS) and by the cross-validated R-squared coefficient (R2cv). The optimal model corresponds to the minimum value of Resid PRESS or, equivalently, to the maximum value of R2cv. This table is represented graphically in the model selection profile plot. The next table contains the regression coefficients and the importance of the predictors in the model. For each response, the importance of a predictor is measured by the standardized regression coefficient = (the corresponding regression coefficient) x (standard deviation of the predictor)/(the standard deviation of the response). This table is presented graphically by the regression coefficient plot.

To calculate the best predictive PLS model, the number of components must be optimised. Model selection gives five options.

Cross-validate all components with leave-one-out method: This method is also called full cross-validation. It is the best strategy for small data sets. If there are many observations, however, this method is quite slow. If there are n observations, then n PLS models are fit. Model i is PLS calculated using the data with observation i omitted.

Cross-validate all components with___xval groups: Allows you to specify the number, j, of cross-validation groups. The value of j should be an integer between 2 and n, the number of observations. The number, j, says how many times the model calculation is repeated. The smaller the number, the faster the calculation, but the less accurate the estimate of goodness-of-prediction. This technique is also called segmented cross-validation.

Cross-validate up to_____components with leave-one-out method: When there are many highly correlated variables (e.g., with spectral data), the optimal number of components is likely to be small compared to the number of predictors. You can limit the number of components to be cross-validated with this option. The number of components you specify should be smaller than the number of variables and the number of observations.

Cross-validate up to___components___xval groups: Allows you to limit the number of components to be cross-validated, as in the preceding option, and to specify the number of cross-validation groups. The number of components you specify should be smaller than the number of variables and the number of observations.

Number of components is: If you know the optimal number of components, you can specify it here.

All of this data is presented graphically:

Regression coefficient plots: If there is just one response, two scatter plots are displayed, regression coefficients versus the predictor ids and predictor importances versus the predictor ids. These plots are especially informative when there are many predictors. The intercept is not on the plot because it often distorts the vertical scale. Try omitting predictors with relatively small importances.

When there are several responses, a plot of the regression coefficients versus the predictor ids is displayed for each response. Plots of the predictor importances are not produced.

Responses plots: If there is just one response, a scatter plot of the calculated responses versus the true response is displayed. Two types of calculated response values are plotted, ordinary fitted response (squares) and cross-validated predicted response (diamonds). Observations where the cross-validated value differs significantly from the ordinary fit are likely to be leverage points. Ideally this plot is close to the straight line, calculated = true. This line is also displayed on the plot.

When there are several responses, a plot is displayed for each.

Residual plots: If there is just one response, two plots are displayed, the ordinary residuals versus the fitted response values and the cross-validated residuals versus the fitted response values (standardized and jackknifed are not available). The line where residual = 0.0 is drawn on each plot. Ideally, residuals distributed symmetrically around this line. These plots reveal nonlinearity, heteroscedasticity, and outliers.

When there are several responses, the ordinary and cross-validated residuals are overlaid, and one such plot is displayed for each response.

Normal residual plots: If there is just one response, two plots are displayed, ordinary residuals versus normal scores (quantiles from standard normal distribution), and cross-validated residual versus normal scores (standardized and jackknifed residuals are not available). This plot is especially useful for checking the normality assumption on the residuals.

When there are several responses, the ordinary residuals are plotted against the normal scores. One plot is displayed for each response. Plots for the cross-validated residuals are not produced.

PLS is a non least squares biased linear regression method that relates a set of predictor variables, **X,** to a set of r response variables, **Y**. A least squares regression is performed on a set of **M** uncorrelated variables, **T**, that are a standardized linear combination of the original predictor variables, **X**. The variables **T** are called the **X** latent variables. PLS models the response(s) as:

Y = TBQ+E

where **B** is a diagonal matrix containing the least squares coefficients from the regression on the latent variables, **Q** contains the weights of the responses, and **E** is the residual (or error) matrix. The latent variables are calculated one at a time, maximizing their covariance.

The PLS components are calculated one at a time.

The bias-variance trade-off is controlled by the number of latent variables, M. The more latent variables included, the larger the variance and the smaller the bias of the estimated regression coefficients. As the number of latent variables increases, the PLS model converges to the OLS model. When there are p latent variables, PLS and OLS are the same. The optimal number of components, M, is determined based on a goodness-of-prediction criterion, e.g. by cross-validation.

Aside: Nonlinear PLS fits a nonlinear regression model by nonlinearizing the inner relationships of the PLS model applying smoothers. The components are calculated one at a time. First the X and Y latent variables are obtained as in the linear PLS model. Next, instead of a linear fit, the two latent variables are related by a nonlinear function estimated by a smoother used in ACE (explanation of this follows). The form of nonlinearity can be analyzed graphically on the latent variable plots.

Partial Least Squares Regression (PLS)

Number of Cross-validation Groups: 133

Regression Coefficients, Predictor Importances

Intercept	v3400	v2960	v2872	v1736	v1712	v1640
68.14	-3.814	26.03	33.88	-25.39	-13.48	18.71

Importance	-0.09682	0.6608	0.8600	-0.6447	-0.3422	0.4750
v1452	v1376	v1252	v1196	v1152	v676	
-102.6	38.54	2.225	22.08	5.138	5.663	
-2.604	0.9783	0.05647	0.5605	0.1304	0.1438	

Optimal Number of Components: 9 **Response Variable: sample**

Goodness-of-Fit and Goodness-of-Prediction

Comp	Residual SS	R2	Resid PRESS	R2cv
1	146940	0.2826	158227	0.2275
2	112581	0.4504	119733	0.4155
3	85109	0.5845	97256	0.5252
4	79748	0.6107	93660	0.5427
5	71014	0.6533	85788	0.5812
6	67408	0.6709	80131	0.6088
7	66527	0.6752	80459	0.6072
8	64557	0.6848	77308	0.6226
9	63894	0.6881	77205	0.6231
10	63770	0.6887	77576	0.6213
11	63433	0.6903	78316	0.6177
12	63432	0.6903	78065	0.6189

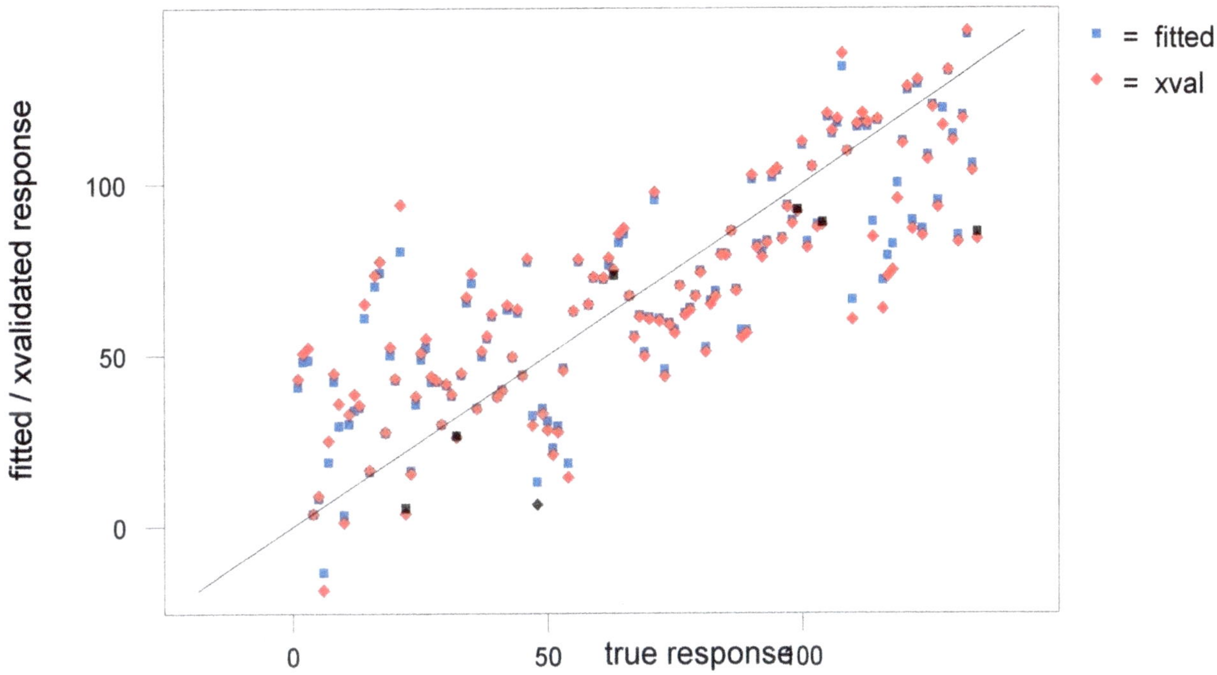

PLS Fit and Prediction

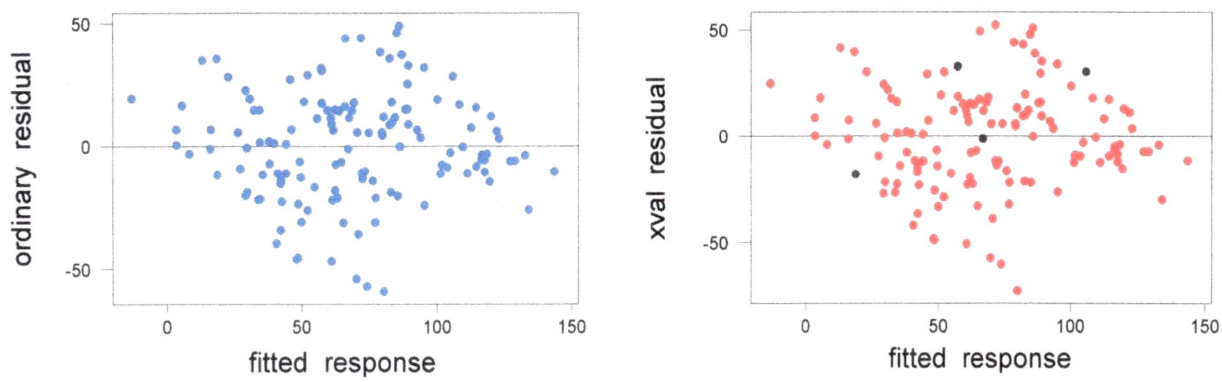

PLS Residual Plots

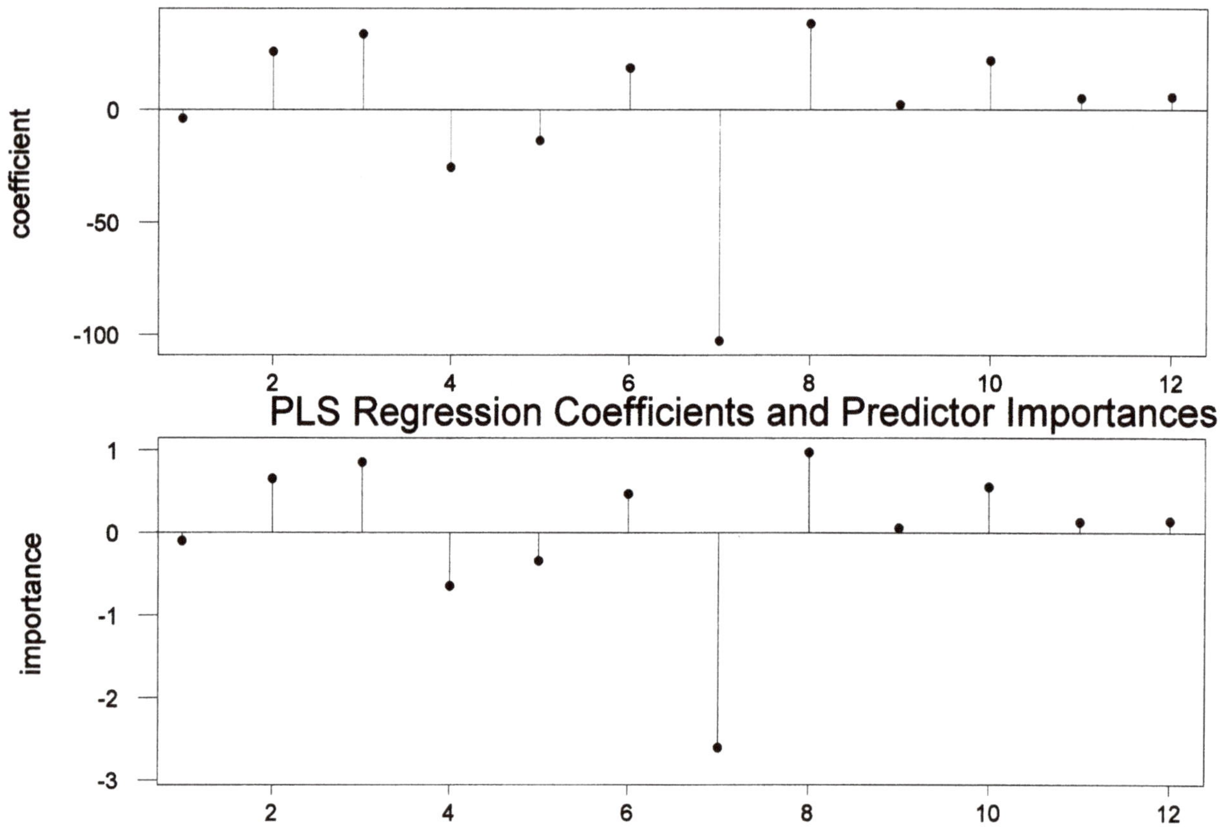
coefficient
0
-50
-100
2
4
6
8
10
12
PLS Regression Coefficients and Predictor Importances
importance
1
0
-1
-2
-3

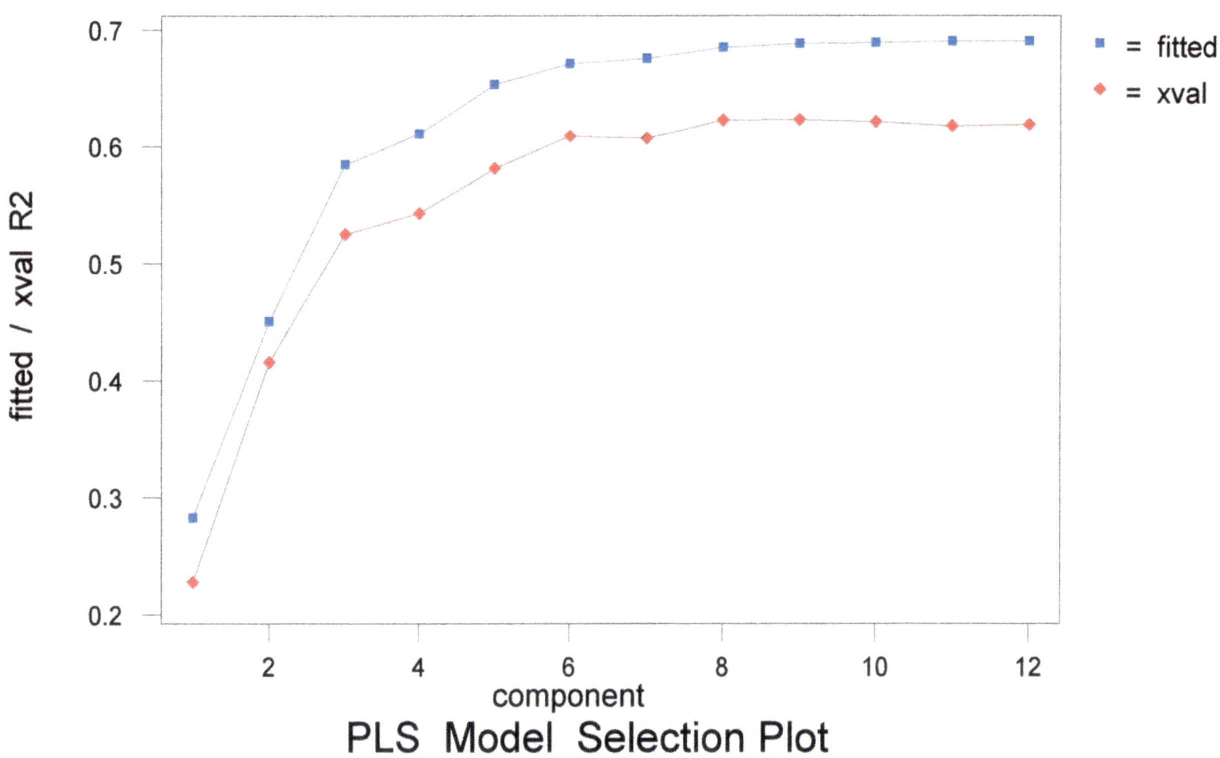
fitted / xval R2
0.7
0.6
0.5
0.4
0.3
0.2
2
4
6
8
10
12
component
= fitted
= xval
PLS Model Selection Plot

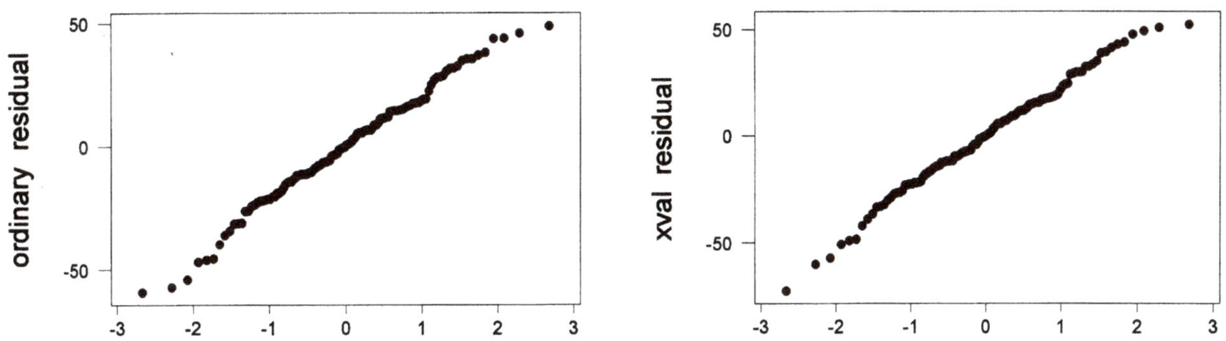

PLS Normal Residual Plots

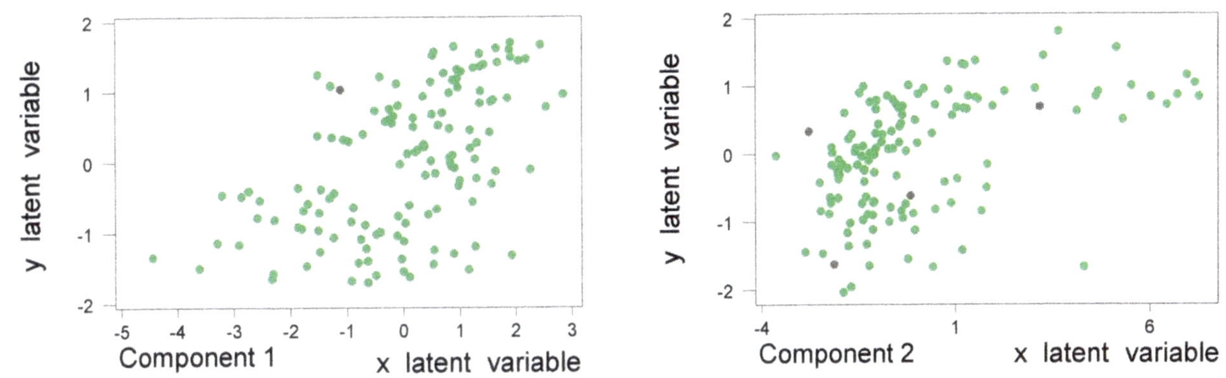

PLS Latent Variable Plots

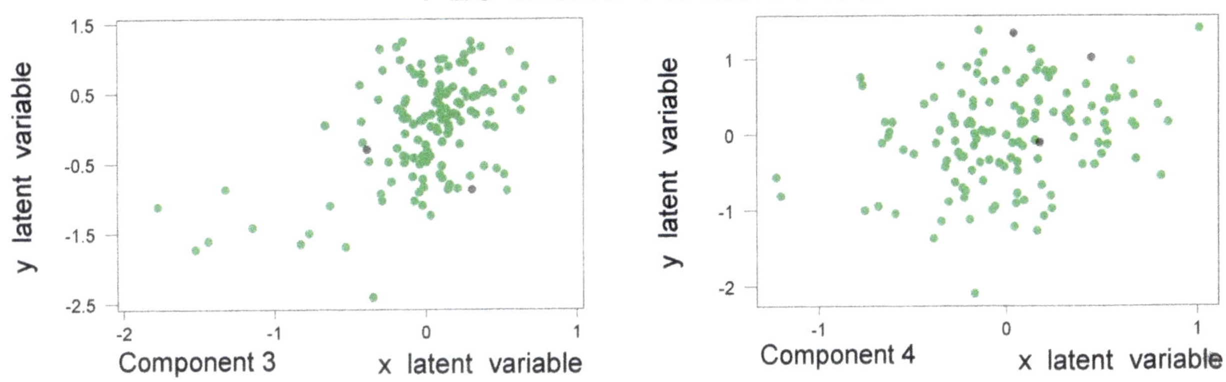

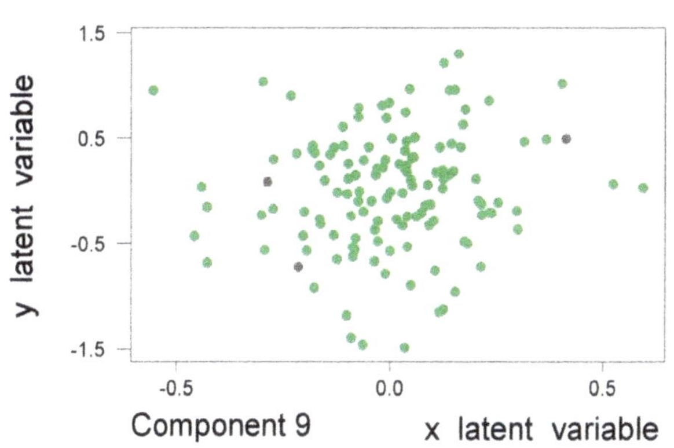

PLS Latent Variable Plots

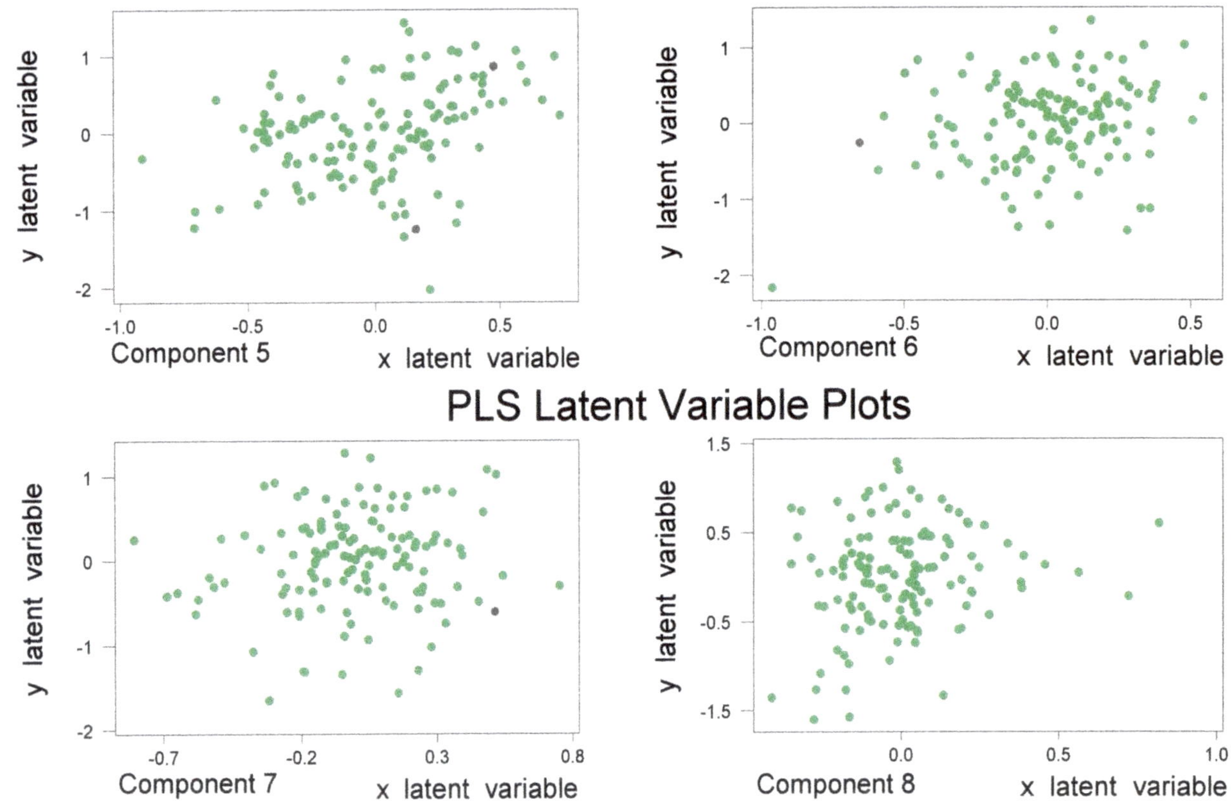

J. Alternating Conditional Expectations (ACE)

ACE calculates a nonlinear nonparametric regression model, where the response is estimated as the sum of functions of the predictors. The only restrictions on these functions is that they must be smooth. This model should be used when the poor performance of the OLS model is due to nonlinearity.

ACE is applicable only if there are enough degrees of freedom, i.e. the data set is well determined, containing many more observations than predictors. This model assumes additive nonlinear predictor - response relationship and identically and independently distributed errors.

Need to specify one or more predictor column for the training set, and one response column.

The numerical output of ACE consists of two tables. The goodness-of-fit and goodness-of-prediction table has model validation parameters, i.e., contains information about how well the calculated model fits the training set data and how well it would predict future data.

The modelling capability (goodness-of-fit) is measured by the residual sum of squares (Resid SS) and by the R-squared coefficient. The predicting capability (goodness-of-prediction) is measured by the residual predictive sum of squares (Resid PRESS) and by the cross-validated R-squared coefficient (R2cv).

The predictor importance table contains the importance of the predictors in the model. It is measured by the standard deviation of the corresponding transform function.

There are no exact formulas available in ACE to calculate cross-validated fits and residuals. These quantities must be estimated by actually leaving out observations and repeating the model calculation several times. If there many objects, these calculations can take a long time.

There are two decisions to make in ACE: how to calculate the goodness-of-prediction of the model and the value of the smoothing parameter.

Cross-validation. **Method:**

No cross-validation: Does not quantify the goodness-of-prediction.

Use__validation groups: Allows the specification of the number of cross-validation groups. This is the number of times the model calculation is repeated. This method is called segmented cross-validation.

Leave-one-out: Repeats the model calculation n times (n is the number of observations). This is the most time consuming option, but the most accurate estimate of the predictive capability. This method is also called full cross-validation.

Smoothing Parameter

Variable span: The smoothing parameter is determined by an adaptive algorithm along the range of the predictors.

Use___% of the observations: Specifies a constant span for the smoothing parameter.

Similarly with all the other techniques, ACE results are also described graphically.

Response plot: Displays a scatter plot of the calculated response versus the true response. Two calculated response values are plotted: ordinary fitted response (squares) and cross-validated predicted response (diamonds). If no cross-validation was done, only the fitted response is plotted. Observations where the cross-validated value differs significantly from the ordinary fit may be leverage points. Ideally this plot is close to the straight line, calculated = true. This line is also displayed on the plot.

Residual plots: Displays two plots, ordinary residuals versus the fitted values and the cross-validated residuals versus the fitted values. If no cross-validation was done, only the plot of the ordinary residuals is displayed. These plots reveal heteroscedasticity and outliers. The horizontal line where residual = 0.0 is displayed on each plot. Ideally, residuals are distributed symmetrically around this line.

Normal residual plots: Displays two plot, ordinary residuals versus the normal scores (quantiles from standard normal distribution), and cross-validated residuals versus the normal scores. If no cross-validation was done, only the plot of the ordinary residuals is displayed. These plots are especially useful for checking the normality assumption of the residuals.

Nonlinear transformations: Displays the transformation function for each predictor variable. In each plot, the transformed values are plotted against the original predictor values. Up to four transformation plots are displayed in one window. These are the most important plots in ACE.
ACE is a nonlinear regression model of the form:

$$y = \sum_j t_j(x_j) + e$$

The functions t_j, called transform functions, are smooth but otherwise unrestricted functions of the predictor variables. They are analogous to the regression coefficients in a linear regression model. The response is modelled as the sum of the transform functions. The ACE functions are obtained by smoothers which are estimated by least squares using an iterative algorithm. The variance of the function t_j, indicates the importance of the predictor variable j in the model. The higher the variance the more important the corresponding variable. The ACE model is defined as a set of point pairs.

In general, the model has no simple analytical form. The transformations can be analysed graphically by plotting each variable against its transformation function. Prediction in ACE is a two step procedure. First the transformation function values corresponding to each predictor are looked up in the function table, usually by calculating a linear interpolation between two values. Second, the p function values are summed to obtain the predicted response value.

Alternating Conditional Expectations Regression (ACE)

Response Variable: sample

Smoothing Parameter (SPAN): 0%

Number of Cross-validation Groups: 133

Goodness-of-Fit and Goodness-of-Prediction

Residual SS	R2	Resid PRESS	R2cv
37960	0.8147	99748	0.5130

Predictor Variable Importances

v3400	v2960	v2872	v1736	v1712	v1640	v1452
0.1438	0.6811	0.8846	0.3398	0.3853	0.5268	2.766
v1376	v1252	v1196	v1152	v676		
1.054	0.3085	1.022	0.2928	0.2386		

K. NONLINEAR PARTIAL LEAST SQUARES (NPLS)

NPLS calculates a nonlinear PLS model for one or more responses, where the inner relationships are nonlinearized by smoothers. This is a biased method that should be used to model nonlinear predictor-response relationship when the data set is underdetermined, i.e. when there are fewer observations than predictors and/or the predictors are highly correlated. In this case, there are not enough degrees of freedom to calculate a least squares nonlinear model. So the only hope is a non least squares method.

There are no exact formulae available in NPLS to calculate cross-validated fit and residuals. These quantities must be estimated by actually leaving out observations and repeating the model calculation several times. Therefore if you have many objects, and the default option for cross-validation using the leave-one-out method, is chosen, the calculation may take a very long time.

Specify one or more predictor columns for the training set, and one or more response columns.

The numerical output consists of one table, the Goodness-of-fit and goodness-of-prediction table, for each response. This table contains model validation parameters, i.e., information about how well each calculated model fits the training set data and how well it would predict future data.

This table contains one row for each potential model that was calculated and cross-validated. If you specified the optimal number of components, then there is only one model. The first column of the table contains the number of components. The modelling capability (goodness-of-fit) is measured by the residual sum of squares (Residual SS) and by the R-squared coefficient (R2). The predicting capability (goodness-of-prediction) is measured by the predictive residual sum of squares (Resid PRESS) and by the cross-validated R-squared coefficient (R2cv). The optimal model corresponds to the minimum value of resid PRESS or, equivalently, to the maximum value of R2cv. This table is presented graphically in the model selection profile plot.

To calculate the best predictive NPLS model, the number of components must be optimized. Model selection gives five options:

1. **Cross-validate all components with leave one out method:** This method is also called full cross-validation. It is the best strategy for small data sets. If there are many observations, however, this method is quite slow. If there are n observations, then n NPLS models are fit. Model i in NPLS is calculated using the data with observation i omitted.

2. **Cross-validate all components with___xval groups:** Allows you to specify the number, j, of cross-validation groups. The value of j should be an integer between 2 and n, the number of observations. The number, j, says how many times the model calculation is repeated. The smaller the number, the faster the calculation, but the less accurate the estimate of goodness-of-prediction. This technique is also called segmented cross-validation.

3. **Cross-validate up to____components with leave-one-out method:** When there are many highly correlated variables (e.g., with spectral data), the optimal number of components is likely to be small compared to the number of predictors. You can limit the number of components to be cross-validated with this option. The number of components you specify should be smaller than the number of variables and the number of observations.

4. **Cross-validate up to___components with___xval groups:** Allows you to limit the number of components to be cross-validated, as in the preceding option, and to specify the number of cross-validation groups. The number of components you specify should be smaller than the number of variables and the number of observations.

5. **Number of components is____:** If you know the optimal number of components, you can specify it here, if this option is selected along with storage or plots that involve cross-validation fits or residuals, these are calculated with five validation groups.

Smoothing parameters

Variable span: The smoothing parameter is determined by an adaptive algorithm along the range of the predictors.

Use__% of the observations: Specifies a constant span for the smoothing parameter.

The results from NPLS are also graphically represented.

Response plots: If there is just one response, a scatter plot of the calculated responses versus the true response is displayed. Two types of calculated response values are plotted, ordinary fitted response (squares) and cross-validated predicted response (diamonds). Observations where the cross-validated value differs significantly from the ordinary fit are likely to be leverage points. Ideally this plot is close to the straight line, calculated = true. This line is also displayed on the plot. When there are several responses, a plot is displayed for each response.

Residual plots: If there is just one response, two plots are displayed, the ordinary residuals versus the fitted response values and the cross-validated residuals versus the fitted response values (standardized and jackknifed residuals are not available). The line where residual = 0.0 is drawn on each plot. Ideally, residuals are distributed symmetrically around this line. When there are several responses, the ordinary and cross-validated residual plots are overlaid, and one such plot is displayed for each response.

Normal residual plots: If there is just one response, two plots are displayed, ordinary residuals versus normal scores (quantiles from standard normal distribution), and cross-validated residual versus normal scores (standardized and jackknifed residuals are not available). This plot is especially useful for checking the normality assumption on the residuals. When there are several responses, the ordinary residuals are plotted against the normal scores. One plot is displayed for each response. Plots for the cross-validated residuals are not produced.

Latent variable plots: One scatter plot for each component is displayed. The Y (response) latent variable is plotted versus the x (predictor) latent variable corresponding to that component. Up to four scatter plots are displayed in one window, each for a different component. This plot is useful for examining the shape of the nonlinearity.

Model selections profile plot: If there is just one response, this option displays a scatter plot of the ordinary fitted R-squared (squares) and cross-validated R-squared (diamonds) versus the number of components. This plot shows the differences among potential models in terms of modelling and predictive capability. When there are several responses, four plots are displayed in one window, each for a different response variable.

Theory

Nonlinear PLS fits a nonlinear regression model by nonlinearizing the inner relationships of the PLS model applying smoothers. The components are calculated one at a time. First the X and Y latent variables are obtained as in the linear PLS model. Next, instead of a linear fit, the two latent variables are related by a nonlinear function estimated by a smoother used in ACE. The form of nonlineaity can be analysed graphically on the latent variable plots.

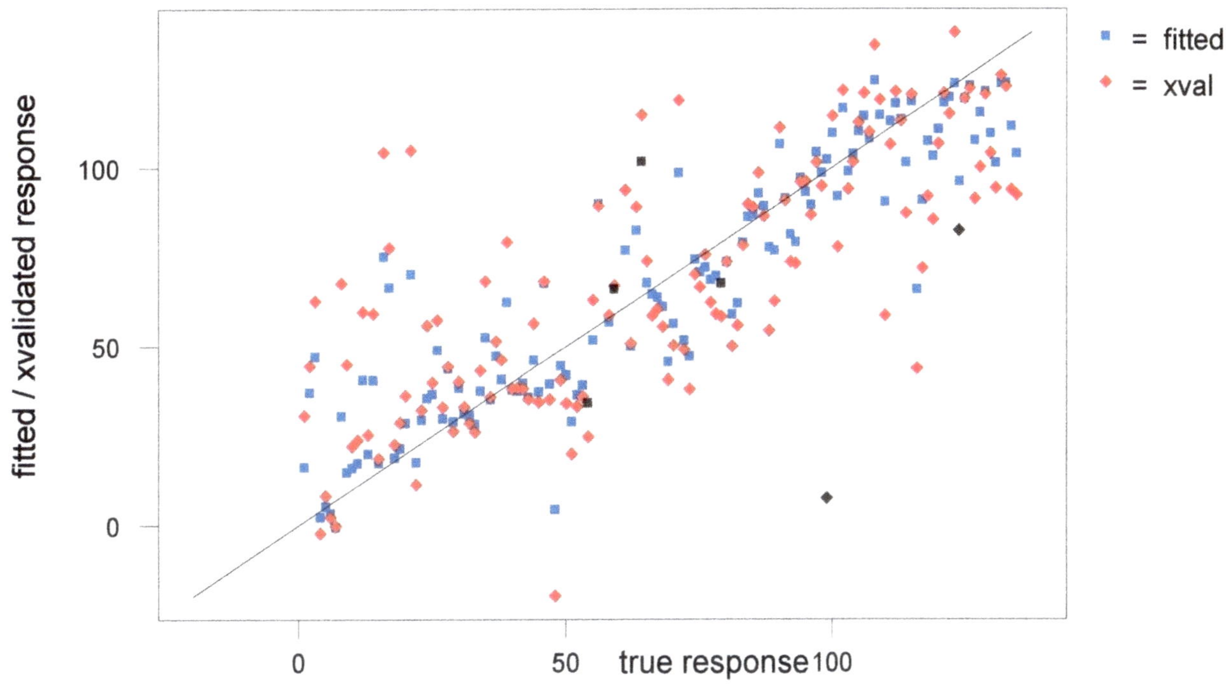

ACE Fit and Prediction

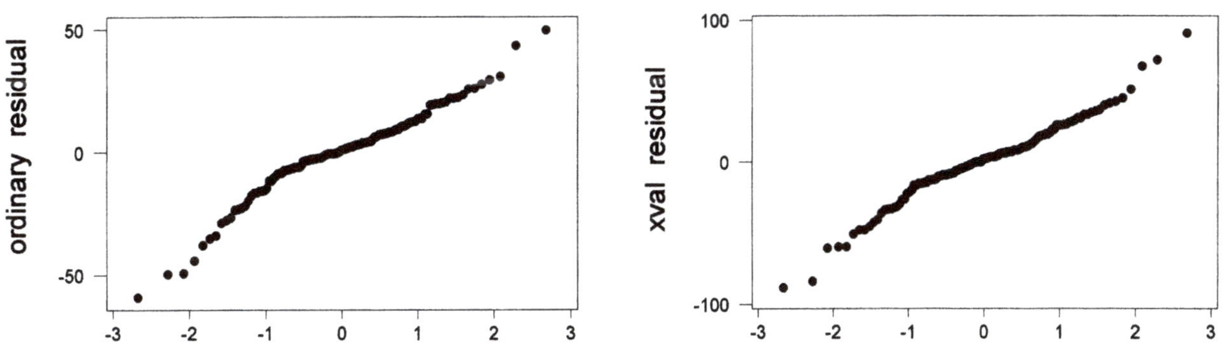

ACE Normal Residual Plots

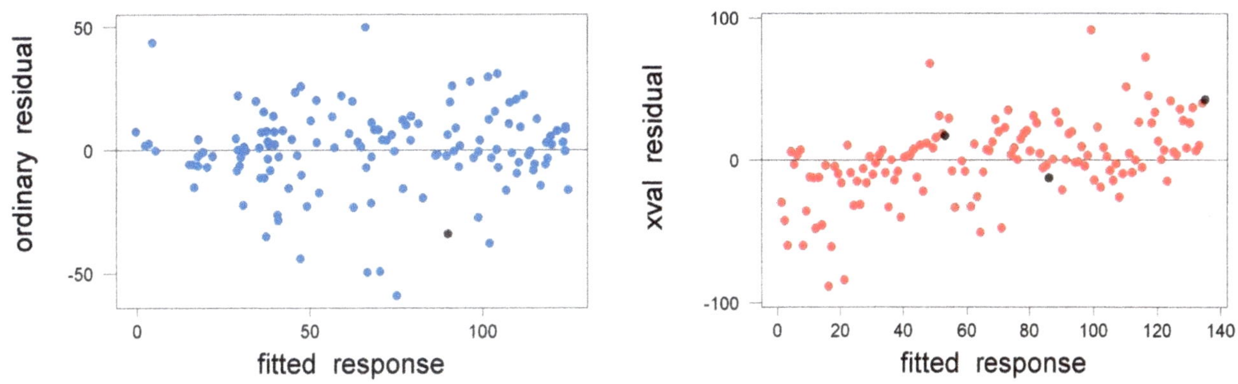

ACE Residual Plots

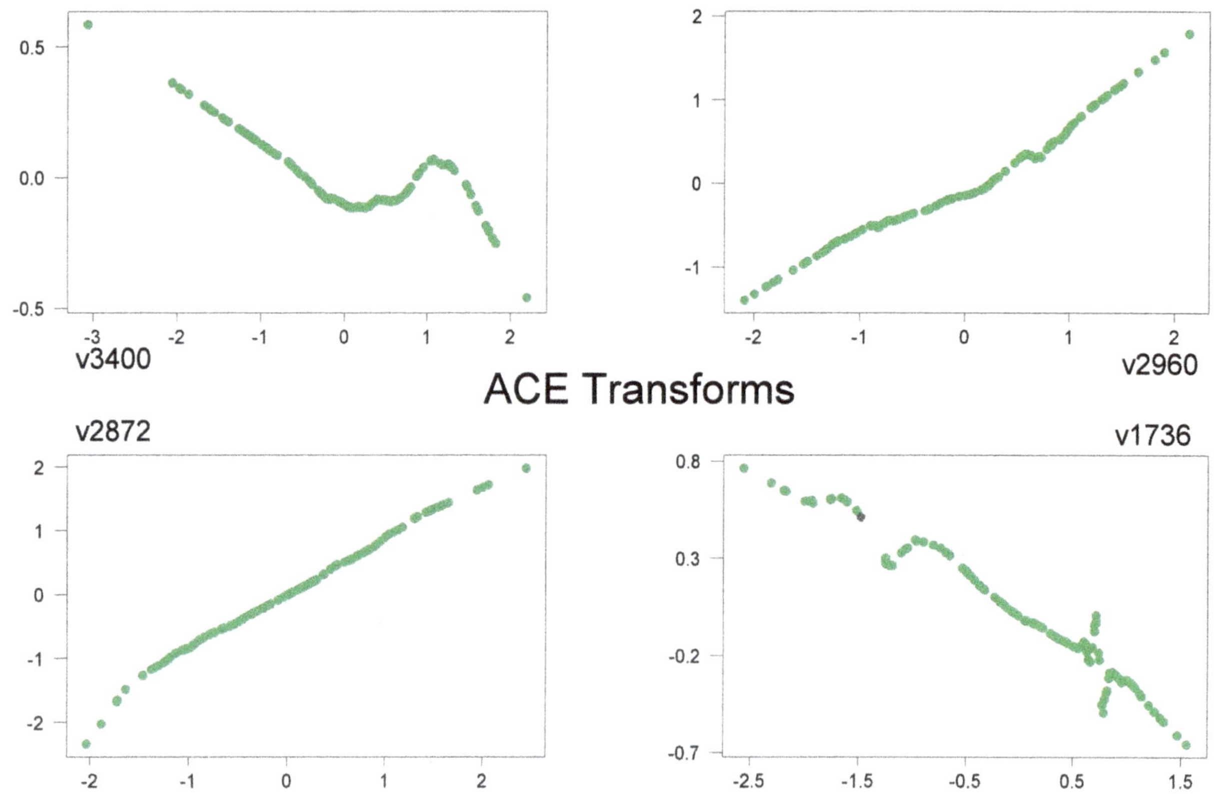
ACE Transforms
v3400
v2960
v2872
v1736

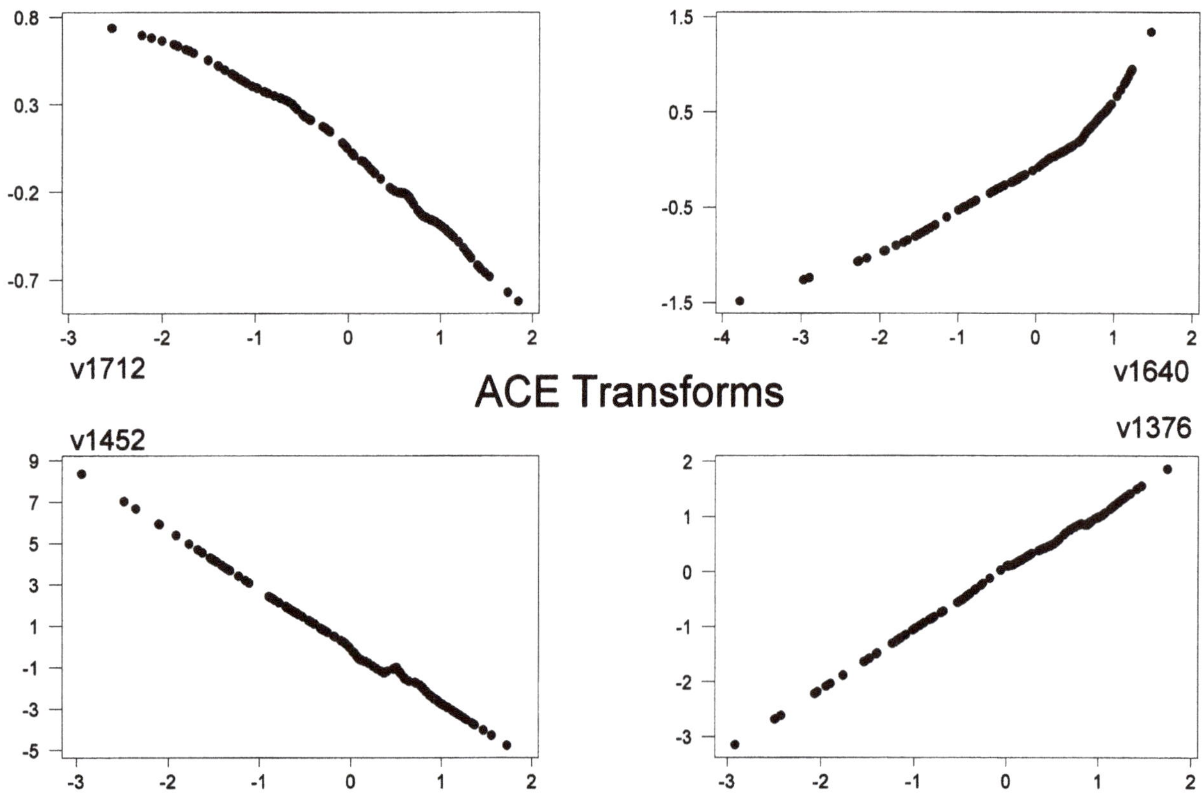
ACE Transforms
v1712
v1640
v1452
v1376

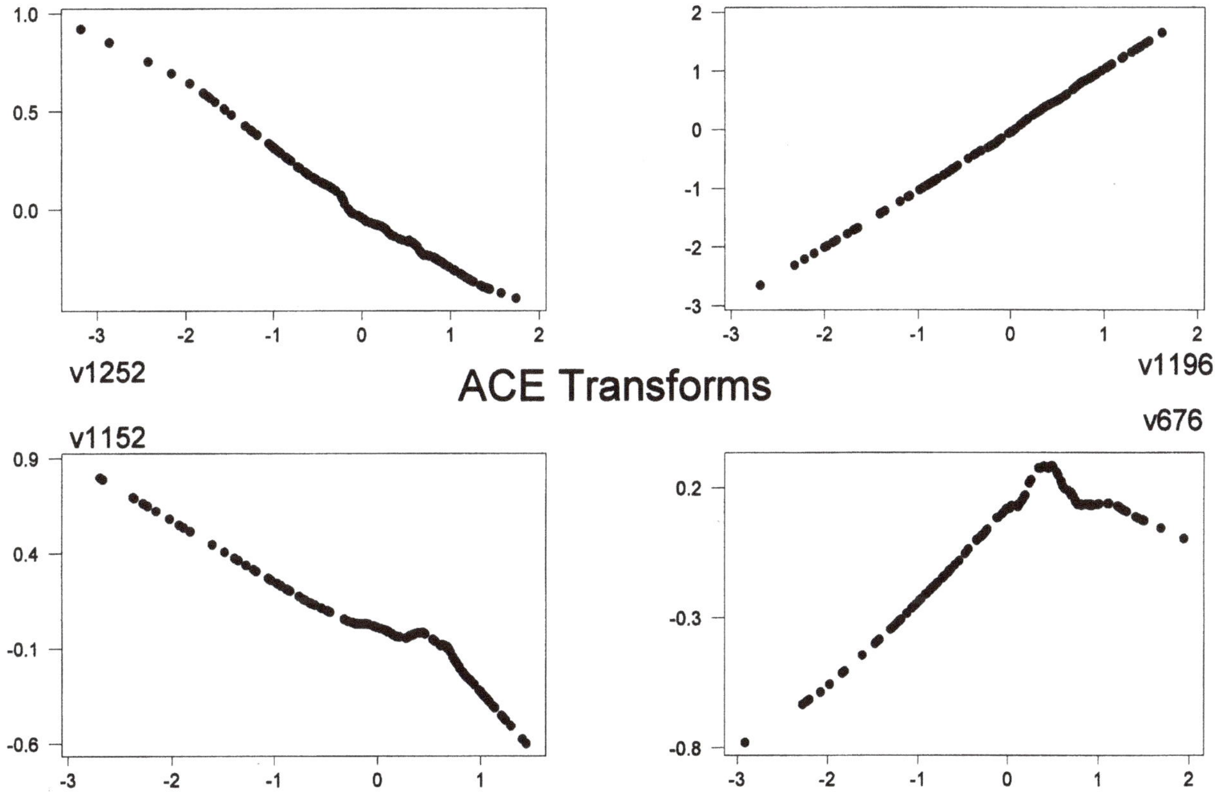

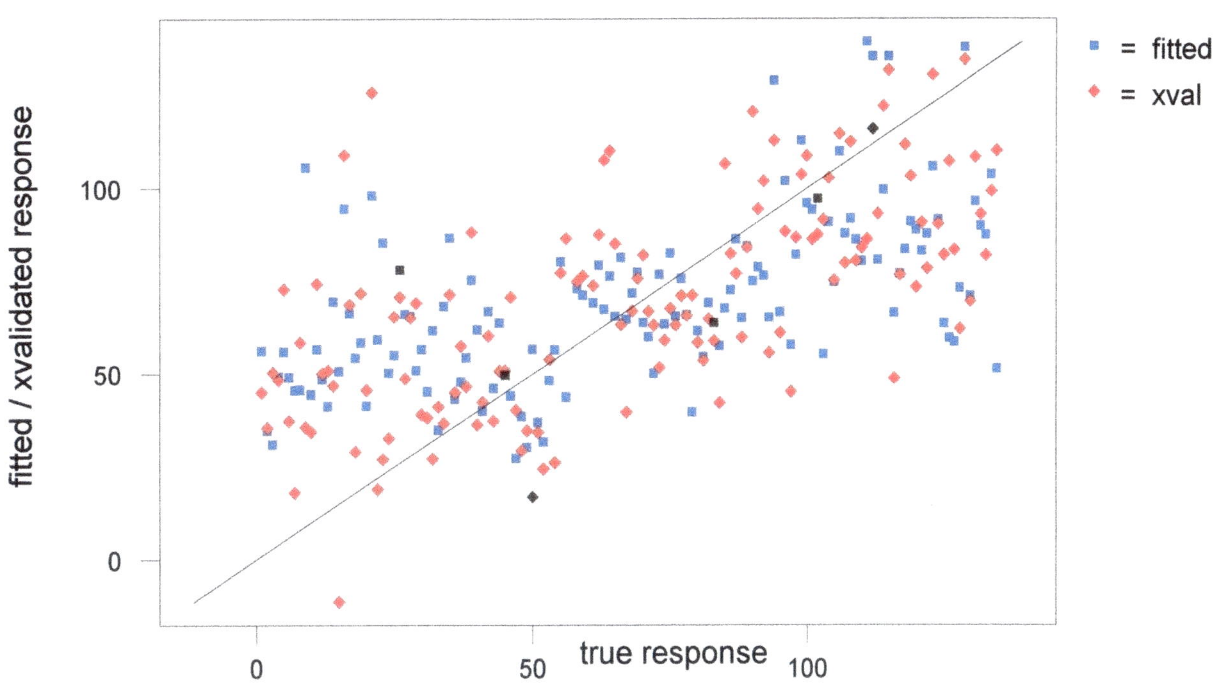

NPLS Fit and Prediction

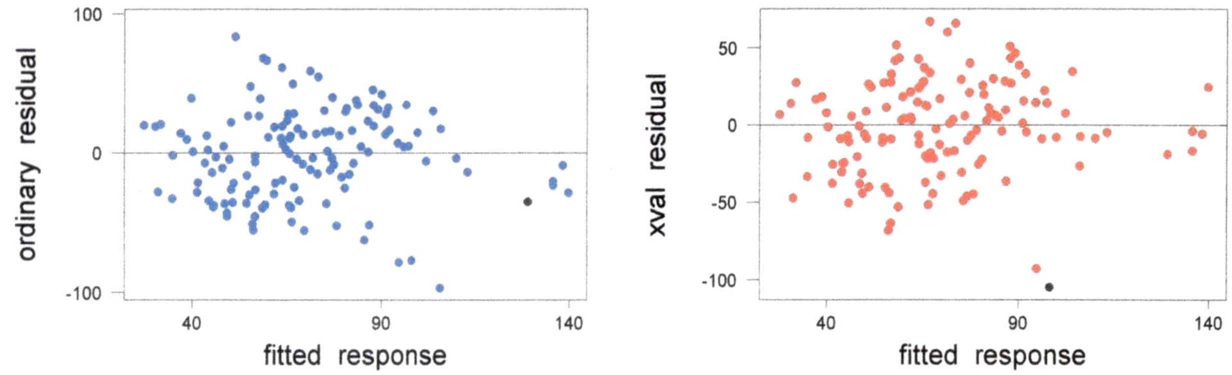

NPLS Residual Plots

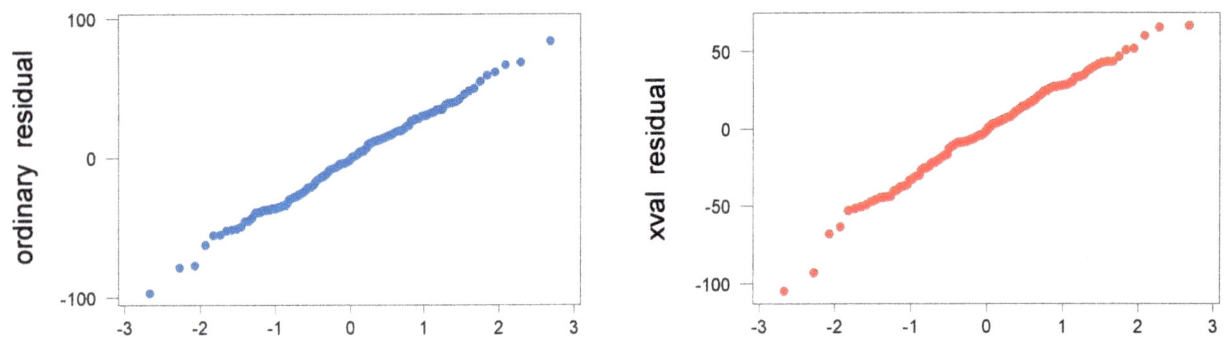

NPLS Normal Residual Plots

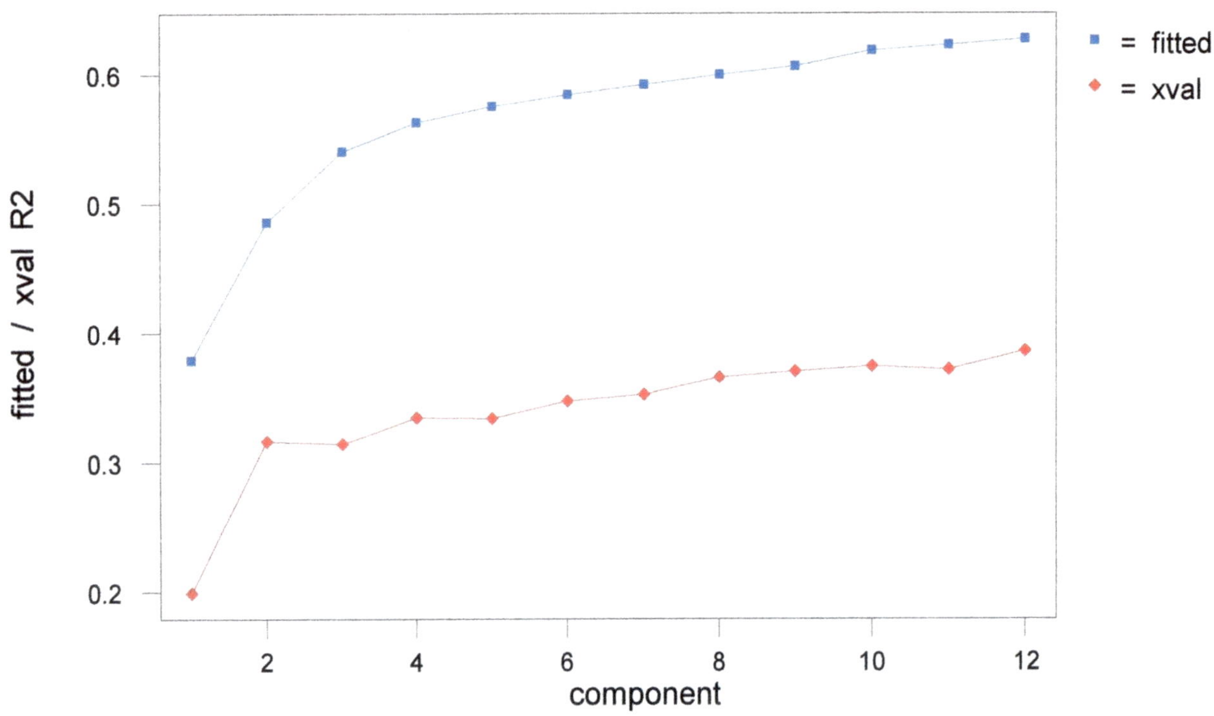

NPLS Model Selection Plot

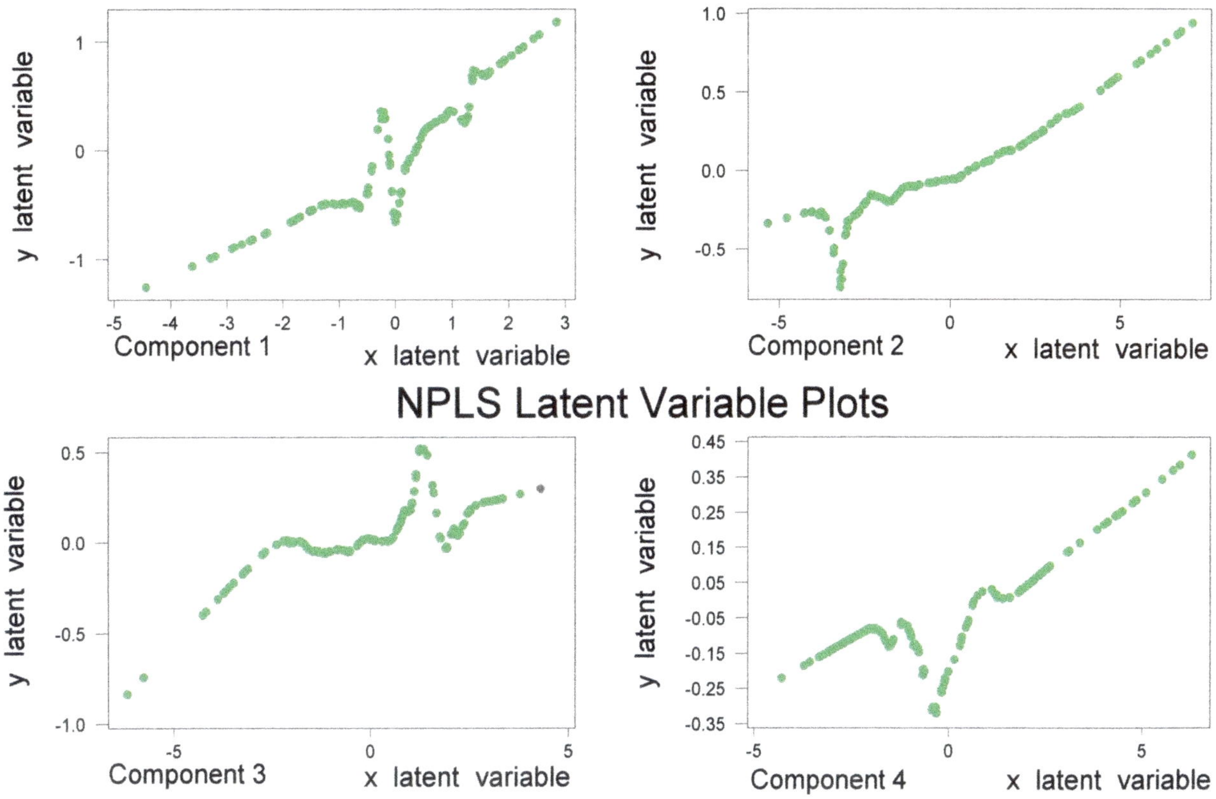
NPLS Latent Variable Plots
Component 1
Component 2
Component 3
Component 4
y latent variable
x latent variable

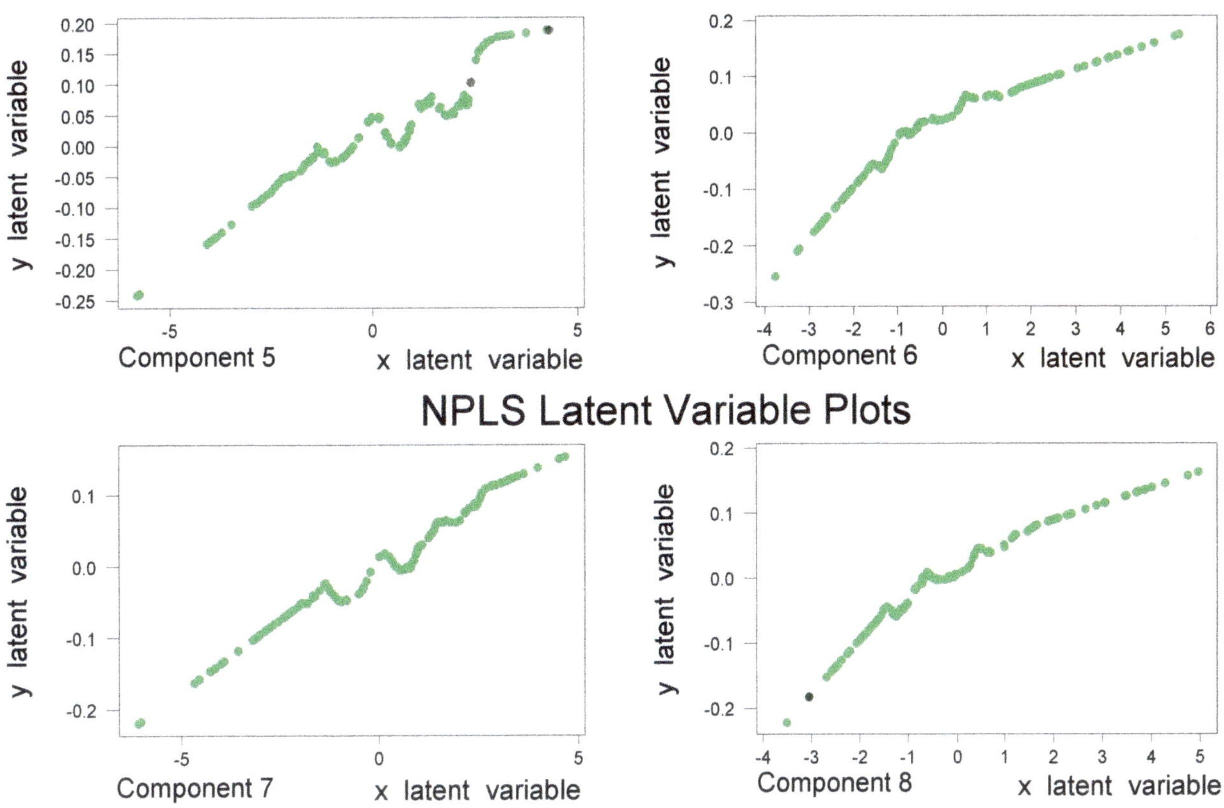
NPLS Latent Variable Plots
Component 5
Component 6
Component 7
Component 8
y latent variable
x latent variable

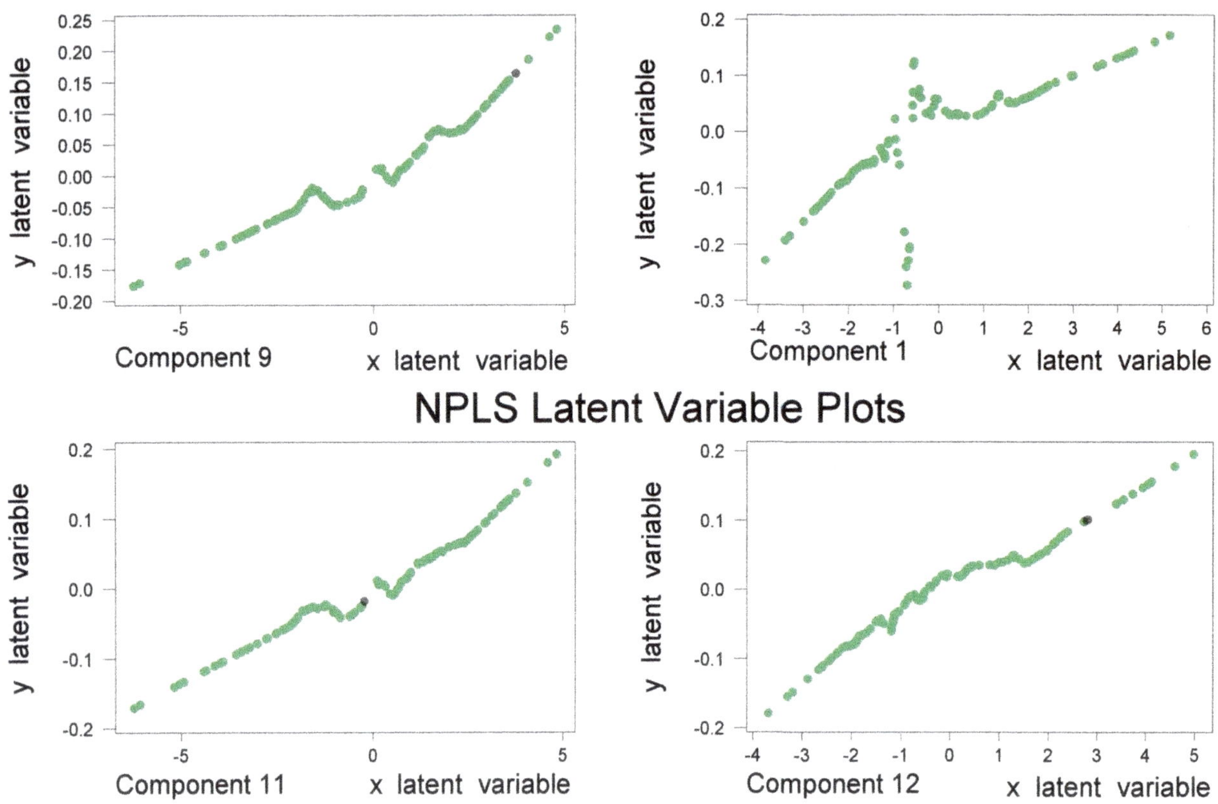
NPLS Latent Variable Plots
Component 9
Component 1
Component 11
Component 12
y latent variable
x latent variable
0.25
0.20
0.15
0.10
0.05
0.00
-0.05
-0.10
-0.15
-0.20
0.2
0.1
0.0
-0.1
-0.2
-0.3
-5
0
5
-4
-3
-2
-1
1
2
3
4
6

PART III

Modern - COO1C - C135C
Archaeological - AROO1C - AR055C
$CHCl_3$

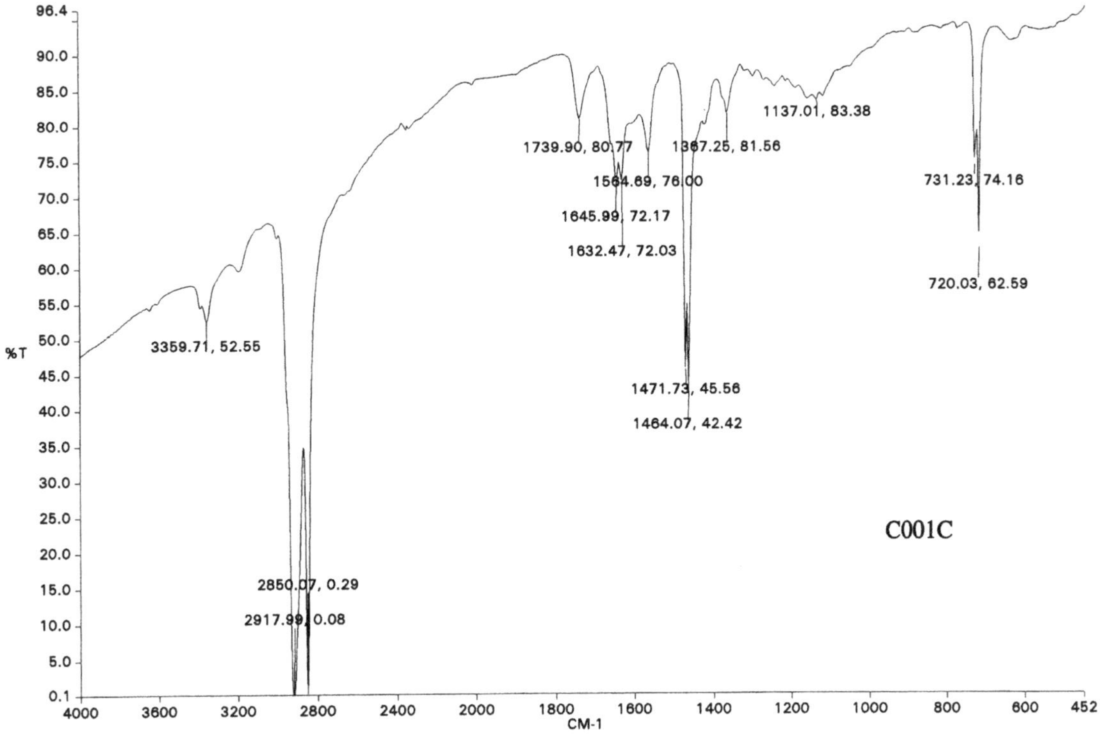

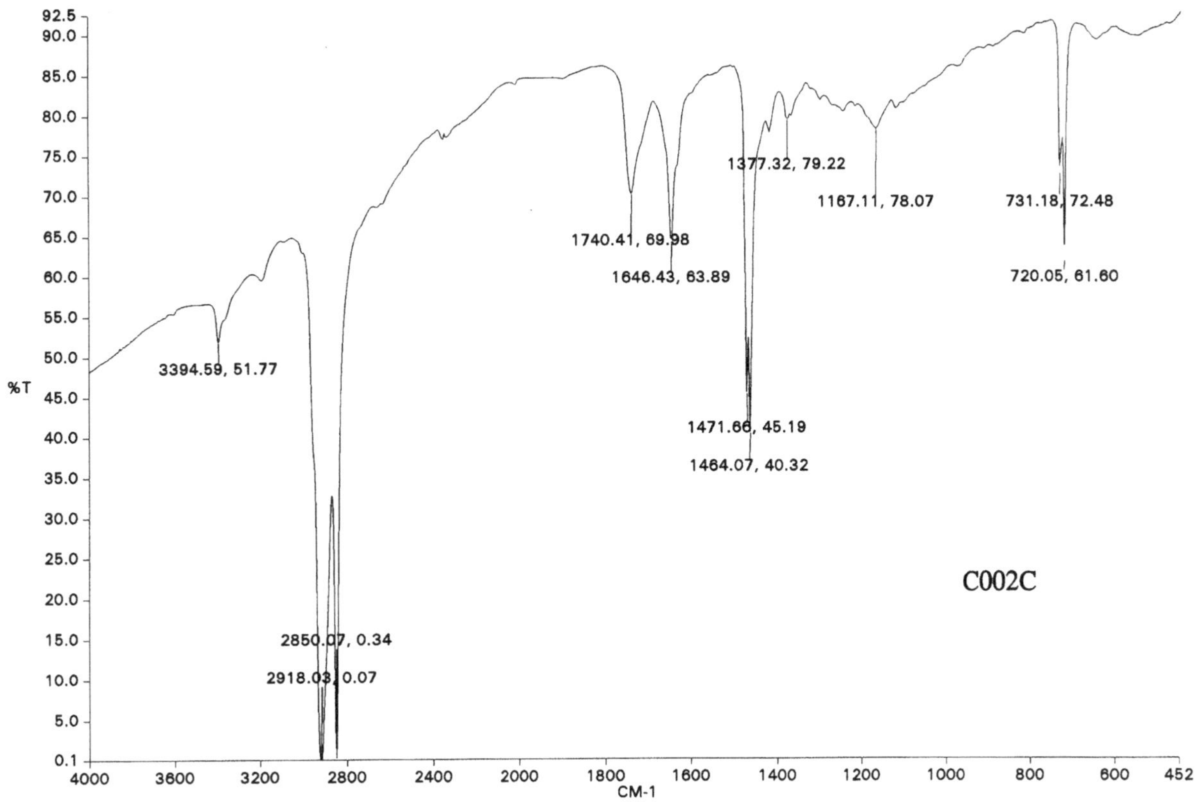

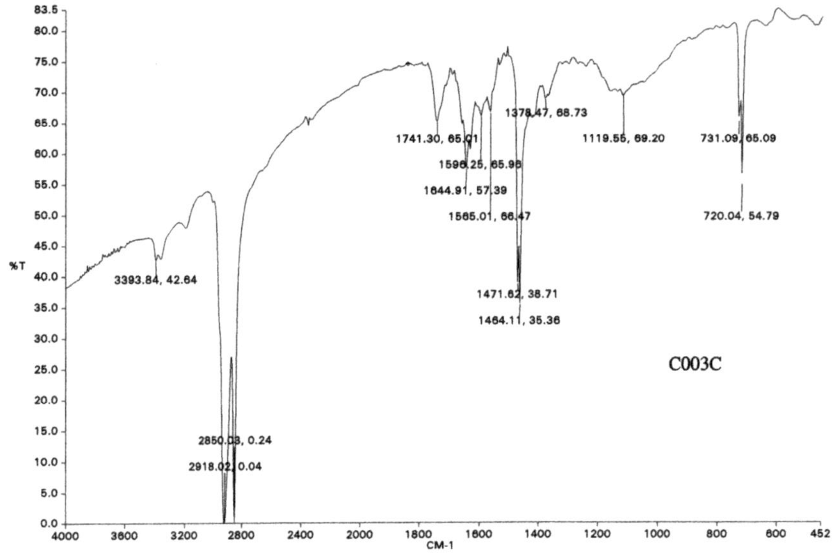
C003C
%T
CM-1
3393.84, 42.64
1741.30, 65.01
1596.25, 65.98
1644.91, 57.39
1565.01, 66.47
1378.47, 68.73
1471.82, 38.71
1464.11, 35.36
1119.55, 69.20
731.09, 65.09
720.04, 54.79
2850.03, 0.24
2918.02, 0.04

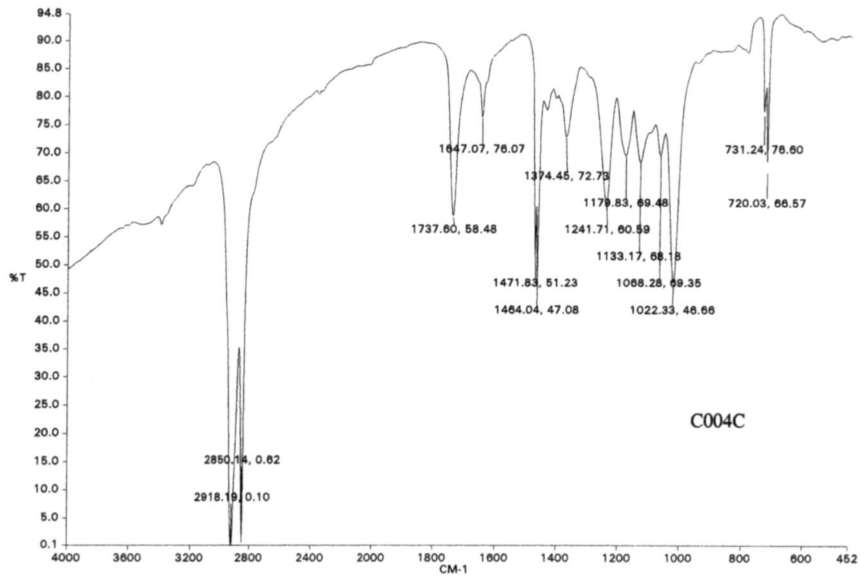
C004C
%T
CM-1
1647.07, 76.07
1737.60, 58.48
1374.45, 72.73
1179.83, 69.48
1241.71, 60.59
1133.17, 68.18
1471.83, 51.23
1464.04, 47.08
1068.28, 69.35
1022.33, 46.66
731.24, 76.60
720.03, 66.57
2850.14, 0.62
2918.19, 0.10

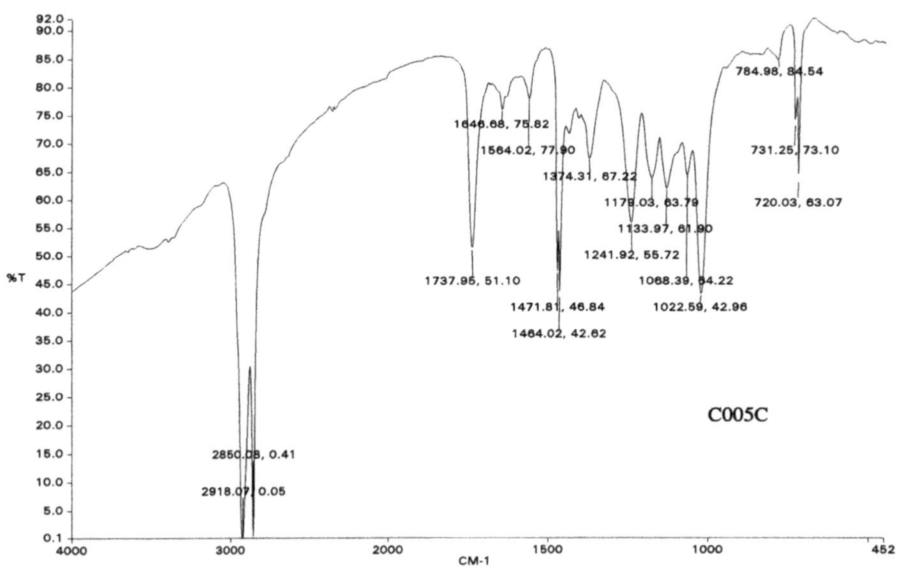
C005C
%T
CM-1
784.98, 84.54
1646.68, 75.82
1564.02, 77.90
1374.31, 67.22
1179.03, 63.79
1133.97, 61.90
1241.92, 55.72
1737.95, 51.10
1068.39, 54.22
1471.81, 46.84
1022.59, 42.96
1464.02, 42.62
731.25, 73.10
720.03, 63.07
2850.08, 0.41
2918.07, 0.05

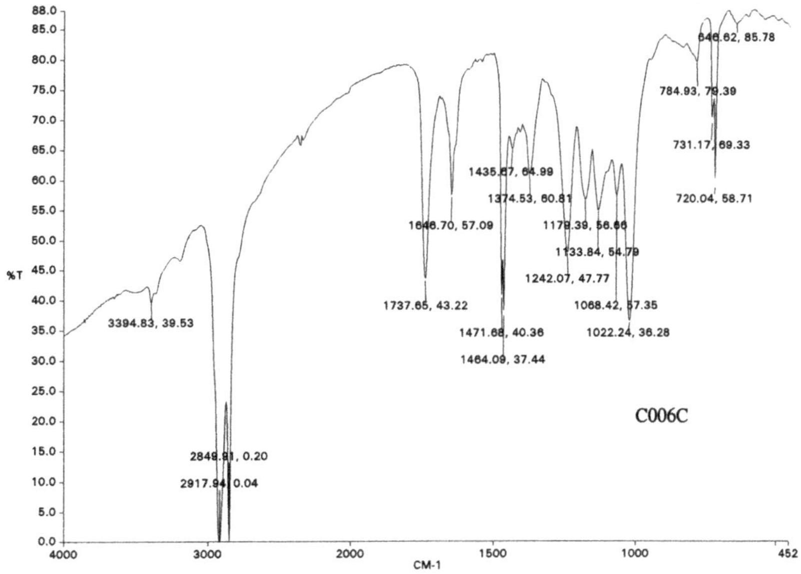
%T
CM-1
3394.83, 39.53
2849.81, 0.20
2917.94, 0.04
1737.65, 43.22
1646.70, 57.09
1471.68, 40.36
1464.09, 37.44
1435.67, 64.99
1374.53, 60.81
1242.07, 47.77
1179.39, 56.66
1133.84, 54.79
1068.42, 37.35
1022.24, 36.28
846.62, 85.78
784.93, 79.39
731.17, 69.33
720.04, 58.71
C006C

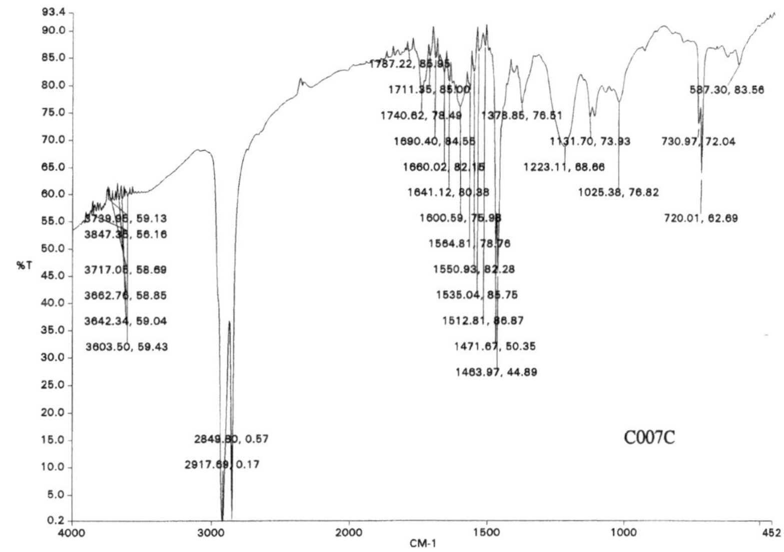
%T
CM-1
3739.96, 59.13
3847.38, 56.16
3717.05, 58.69
3662.76, 58.85
3642.34, 59.04
3603.50, 59.43
2849.80, 0.57
2917.69, 0.17
1787.22, 85.95
1711.35, 85.00
1740.62, 78.49
1690.40, 84.55
1660.02, 82.15
1641.12, 80.38
1600.59, 75.98
1564.81, 78.76
1550.93, 82.28
1535.04, 85.75
1512.81, 86.87
1471.67, 50.35
1463.97, 44.89
1378.85, 76.51
1131.70, 73.93
1223.11, 68.66
1025.38, 76.82
587.30, 83.56
730.97, 72.04
720.01, 62.69
C007C

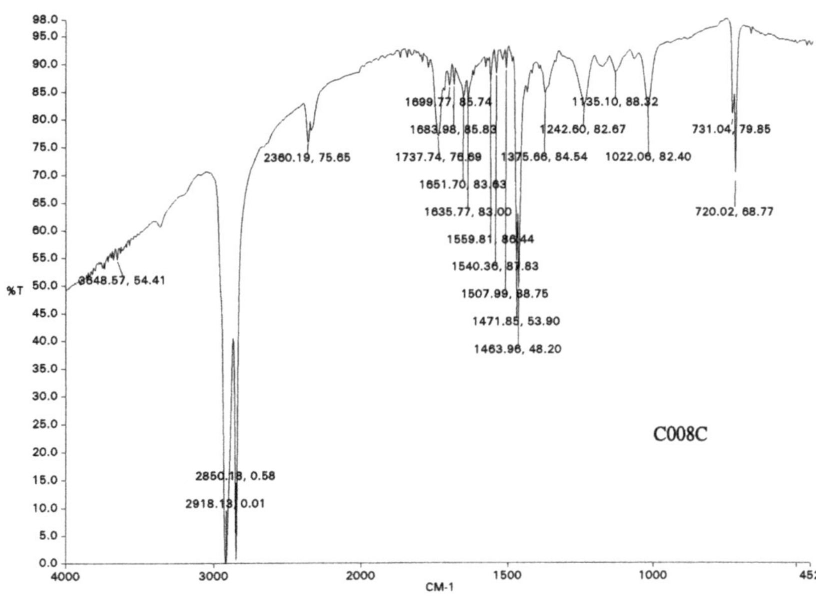
%T
CM-1
3648.57, 54.41
2360.19, 75.65
2850.18, 0.58
2918.13, 0.01
1699.77, 85.74
1683.98, 85.83
1737.74, 76.69
1651.70, 83.63
1635.77, 83.00
1559.81, 86.44
1540.36, 87.83
1507.99, 88.75
1471.85, 53.90
1463.96, 48.20
1375.66, 84.54
1242.60, 82.67
1135.10, 88.32
1022.06, 82.40
731.04, 79.85
720.02, 68.77
C008C

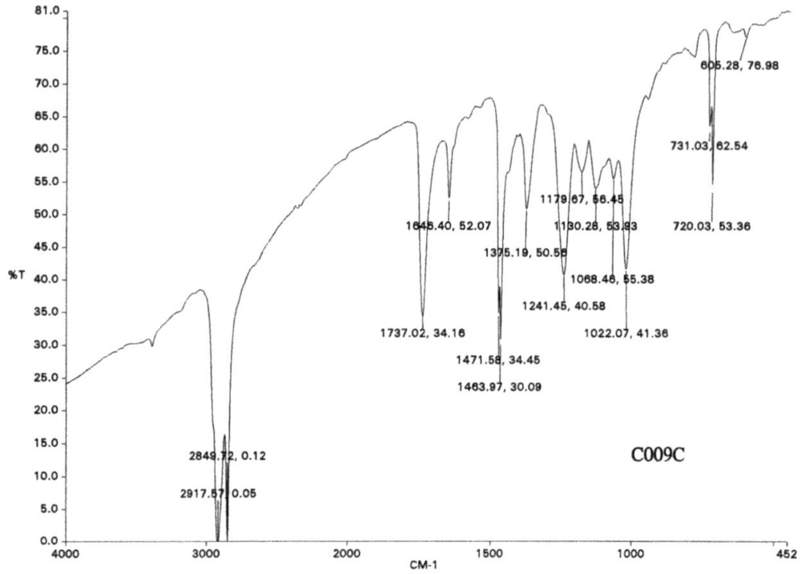

%T
CM-1
2849.72, 0.12
2917.57, 0.05
1737.02, 34.16
1645.40, 52.07
1375.19, 50.56
1471.58, 34.45
1463.97, 30.09
1241.45, 40.58
1179.67, 56.45
1130.28, 53.93
1068.46, 55.38
1022.07, 41.36
731.03, 62.54
720.03, 53.36
605.28, 76.98
C009C

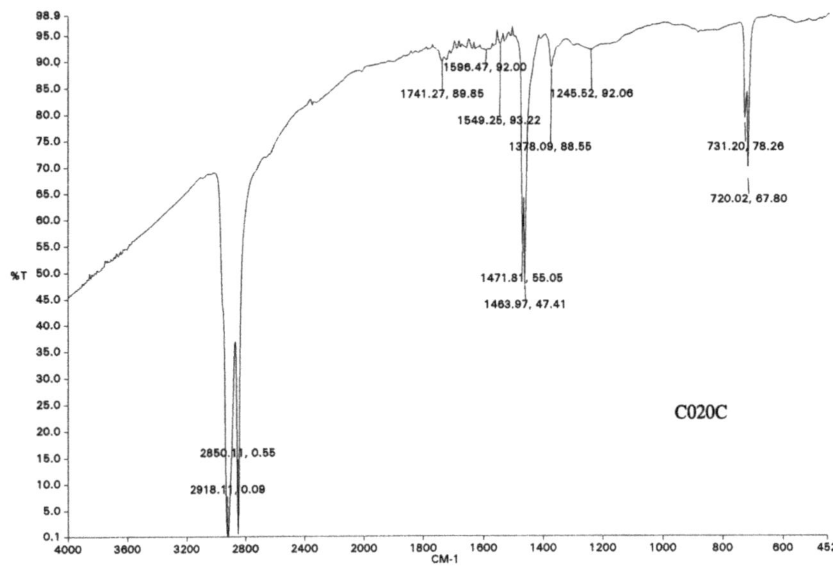

%T
CM-1
1596.47, 92.00
1741.27, 89.85
1549.25, 93.22
1245.52, 92.06
1378.09, 88.55
731.20, 78.26
720.02, 67.80
1471.81, 55.05
1463.97, 47.41
2850.11, 0.55
2918.11, 0.09
C020C

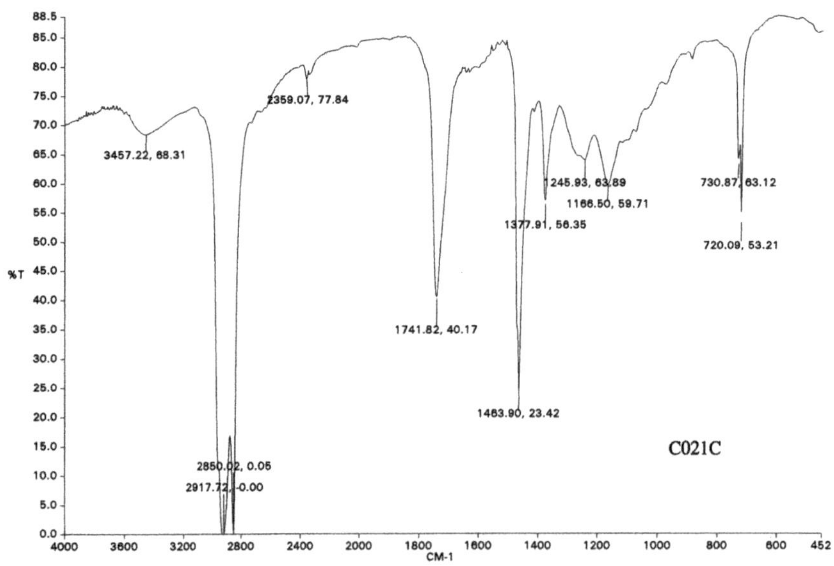

%T
CM-1
2359.07, 77.84
3457.22, 68.31
1245.93, 63.89
1166.50, 59.71
1377.91, 56.35
730.87, 63.12
720.09, 53.21
1741.82, 40.17
1463.90, 23.42
2850.02, 0.05
2917.72, -0.00
C021C

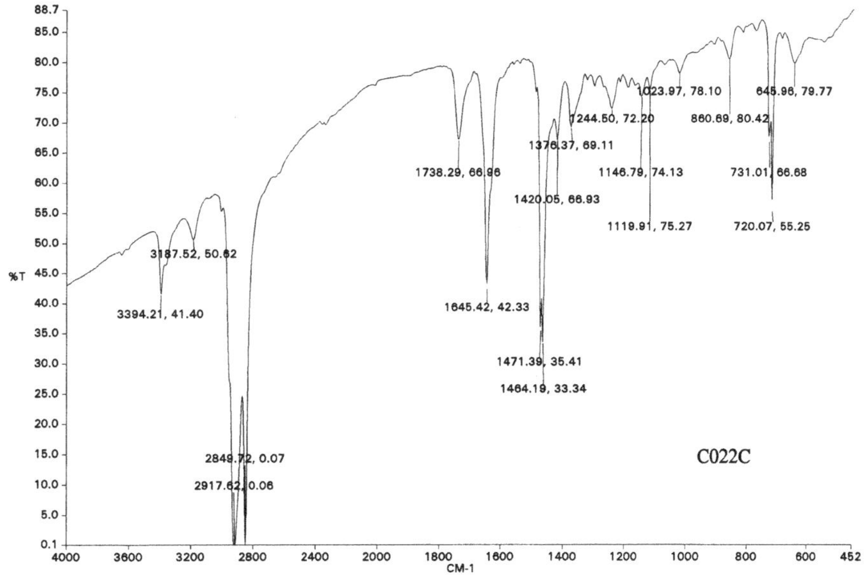
%T
CM-1
3394.21, 41.40
3187.52, 50.82
2849.72, 0.07
2917.62, 0.06
1738.29, 66.96
1645.42, 42.33
1471.39, 35.41
1464.19, 33.34
1420.05, 66.93
1376.37, 69.11
1244.50, 72.20
1146.79, 74.13
1119.91, 75.27
1023.97, 78.10
860.69, 80.42
731.01, 66.68
720.07, 55.25
645.96, 79.77
C022C

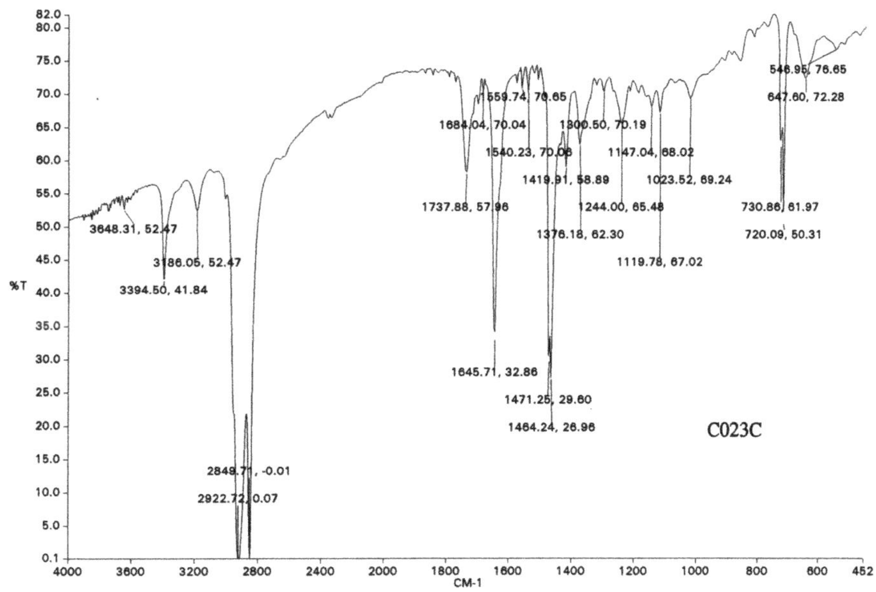
%T
CM-1
3648.31, 52.47
3186.05, 52.47
3394.50, 41.84
2849.71, -0.01
2922.72, 0.07
1684.04, 70.04
1559.74, 70.65
1540.23, 70.06
1737.88, 57.96
1645.71, 32.86
1471.25, 29.60
1464.24, 26.96
1419.91, 58.89
1376.18, 62.30
1300.50, 70.19
1244.00, 65.48
1147.04, 68.02
1119.78, 67.02
1023.52, 69.24
730.86, 61.97
720.09, 50.31
546.95, 76.65
647.60, 72.28
C023C

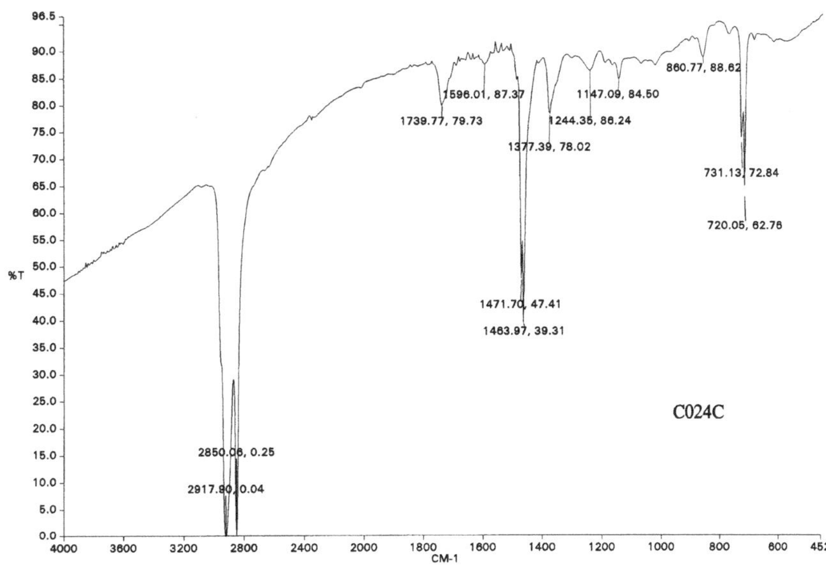
%T
CM-1
1596.01, 87.37
1739.77, 79.73
1377.39, 78.02
1244.35, 86.24
1147.09, 84.50
860.77, 88.62
731.13, 72.84
720.05, 62.76
1471.70, 47.41
1463.97, 39.31
2850.08, 0.25
2917.90, 0.04
C024C

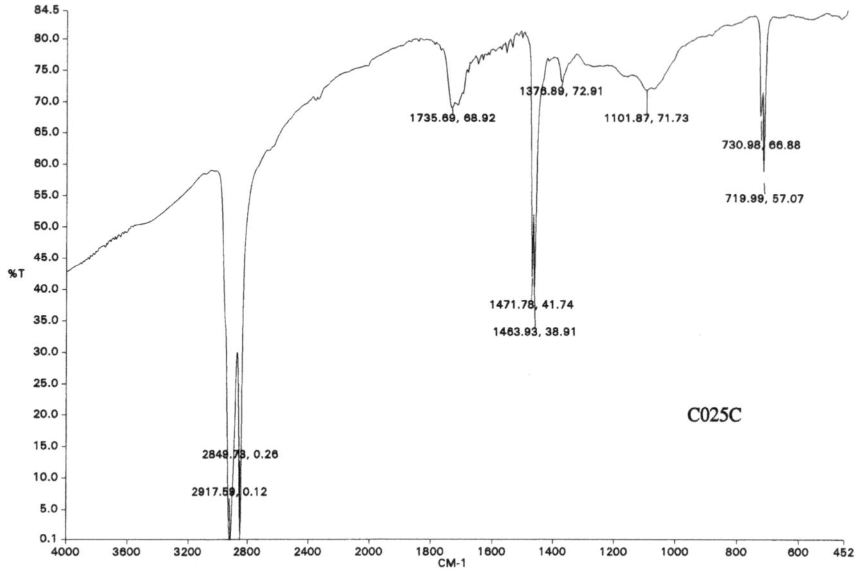
%T
CM-1
1735.69, 68.92
1378.89, 72.91
1101.87, 71.73
730.98, 66.88
719.99, 57.07
1471.78, 41.74
1463.93, 38.91
2849.78, 0.26
2917.59, 0.12
C025C

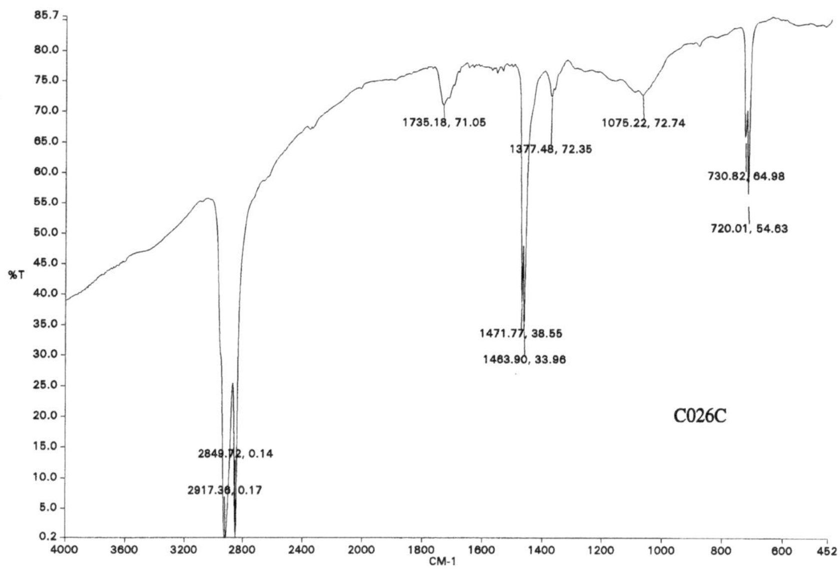
%T
CM-1
1735.18, 71.05
1377.48, 72.35
1075.22, 72.74
730.82, 64.98
720.01, 54.63
1471.77, 38.55
1463.90, 33.96
2849.72, 0.14
2917.36, 0.17
C026C

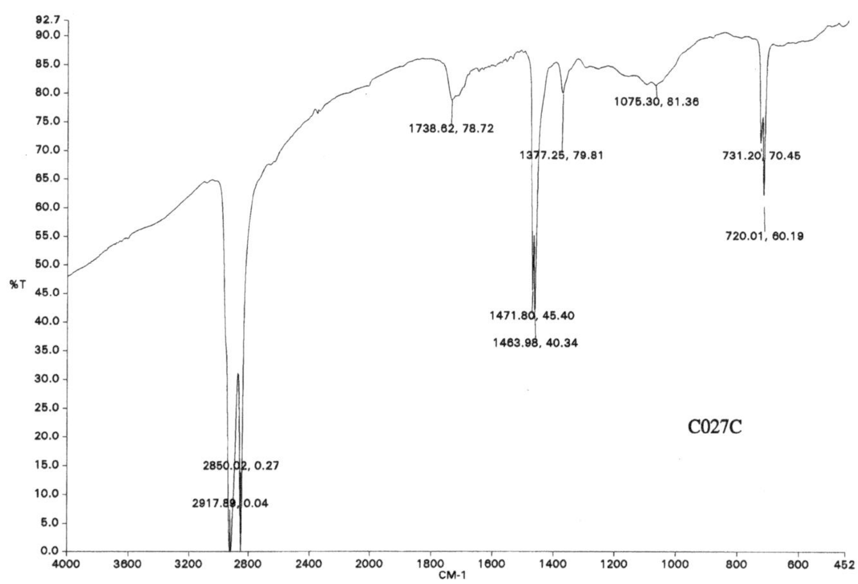
%T
CM-1
1075.30, 81.36
1738.62, 78.72
1377.25, 79.81
731.20, 70.45
720.01, 60.19
1471.80, 45.40
1463.98, 40.34
2850.02, 0.27
2917.89, 0.04
C027C

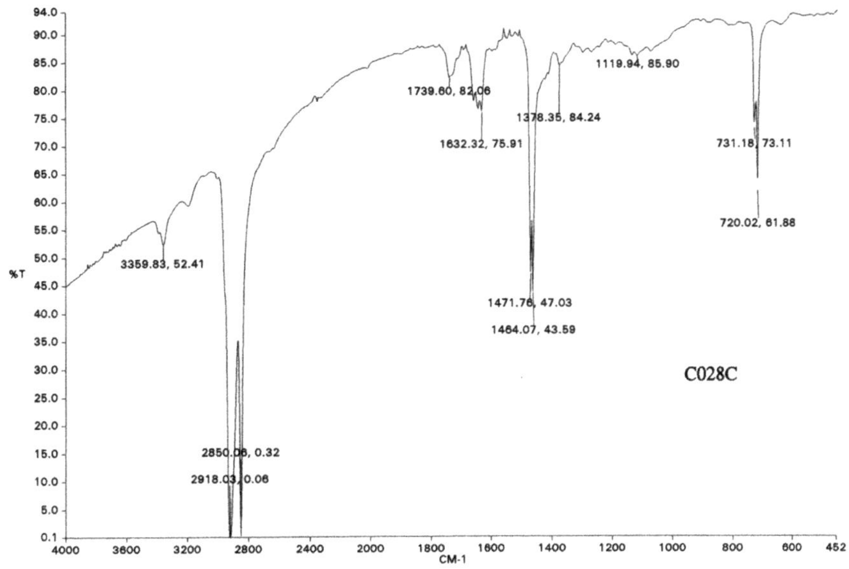
C028C
%T
CM-1
1739.60, 82.06
1632.32, 75.91
1378.35, 84.24
1119.94, 85.90
731.18, 73.11
720.02, 61.88
3359.83, 52.41
1471.76, 47.03
1464.07, 43.59
2850.06, 0.32
2918.03, 0.06

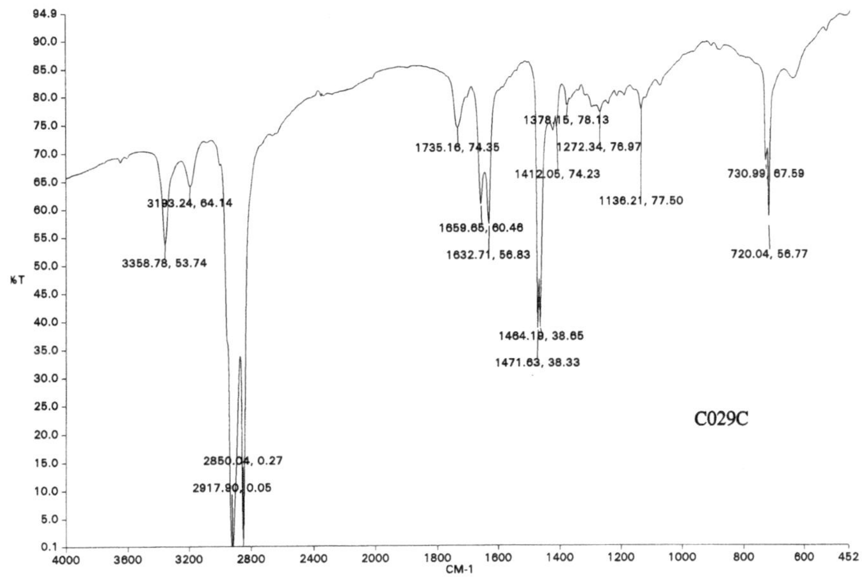
C029C
%T
CM-1
1378.15, 78.13
1735.16, 74.35
1272.34, 76.97
1412.05, 74.23
730.99, 67.59
3193.24, 64.14
1136.21, 77.50
1659.65, 60.46
1632.71, 56.83
720.04, 56.77
3358.78, 53.74
1464.19, 38.65
1471.63, 38.33
2850.04, 0.27
2917.90, 0.05

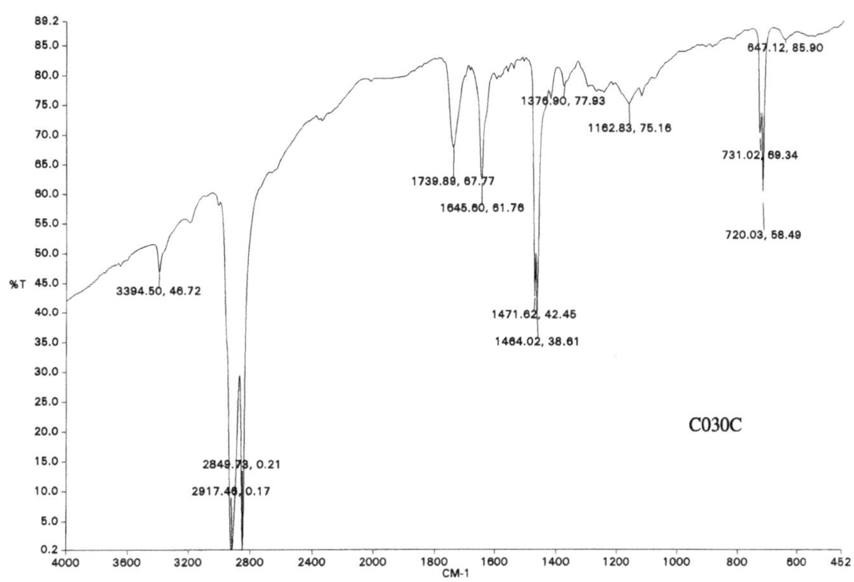
C030C
%T
CM-1
647.12, 85.90
1376.90, 77.93
1162.83, 75.16
731.02, 69.34
1739.89, 67.77
1645.60, 61.76
720.03, 58.49
3394.50, 46.72
1471.62, 42.45
1464.02, 38.61
2849.73, 0.21
2917.46, 0.17

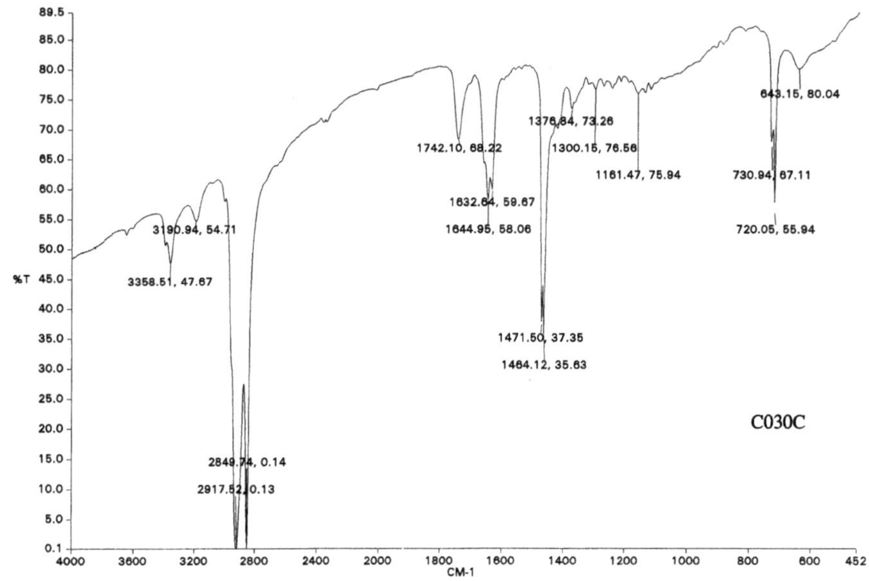
%T
CM-1
643.15, 80.04
1376.84, 73.26
1742.10, 68.22
1300.15, 76.56
1161.47, 75.94
730.94, 67.11
1632.64, 59.67
1644.95, 58.06
3190.94, 54.71
720.05, 55.94
3358.51, 47.67
1471.50, 37.35
1464.12, 35.63
C030C
2849.74, 0.14
2917.52, 0.13

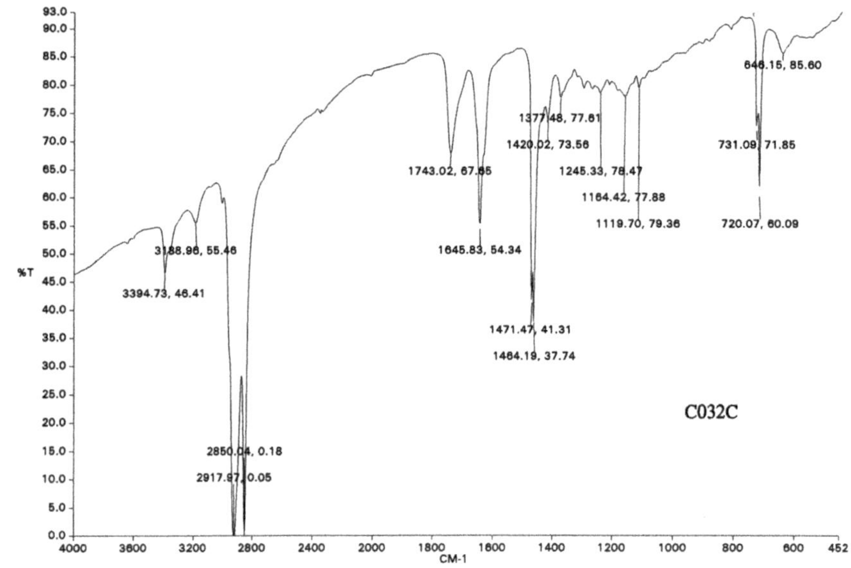
%T
CM-1
646.15, 85.60
1377.48, 77.61
1420.02, 73.56
731.09, 71.85
1743.02, 67.85
1245.33, 78.47
1164.42, 77.88
1119.70, 79.36
720.07, 60.09
1645.83, 54.34
3188.96, 55.46
3394.73, 46.41
1471.47, 41.31
1464.19, 37.74
C032C
2850.04, 0.18
2917.97, 0.05

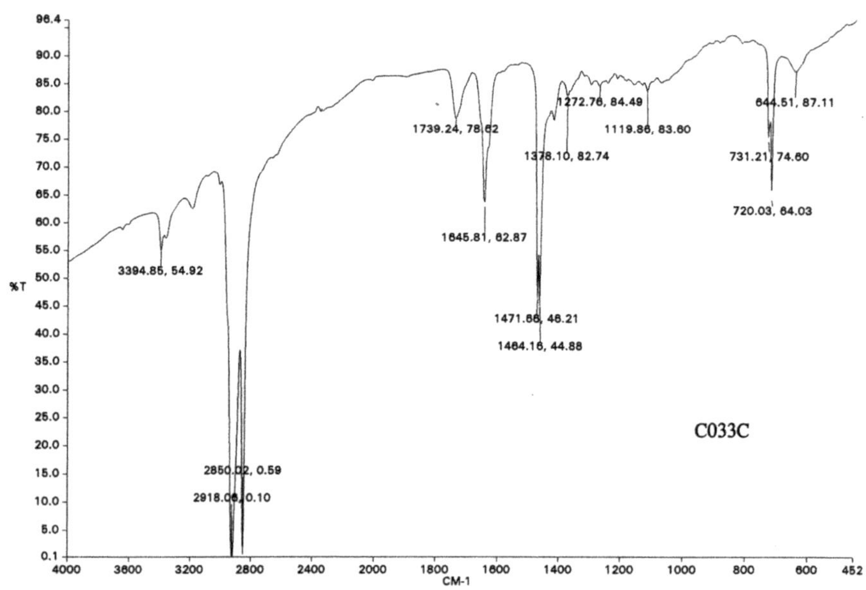
%T
CM-1
1272.76, 84.49
644.51, 87.11
1739.24, 78.62
1119.86, 83.60
1378.10, 82.74
731.21, 74.60
720.03, 64.03
1645.81, 62.87
3394.85, 54.92
1471.86, 46.21
1464.16, 44.88
C033C
2850.02, 0.59
2918.06, 0.10

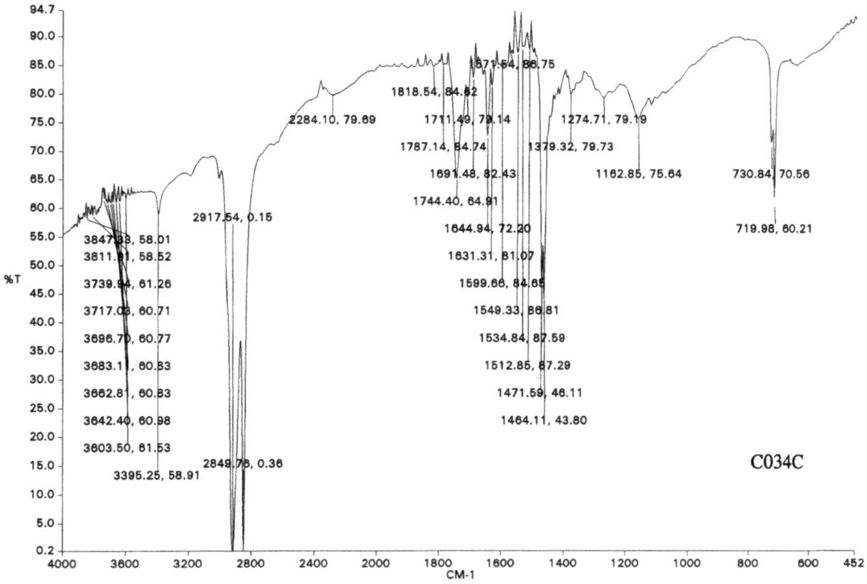
C034C
%T
CM-1
2284.10, 79.69
2917.54, 0.15
2849.76, 0.36
3395.25, 58.91
1818.54, 84.82
1711.49, 79.14
1787.14, 84.74
1691.48, 82.43
1744.40, 64.91
1644.94, 72.20
1631.31, 81.07
1599.66, 84.65
1549.33, 86.81
1534.84, 87.59
1512.85, 87.29
1471.59, 46.11
1464.11, 43.80
1274.71, 79.19
1379.32, 79.73
1162.85, 75.64
730.84, 70.56
719.98, 60.21

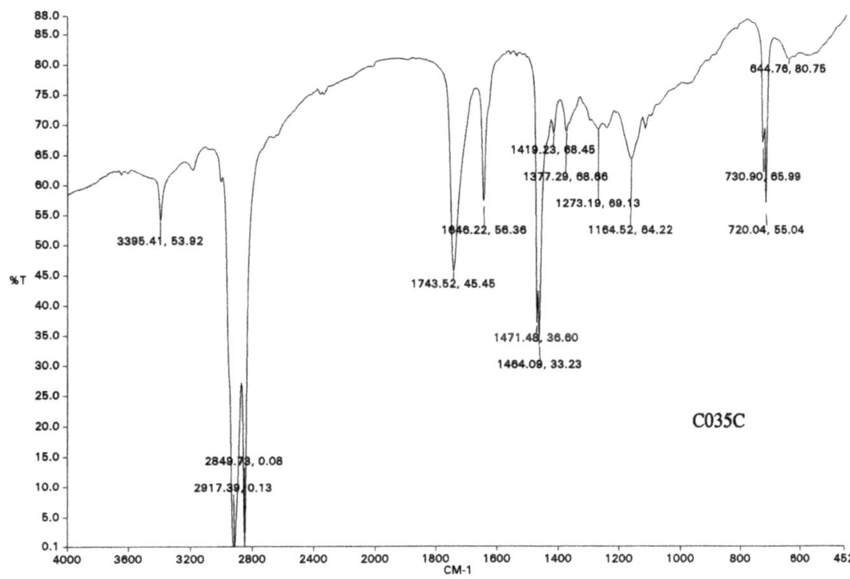
C035C
%T
CM-1
3395.41, 53.92
2849.78, 0.08
2917.39, 0.13
1646.22, 56.36
1743.52, 45.45
1419.23, 68.45
1377.29, 68.66
1273.19, 69.13
1164.52, 64.22
1471.48, 36.60
1464.09, 33.23
644.76, 80.75
730.90, 65.99
720.04, 55.04

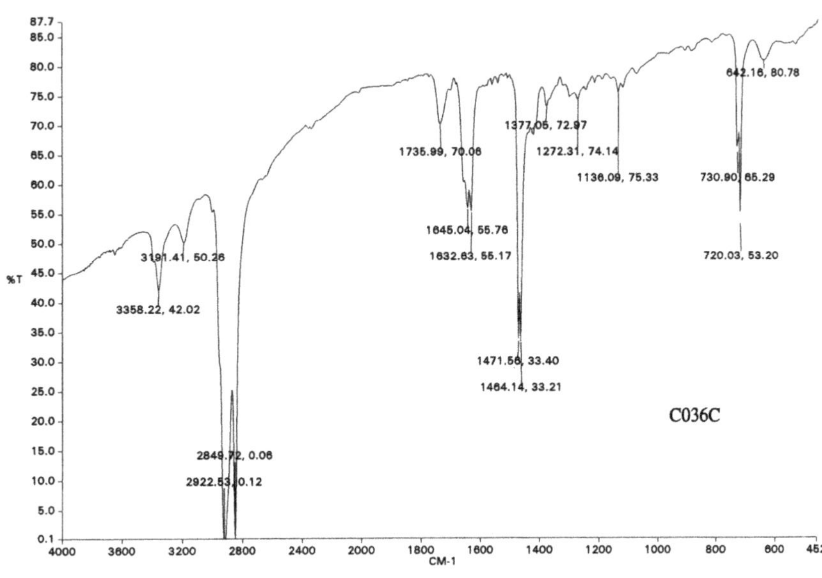
C036C
%T
CM-1
3191.41, 50.26
3358.22, 42.02
2849.72, 0.06
2922.53, 0.12
1735.99, 70.06
1645.04, 55.76
1632.63, 55.17
1377.05, 72.97
1272.31, 74.14
1136.09, 75.33
1471.56, 33.40
1464.14, 33.21
642.16, 80.78
730.90, 65.29
720.03, 53.20

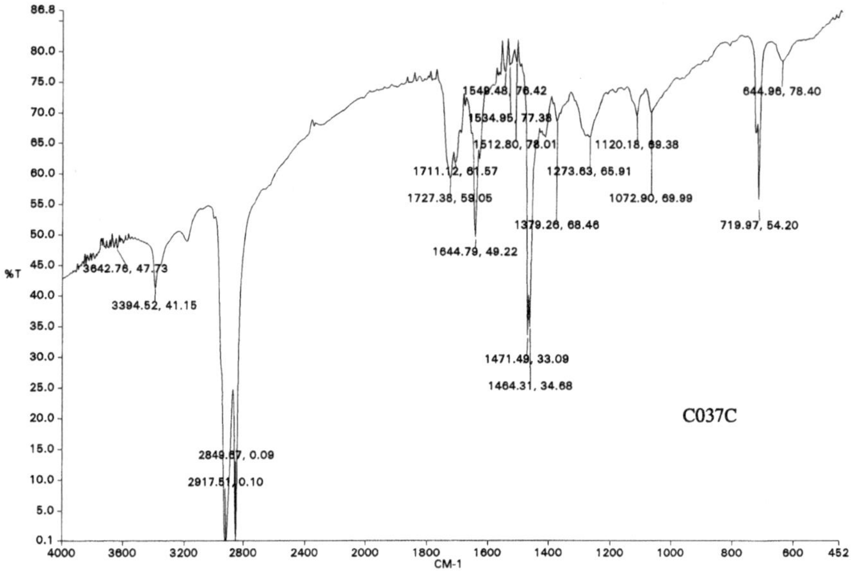
%T
CM-1
3642.76, 47.73
3394.52, 41.15
2849.87, 0.09
2917.51, 0.10
1549.48, 76.42
1534.95, 77.38
1512.80, 78.01
1711.12, 61.57
1727.38, 59.05
1644.79, 49.22
1379.26, 68.46
1273.63, 65.91
1120.18, 69.38
1072.90, 69.99
1471.49, 33.09
1464.31, 34.68
644.96, 78.40
719.97, 54.20
C037C

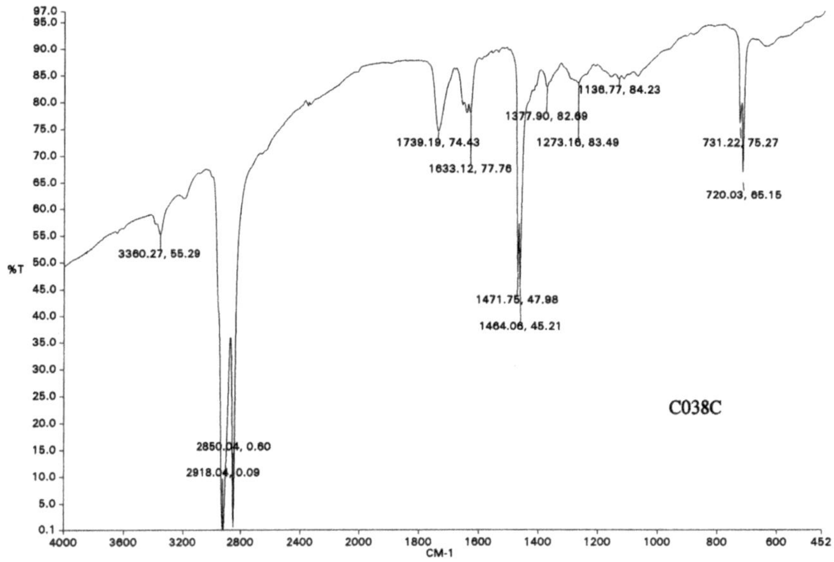
%T
CM-1
3360.27, 55.29
2850.04, 0.60
2918.04, 0.09
1739.19, 74.43
1633.12, 77.76
1377.90, 82.69
1273.16, 83.49
1136.77, 84.23
1471.75, 47.98
1464.06, 45.21
731.22, 75.27
720.03, 65.15
C038C

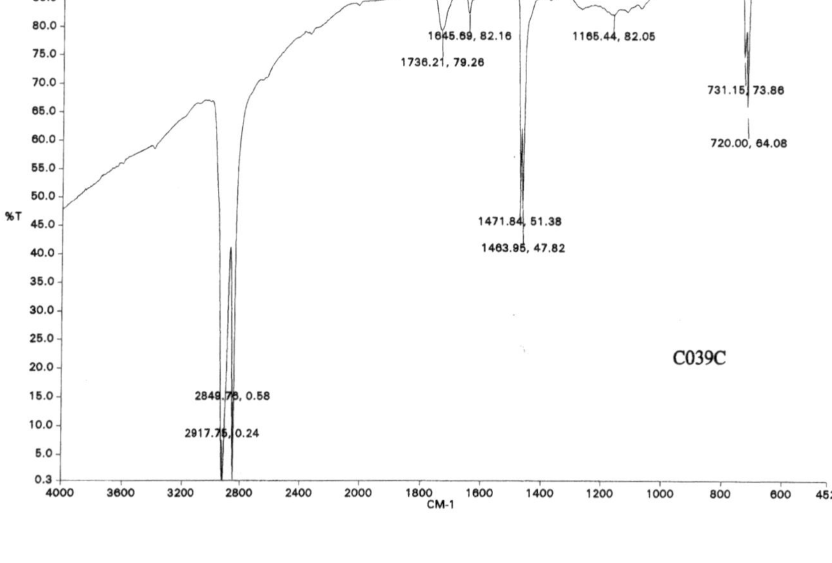
%T
CM-1
1645.69, 82.16
1736.21, 79.26
1165.44, 82.05
731.15, 73.86
720.00, 64.08
1471.84, 51.38
1463.95, 47.82
2849.76, 0.58
2917.76, 0.24
C039C

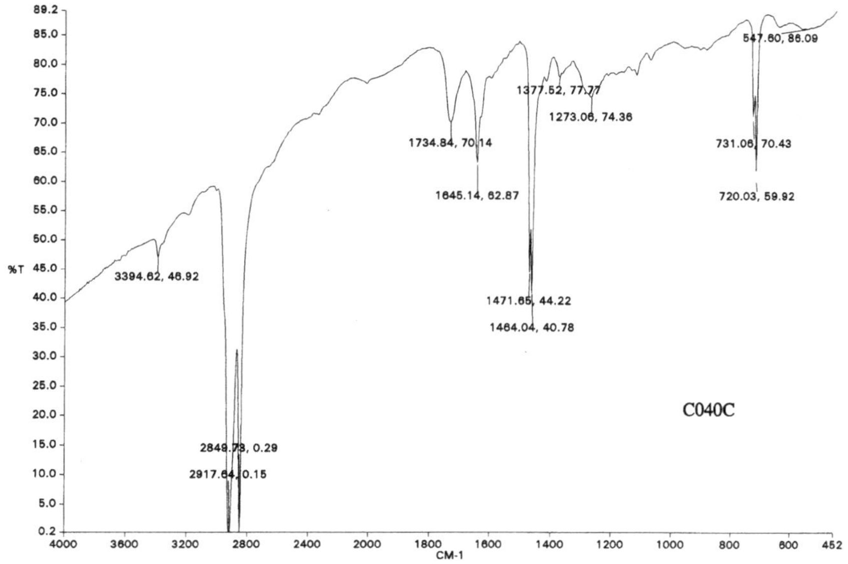
C040C
%T
CM-1
3394.62, 46.92
2849.73, 0.29
2917.64, 0.15
1734.84, 70.14
1645.14, 62.87
1471.65, 44.22
1464.04, 40.78
1377.52, 77.77
1273.06, 74.36
547.60, 86.09
731.06, 70.43
720.03, 59.92

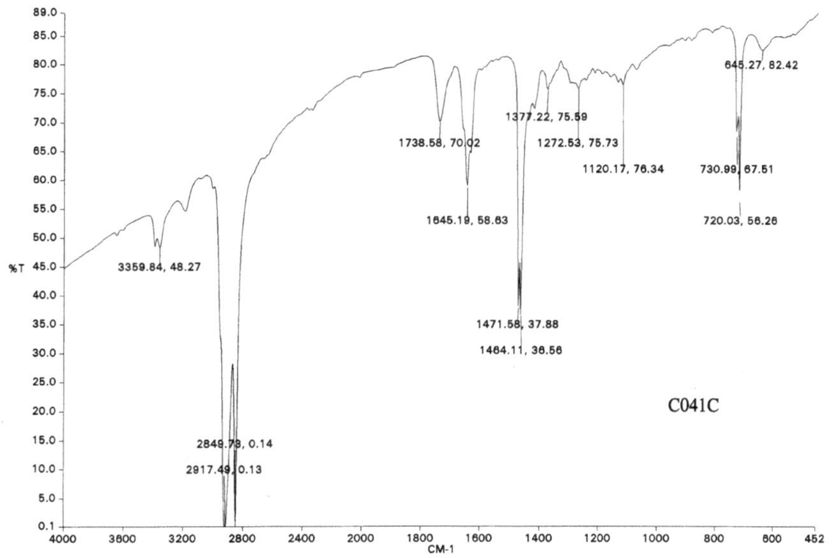
C041C
%T
CM-1
3359.84, 48.27
2849.73, 0.14
2917.49, 0.13
1738.58, 70.02
1645.19, 58.63
1471.58, 37.88
1464.11, 36.56
1377.22, 75.59
1272.53, 75.73
1120.17, 76.34
645.27, 82.42
730.99, 67.51
720.03, 56.26

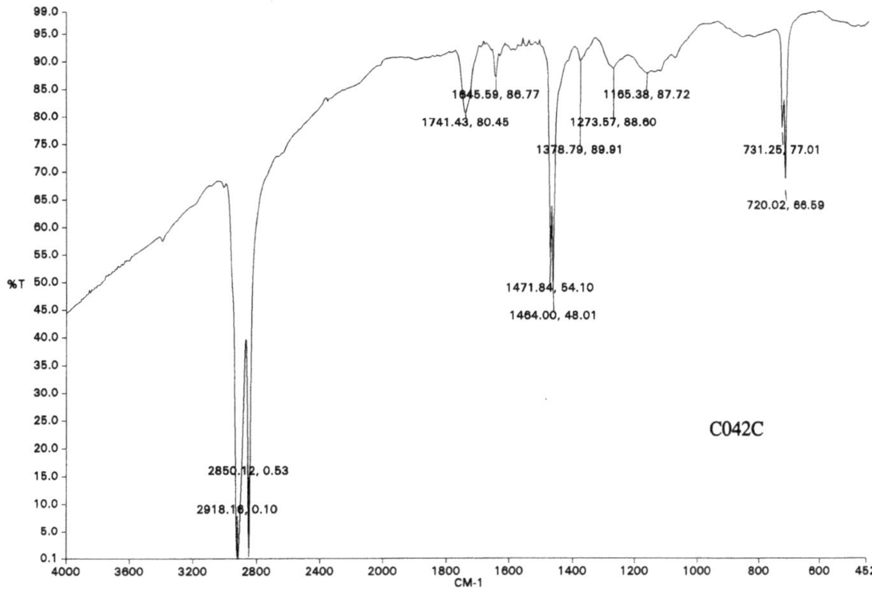
C042C
%T
CM-1
2850.12, 0.53
2918.16, 0.10
1645.59, 86.77
1741.43, 80.45
1165.38, 87.72
1273.57, 88.60
1378.79, 89.91
1471.84, 54.10
1464.00, 48.01
731.25, 77.01
720.02, 66.59

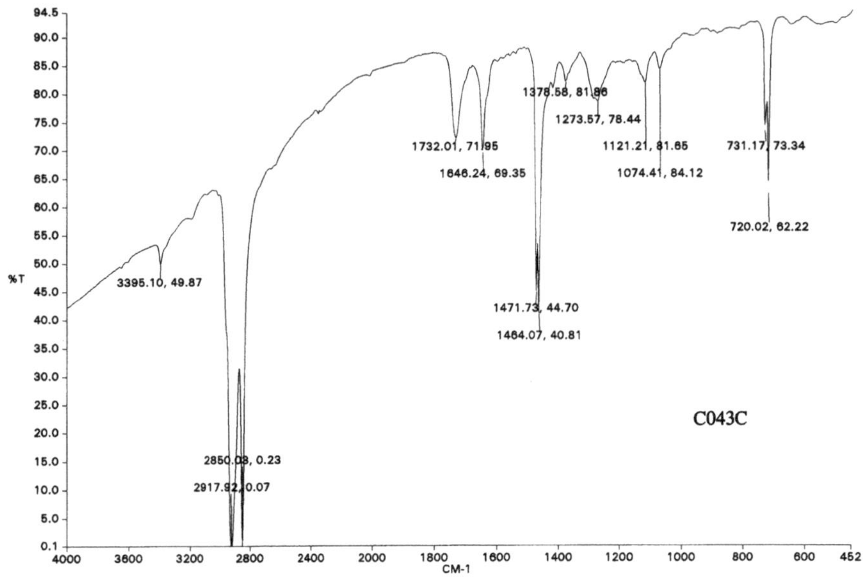
%T
CM-1
3395.10, 49.87
2850.03, 0.23
2917.92, 0.07
1732.01, 71.95
1646.24, 69.35
1378.58, 81.86
1273.57, 78.44
1471.73, 44.70
1464.07, 40.81
1121.21, 81.65
1074.41, 84.12
731.17, 73.34
720.02, 62.22
C043C

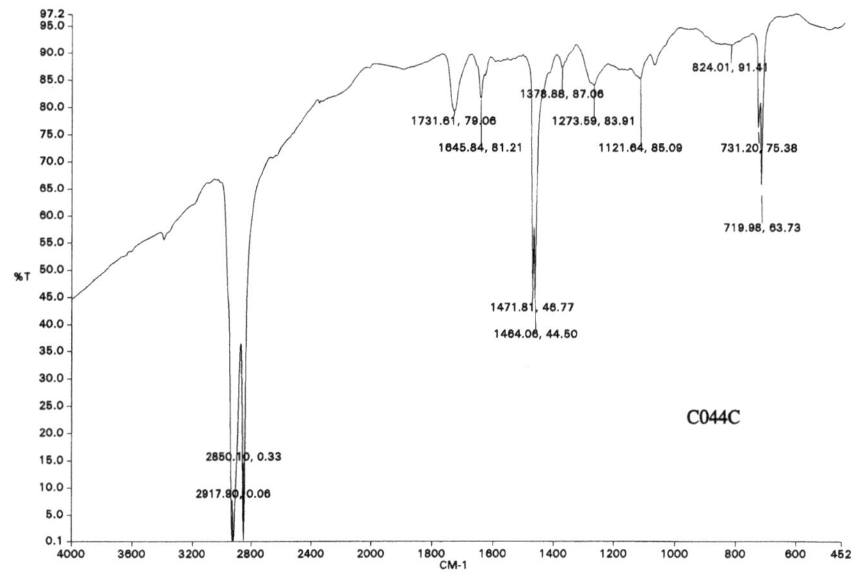
%T
CM-1
2850.10, 0.33
2917.90, 0.06
1731.61, 79.06
1645.84, 81.21
1378.88, 87.06
1273.59, 83.91
1471.81, 46.77
1464.06, 44.50
1121.64, 85.09
824.01, 91.41
731.20, 75.38
719.96, 63.73
C044C

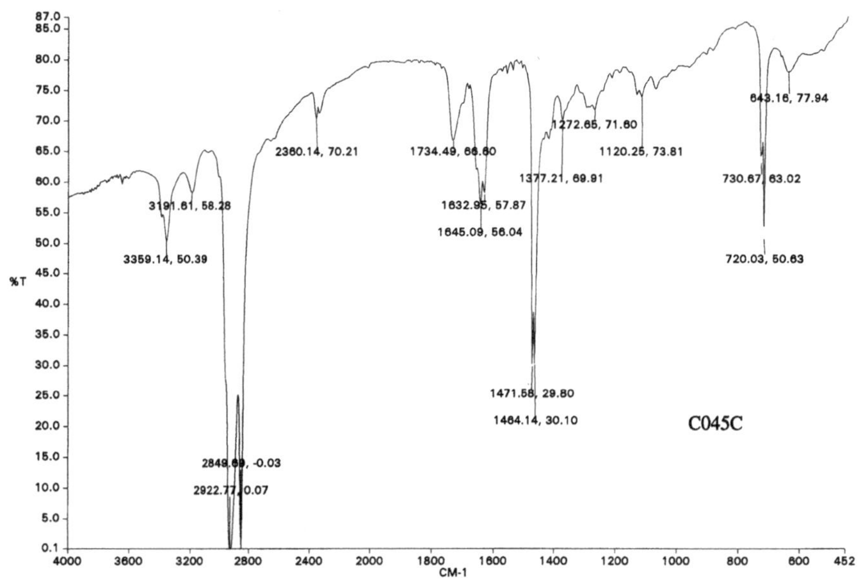
%T
CM-1
3191.61, 58.28
3359.14, 50.39
2360.14, 70.21
2849.69, -0.03
2922.77, 0.07
1734.49, 66.60
1632.95, 57.87
1645.09, 56.04
1272.65, 71.60
1377.21, 69.91
1471.58, 29.80
1464.14, 30.10
1120.25, 73.81
643.16, 77.94
730.67, 63.02
720.03, 50.63
C045C

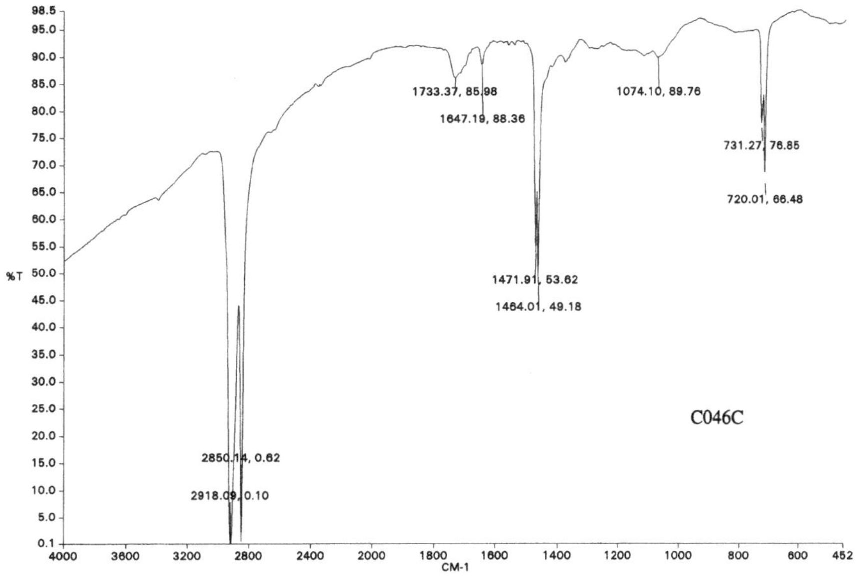
%T
CM-1
1733.37, 85.98
1647.19, 88.36
1074.10, 89.76
731.27, 76.85
720.01, 66.48
1471.91, 53.62
1464.01, 49.18
2850.14, 0.62
2918.09, 0.10
C046C

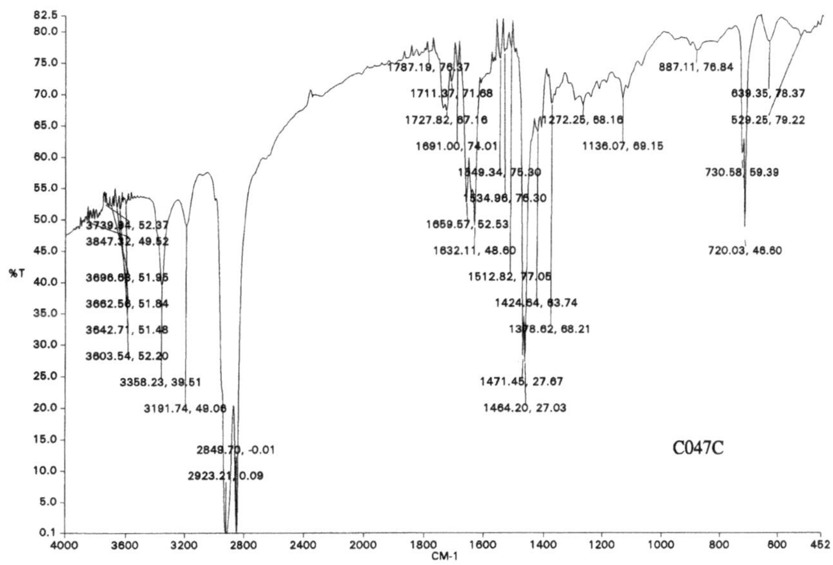
%T
CM-1
1787.19, 76.37
1711.37, 71.68
1727.82, 67.16
1691.00, 74.01
1549.34, 75.30
1534.96, 76.30
1659.57, 52.53
1632.11, 48.60
1512.82, 77.05
1424.84, 63.74
1378.62, 68.21
1471.45, 27.67
1464.20, 27.03
1272.25, 68.16
1136.07, 69.15
887.11, 76.84
639.35, 78.37
529.25, 79.22
730.58, 59.39
720.03, 46.60
3739.94, 52.37
3847.32, 49.52
3696.68, 51.95
3662.56, 51.84
3642.71, 51.48
3603.54, 52.20
3358.23, 39.51
3191.74, 49.06
2849.70, -0.01
2923.21, 0.09
C047C

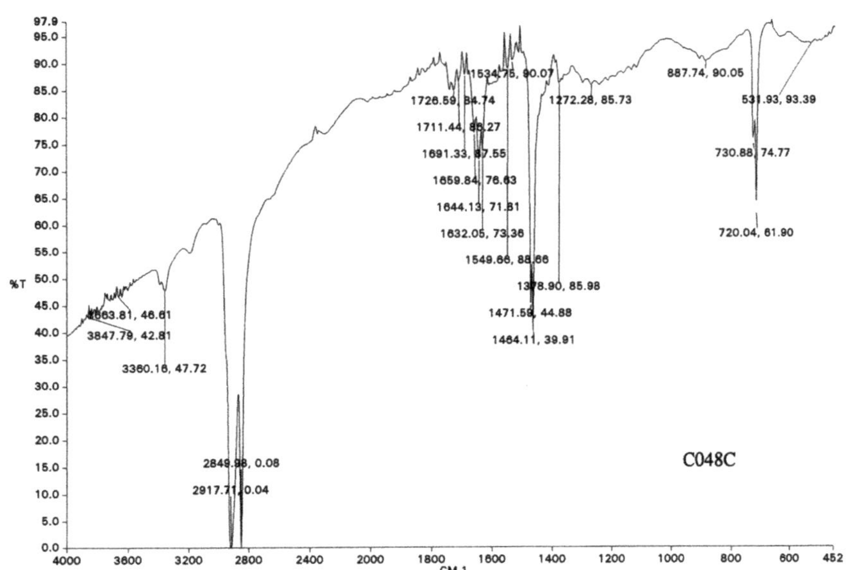
%T
CM-1
1534.75, 90.07
1726.59, 84.74
1711.44, 85.27
1691.33, 87.55
1659.84, 76.63
1644.13, 71.81
1632.05, 73.36
1549.66, 88.66
1378.90, 85.98
1471.59, 44.88
1464.11, 39.91
1272.28, 85.73
887.74, 90.05
531.93, 93.39
730.88, 74.77
720.04, 61.90
3863.81, 46.61
3847.79, 42.81
3360.16, 47.72
2849.98, 0.08
2917.71, 0.04
C048C

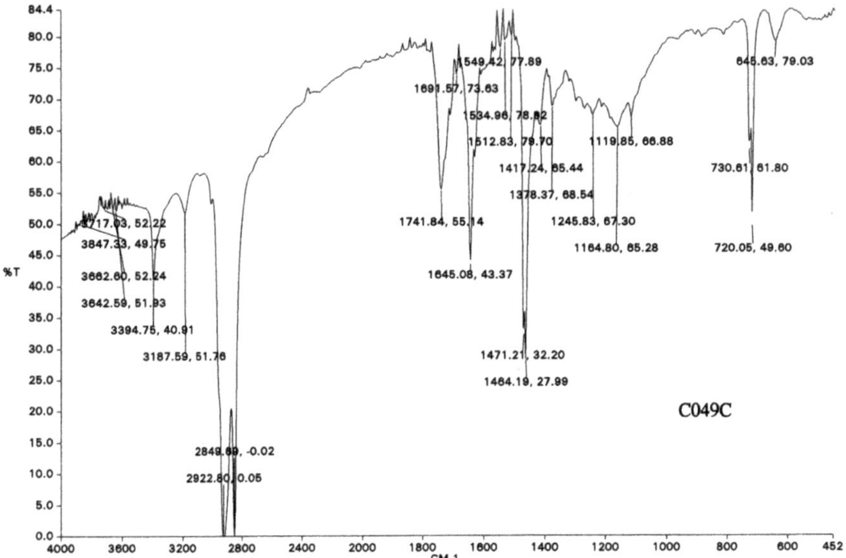
C049C
%T
CM-1

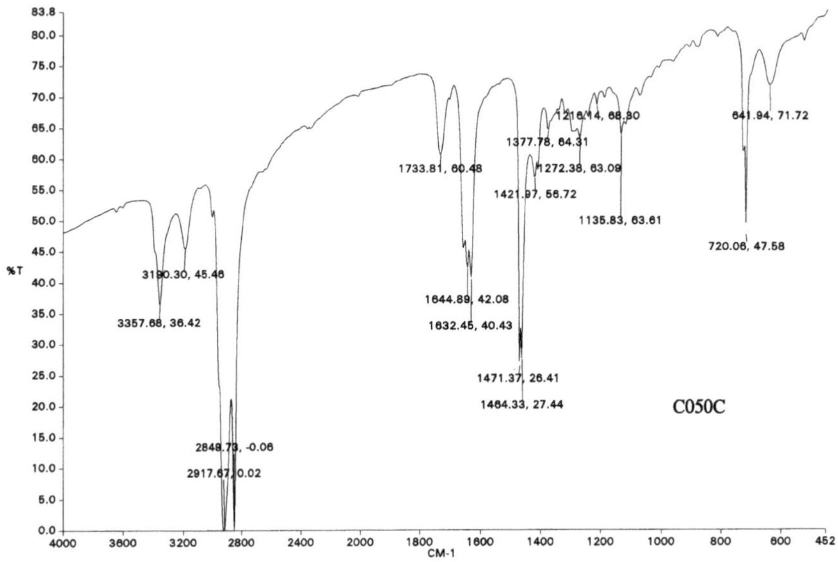
C050C
%T
CM-1

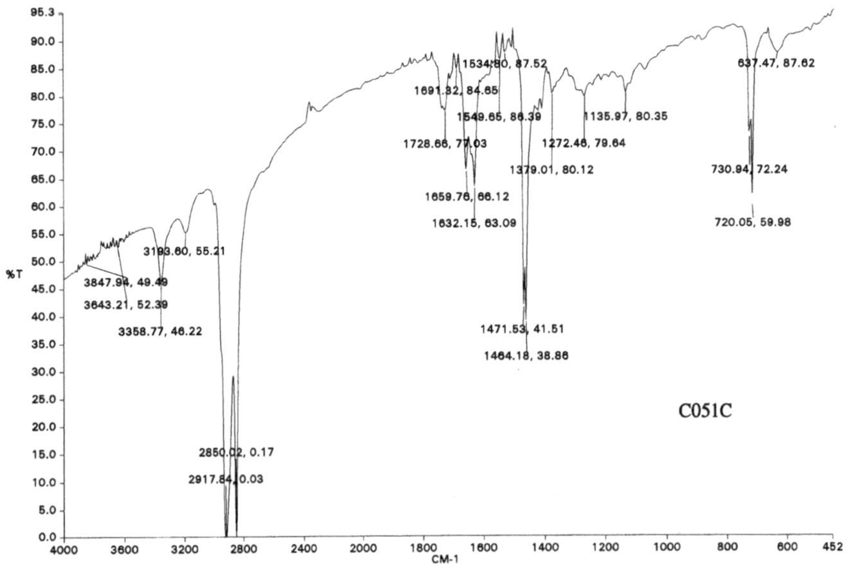
C051C
%T
CM-1

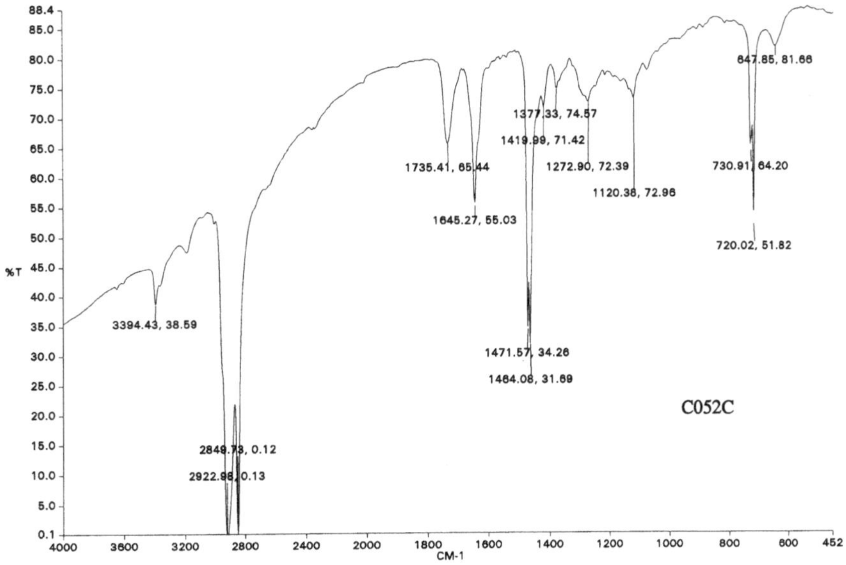
%T
CM-1
647.85, 81.66
1377.33, 74.57
1419.99, 71.42
1735.41, 65.44
1272.90, 72.39
730.91, 64.20
1120.38, 72.96
1645.27, 55.03
720.02, 51.82
3394.43, 38.59
1471.57, 34.26
1464.06, 31.69
C052C
2849.73, 0.12
2922.96, 0.13

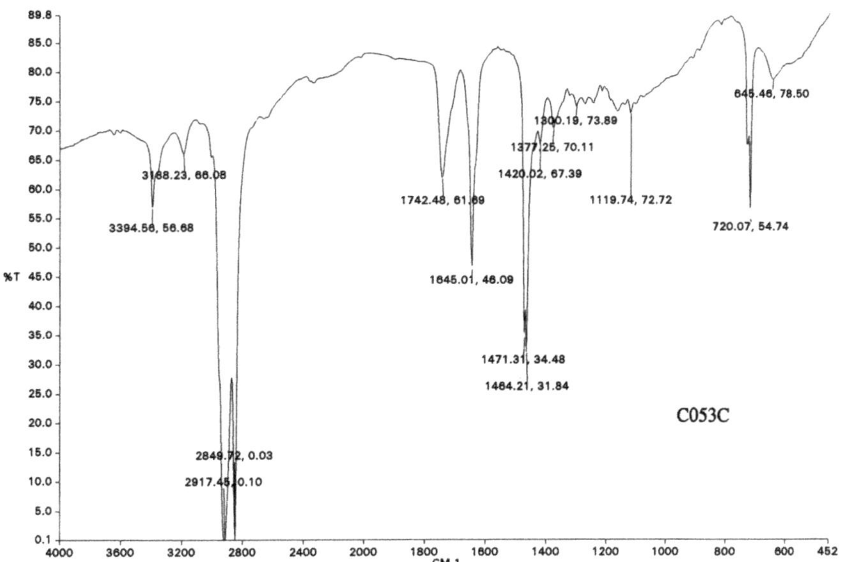
%T
CM-1
645.46, 78.50
1300.19, 73.89
1377.25, 70.11
3168.23, 66.08
1420.02, 67.39
1742.48, 61.69
1119.74, 72.72
3394.56, 56.68
720.07, 54.74
1645.01, 46.09
1471.31, 34.48
1464.21, 31.84
C053C
2849.72, 0.03
2917.45, 0.10

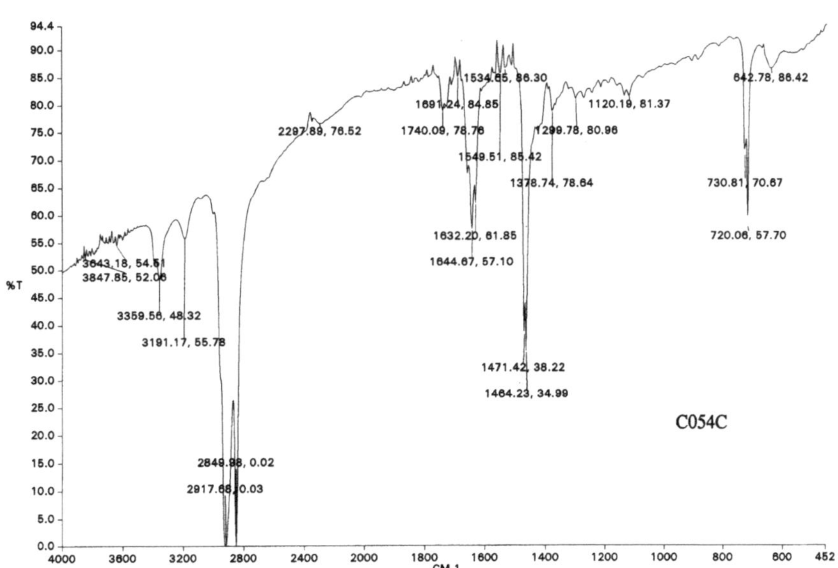
%T
CM-1
1534.65, 86.30
642.78, 86.42
1691.24, 84.85
1120.19, 81.37
2297.89, 76.52
1740.09, 76.76
1299.78, 80.96
1549.51, 85.42
1378.74, 78.64
730.81, 70.67
1632.20, 61.85
720.06, 57.70
1644.67, 57.10
3643.18, 54.51
3847.85, 52.06
3359.56, 48.32
3191.17, 55.78
1471.42, 38.22
1464.23, 34.99
C054C
2849.98, 0.02
2917.66, 0.03

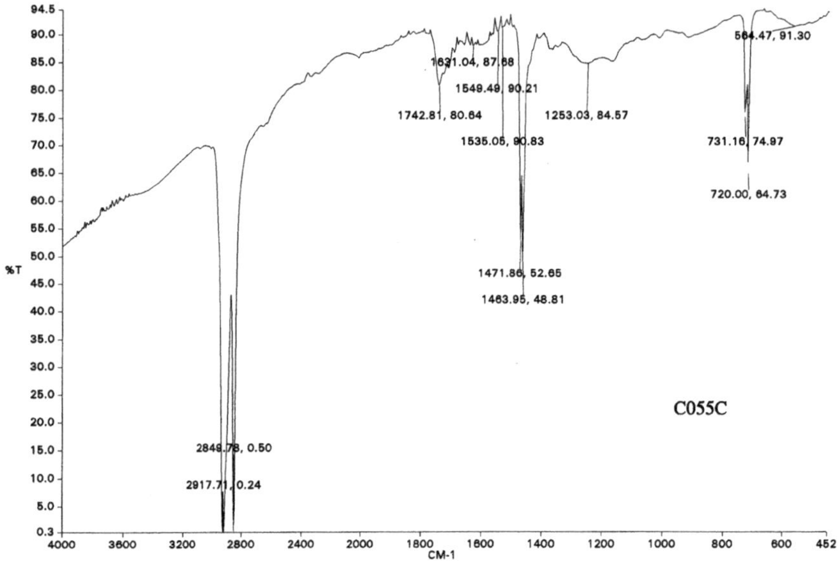

%T
CM-1
1621.04, 87.88
1549.49, 90.21
1742.81, 80.64
1253.03, 84.57
1535.05, 90.83
564.47, 91.30
731.16, 74.97
720.00, 64.73
1471.86, 52.65
1463.95, 48.81
C055C
2849.78, 0.50
2917.71, 0.24

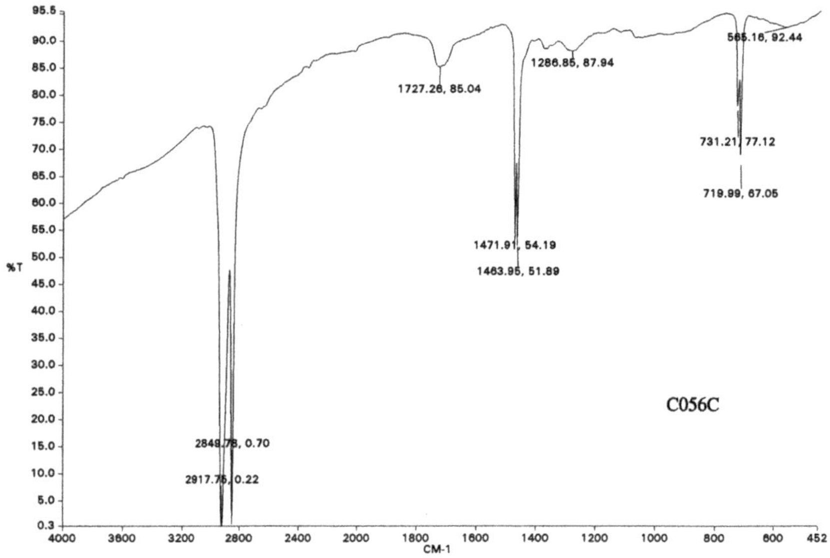

%T
CM-1
1727.26, 85.04
1286.85, 87.94
565.16, 92.44
731.21, 77.12
719.99, 67.05
1471.91, 54.19
1463.95, 51.89
C056C
2849.78, 0.70
2917.75, 0.22

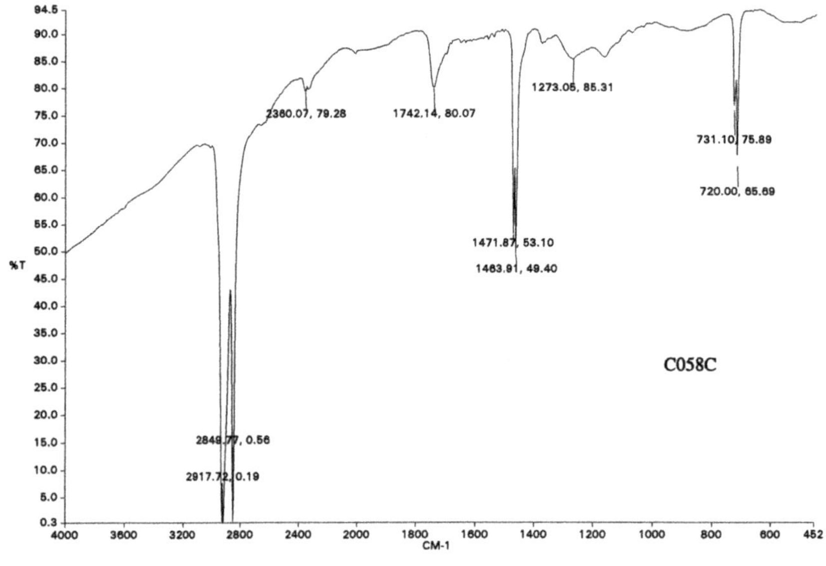

%T
CM-1
1273.05, 85.31
2360.07, 79.28
1742.14, 80.07
731.10, 75.89
720.00, 65.69
1471.87, 53.10
1463.91, 49.40
C058C
2849.77, 0.56
2917.72, 0.19

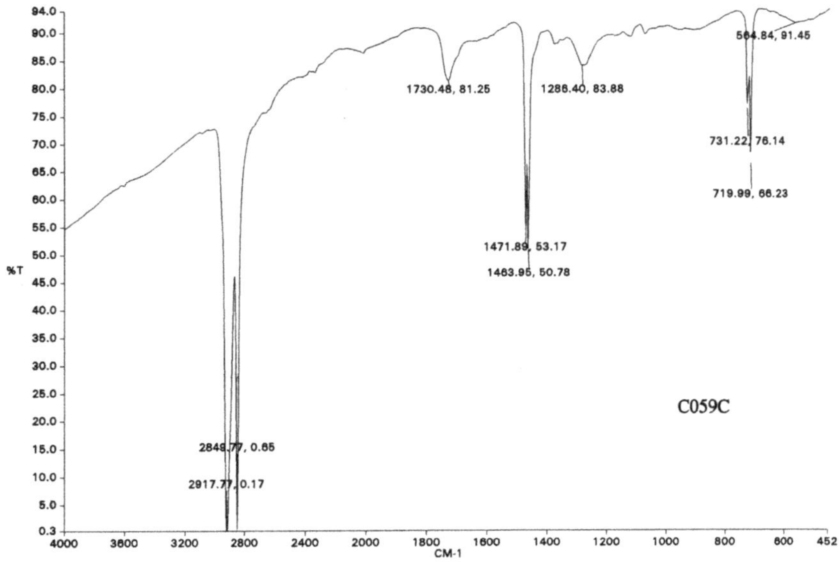
%T
CM-1
564.84, 91.45
1730.48, 81.25
1286.40, 83.88
731.22, 76.14
719.99, 66.23
1471.89, 53.17
1463.95, 50.78
C059C
2849.77, 0.65
2917.77, 0.17

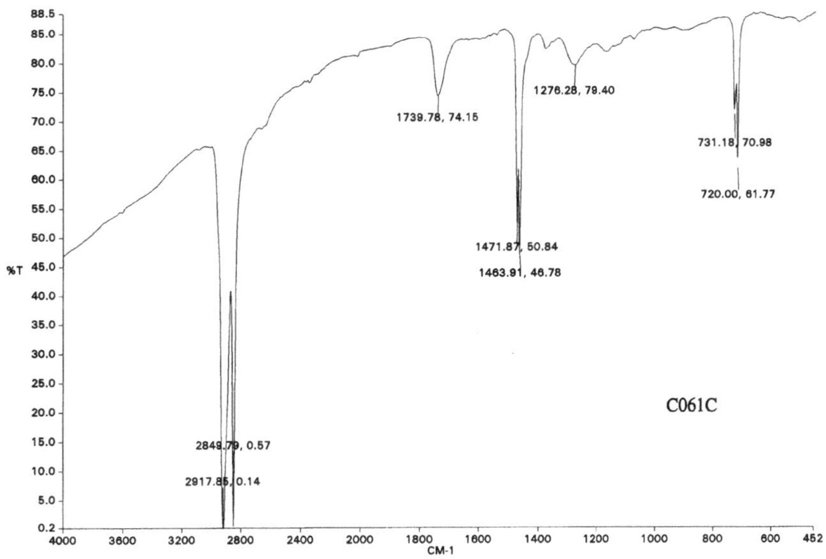
%T
CM-1
1276.28, 79.40
1739.78, 74.15
731.18, 70.98
720.00, 61.77
1471.87, 50.84
1463.91, 46.78
C061C
2849.79, 0.57
2917.85, 0.14

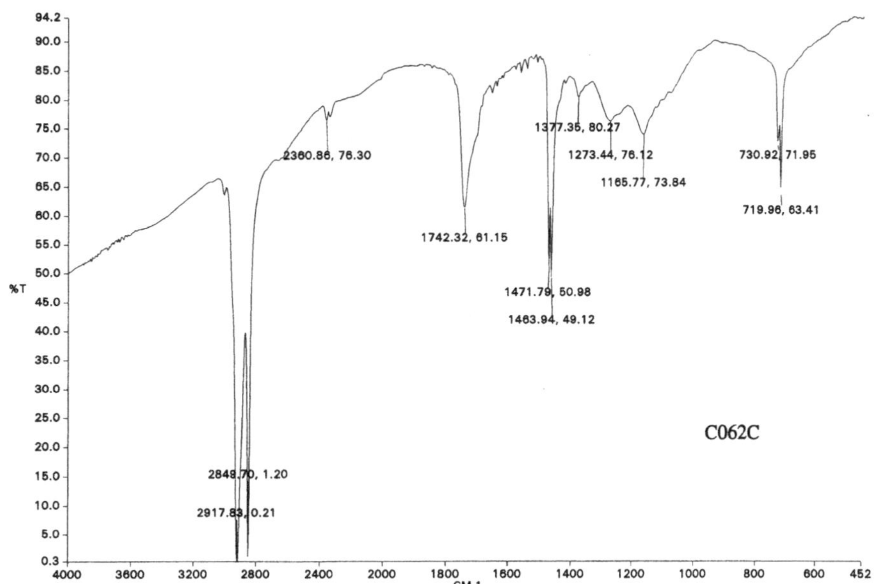
%T
CM-1
1377.35, 80.27
2360.86, 76.30
1273.44, 76.12
730.92, 71.95
1165.77, 73.84
719.96, 63.41
1742.32, 61.15
1471.79, 50.98
1463.94, 49.12
C062C
2849.70, 1.20
2917.83, 0.21

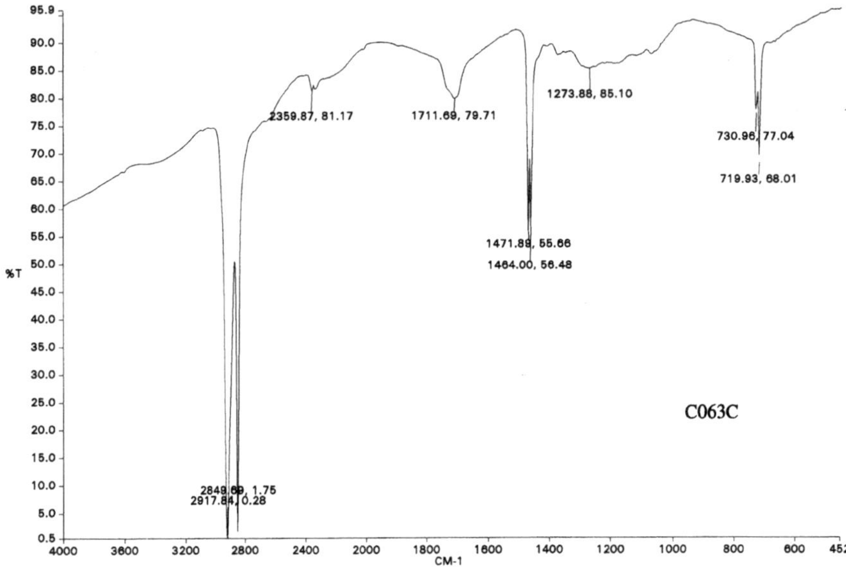
2359.87, 81.17
1711.69, 79.71
1273.88, 85.10
730.96, 77.04
719.93, 68.01
1471.89, 55.66
1464.00, 56.48
%T
C063C
2849.69, 1.75
2917.84, 0.28
CM-1

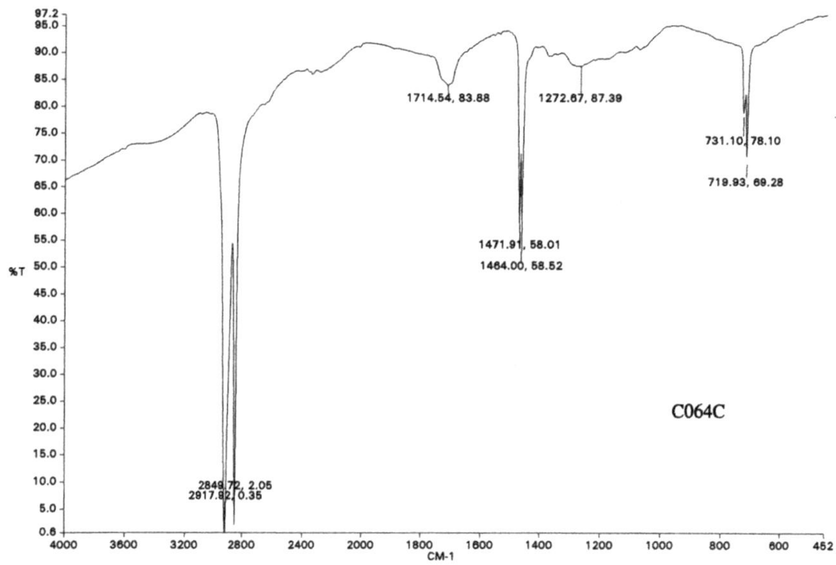
1714.54, 83.88
1272.67, 87.39
731.10, 78.10
719.93, 69.28
1471.91, 58.01
1464.00, 58.52
%T
C064C
2849.72, 2.05
2917.82, 0.35
CM-1

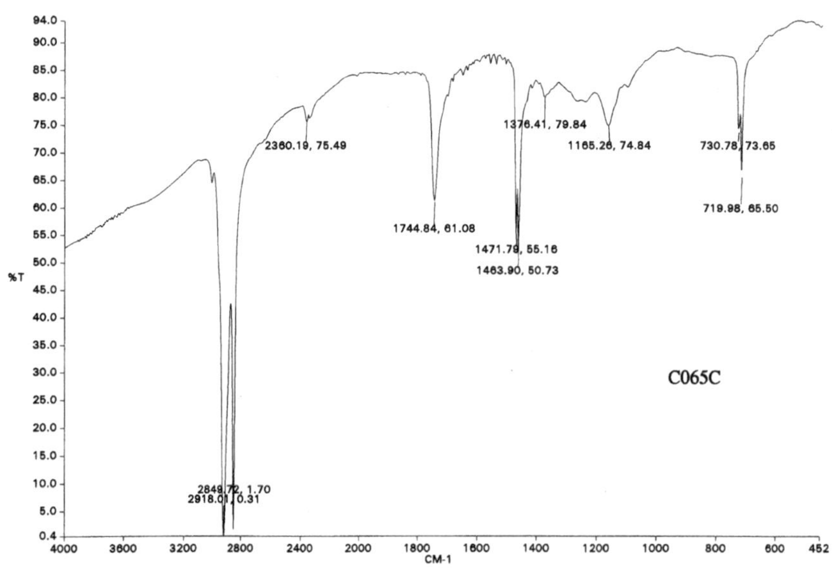
1376.41, 79.84
2360.19, 75.49
1165.26, 74.84
730.78, 73.65
719.98, 65.50
1744.84, 61.08
1471.79, 55.16
1463.90, 50.73
%T
C065C
2849.72, 1.70
2918.01, 0.31
CM-1

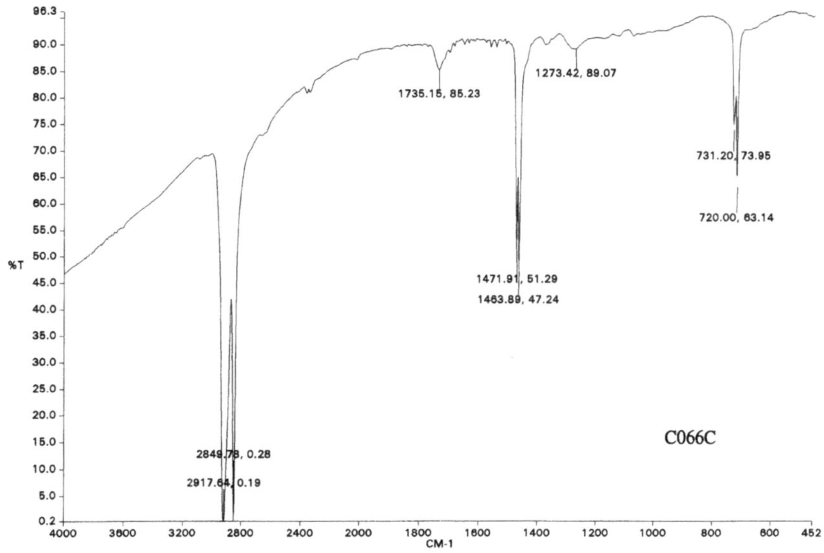
1735.15, 85.23
1273.42, 89.07
731.20, 73.95
720.00, 63.14
1471.91, 51.29
1463.89, 47.24
C066C
2849.78, 0.28
2917.64, 0.19
%T
CM-1

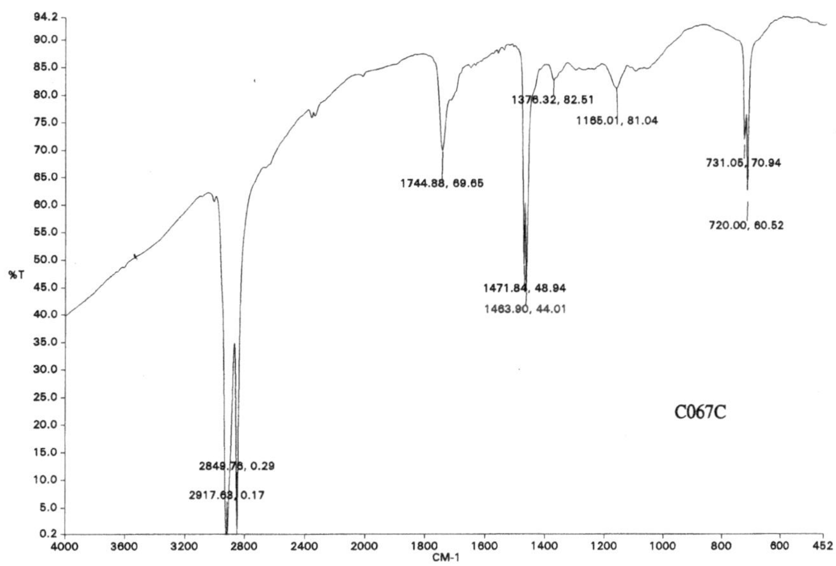
1376.32, 82.51
1165.01, 81.04
731.05, 70.94
1744.88, 69.65
720.00, 60.52
1471.84, 48.94
1463.90, 44.01
C067C
2849.78, 0.29
2917.68, 0.17
%T
CM-1

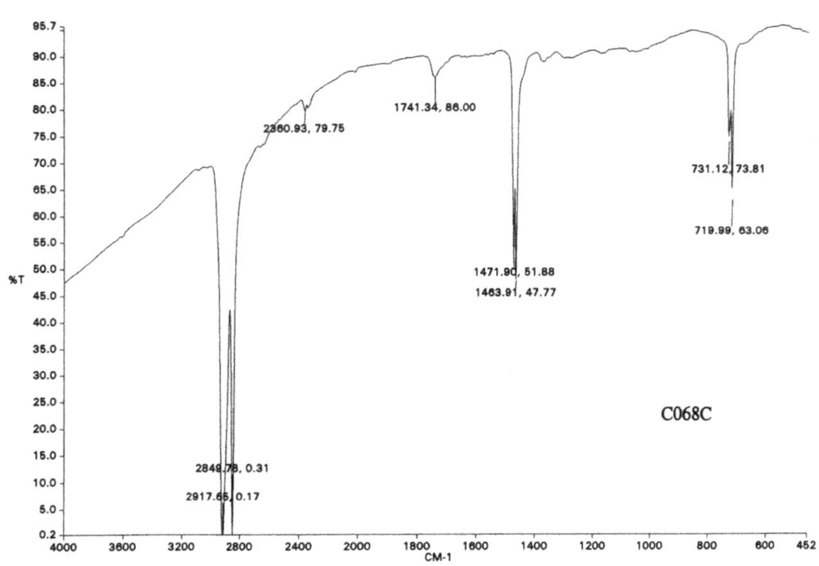
1741.34, 86.00
2360.93, 79.75
731.12, 73.81
719.99, 63.06
1471.90, 51.88
1463.91, 47.77
C068C
2849.78, 0.31
2917.65, 0.17
%T
CM-1

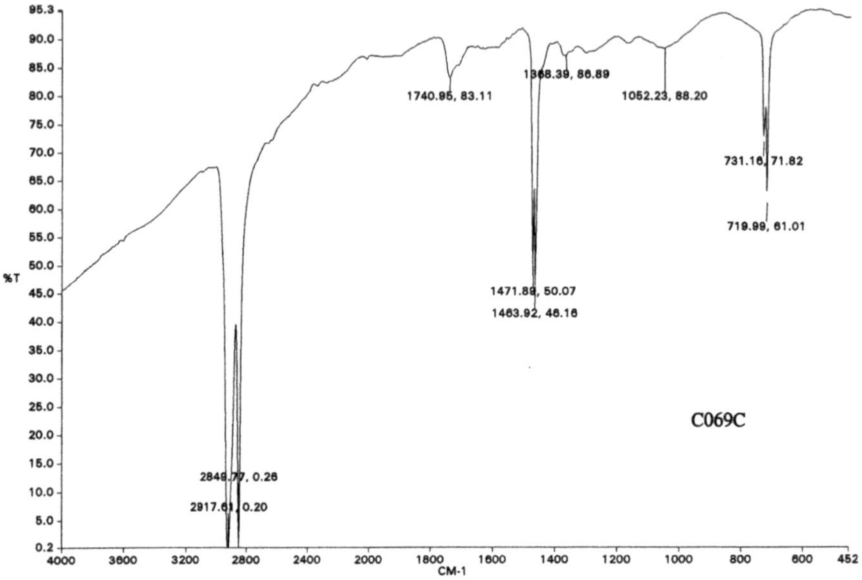
C069C
%T
CM-1
1740.95, 83.11
1368.39, 86.89
1052.23, 88.20
731.16, 71.82
719.99, 61.01
1471.89, 50.07
1463.92, 46.16
2849.77, 0.26
2917.61, 0.20

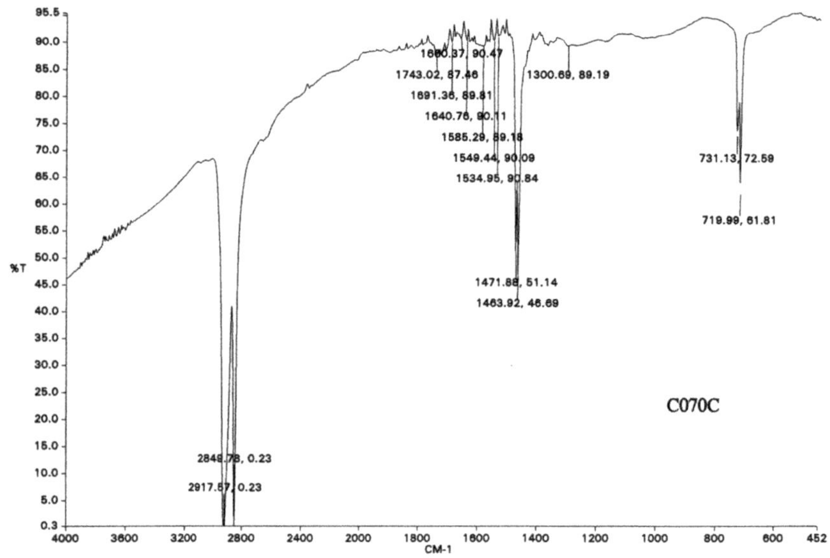
C070C
%T
CM-1
1743.02, 87.46
1300.69, 89.19
731.13, 72.59
719.99, 61.81
1471.88, 51.14
1463.92, 46.69
2849.78, 0.23
2917.57, 0.23

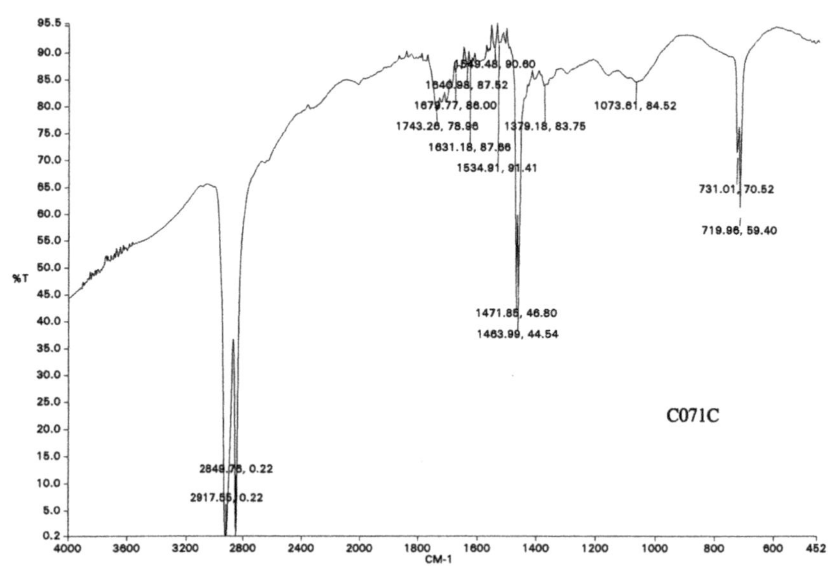
C071C
%T
CM-1
1073.61, 84.52
1743.26, 78.96
1379.18, 83.75
1631.18, 87.86
1534.91, 91.41
731.01, 70.52
719.96, 59.40
1471.85, 46.80
1463.99, 44.54
2849.76, 0.22
2917.55, 0.22

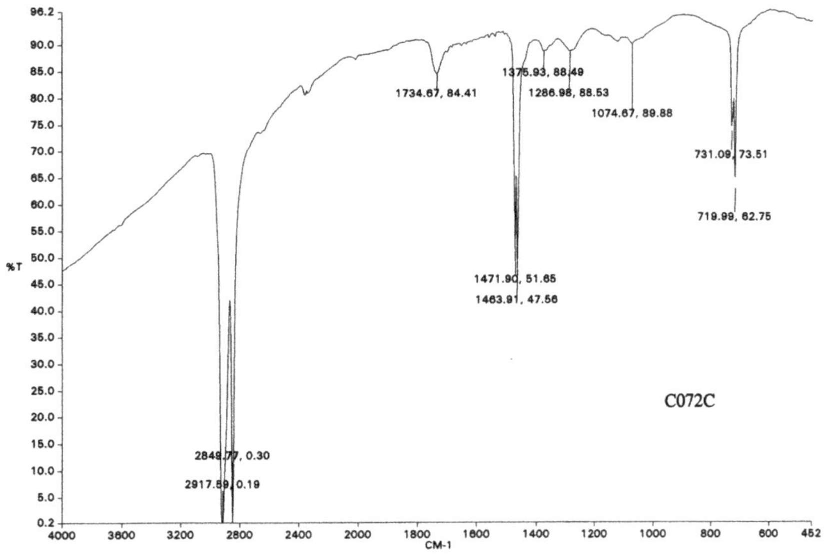
%T
CM-1
1734.67, 84.41
1375.93, 88.49
1286.98, 88.53
1074.67, 89.88
731.09, 73.51
719.99, 62.75
1471.90, 51.65
1463.91, 47.56
2849.77, 0.30
2917.59, 0.19
C072C

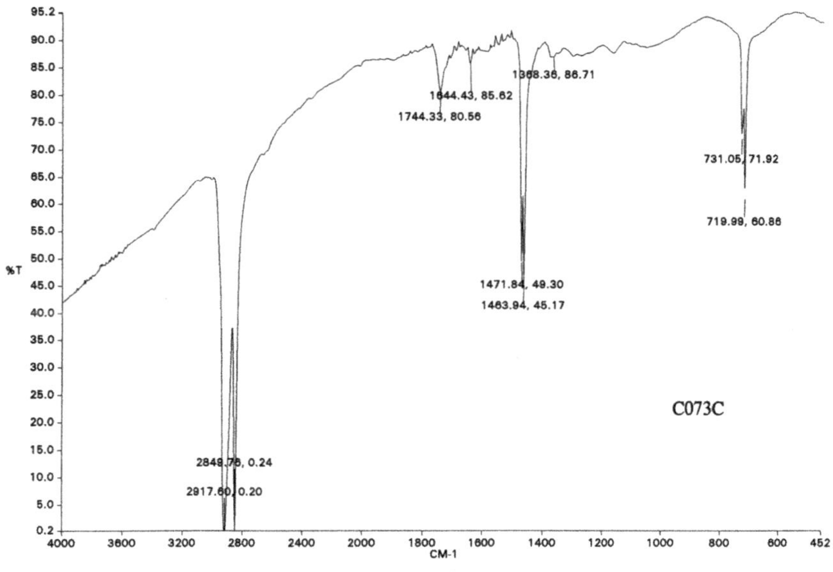
%T
CM-1
1368.36, 86.71
1644.43, 85.62
1744.33, 80.56
731.05, 71.92
719.99, 60.86
1471.84, 49.30
1463.94, 45.17
2849.76, 0.24
2917.60, 0.20
C073C

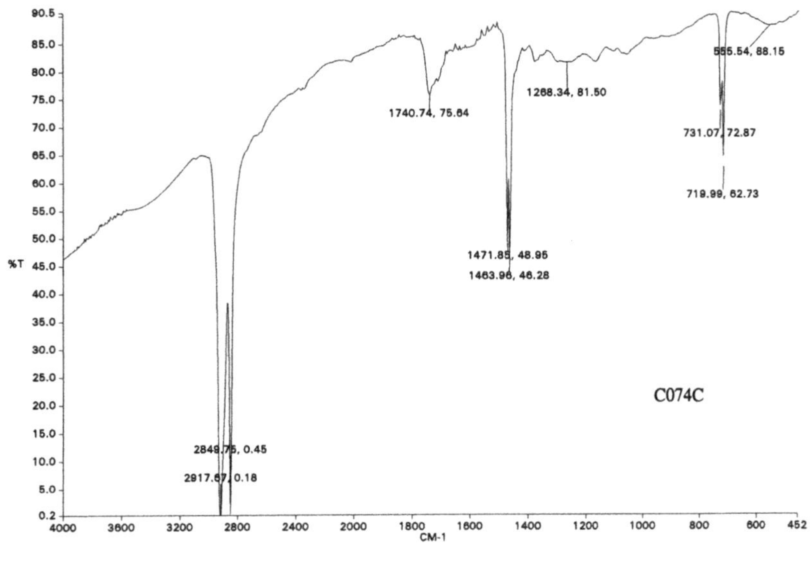
%T
CM-1
555.54, 88.15
1268.34, 81.50
1740.74, 75.64
731.07, 72.87
719.99, 62.73
1471.85, 48.95
1463.96, 46.28
2849.75, 0.45
2917.57, 0.18
C074C

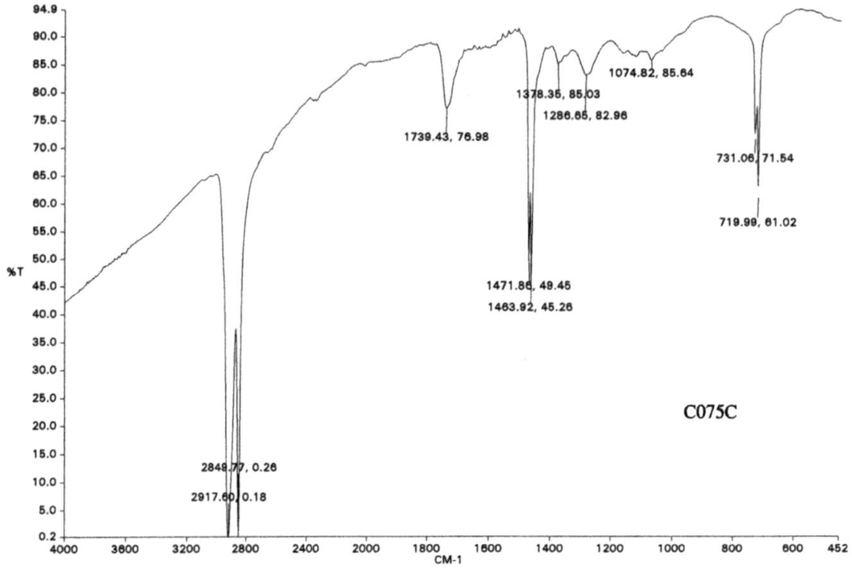
1739.43, 76.98
1378.35, 85.03
1286.65, 82.96
1074.82, 85.64
731.06, 71.54
719.99, 61.02
1471.86, 49.45
1463.92, 45.26
2848.77, 0.26
2917.60, 0.18
%T
CM-1
C075C

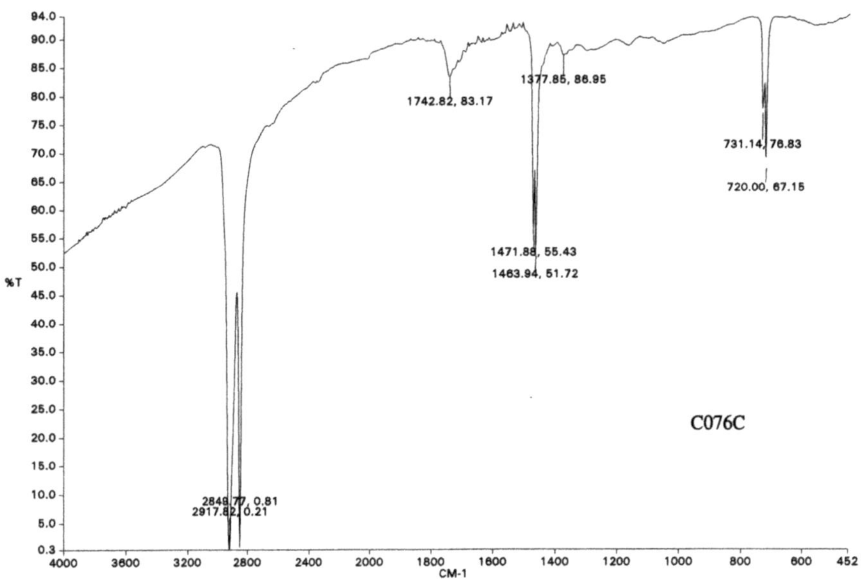
1742.82, 83.17
1377.85, 86.95
731.14, 76.83
720.00, 67.15
1471.88, 55.43
1463.94, 51.72
2848.77, 0.81
2917.82, 0.21
%T
CM-1
C076C

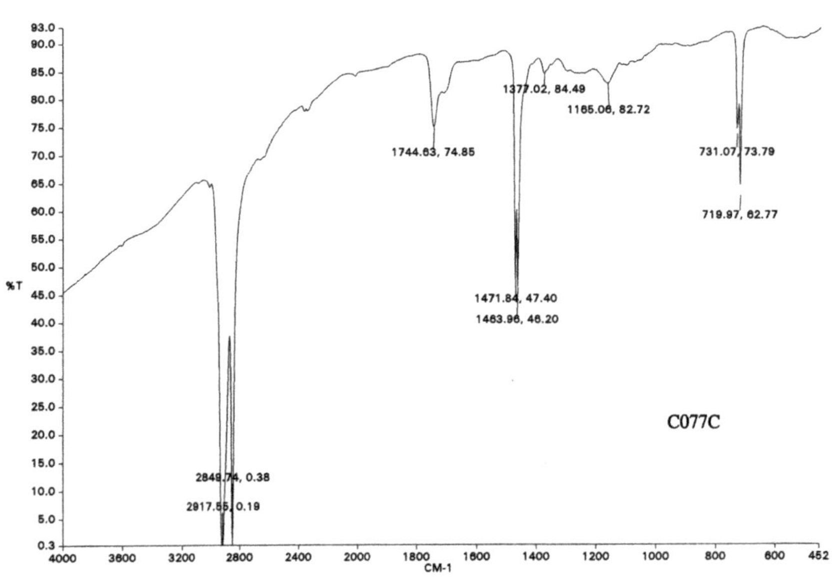
1744.63, 74.85
1377.02, 84.49
1165.06, 82.72
731.07, 73.79
719.97, 62.77
1471.84, 47.40
1463.96, 46.20
2848.74, 0.38
2917.55, 0.19
%T
CM-1
C077C

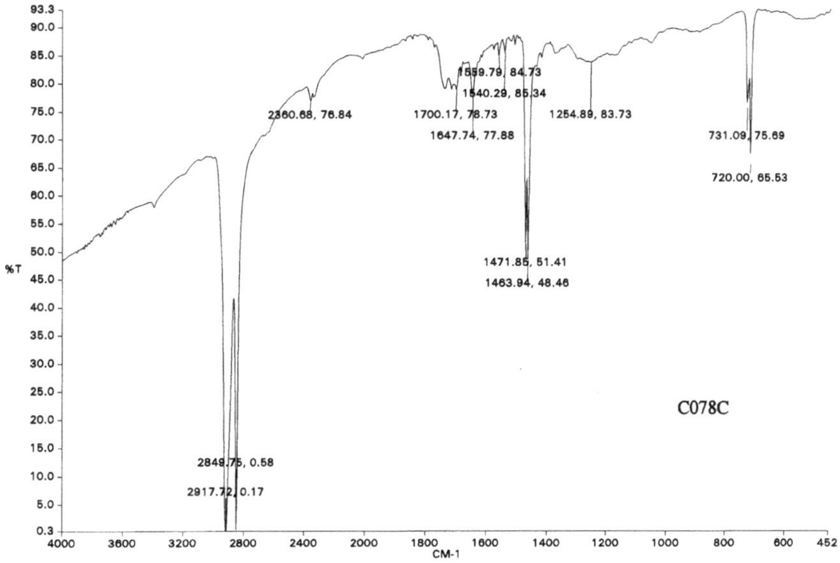

%T
CM-1
2360.68, 76.84
1559.79, 84.73
1540.29, 85.34
1700.17, 78.73
1647.74, 77.88
1254.89, 83.73
731.09, 75.69
720.00, 65.53
1471.85, 51.41
1463.94, 48.46
2849.75, 0.58
2917.72, 0.17
C078C

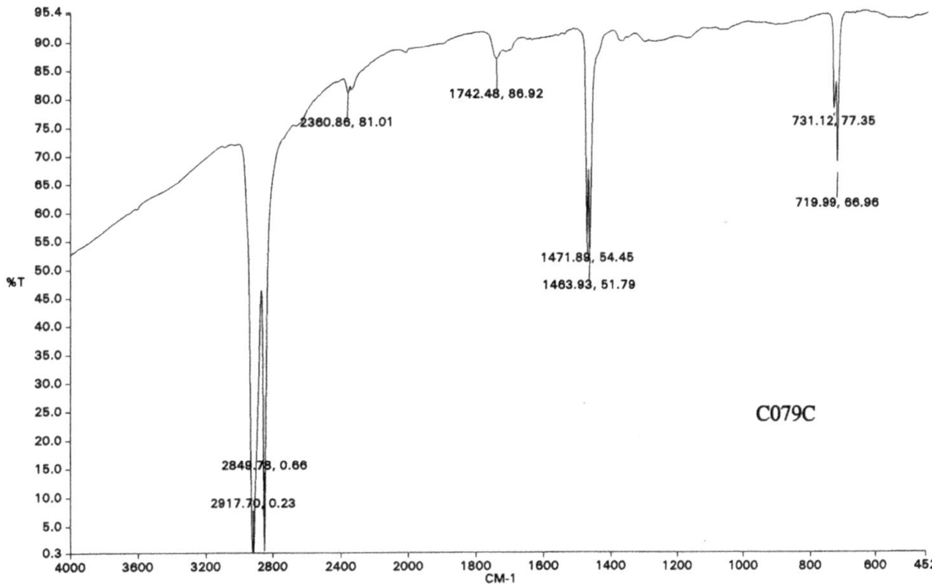

%T
CM-1
1742.48, 86.92
2360.86, 81.01
731.12, 77.35
719.99, 66.96
1471.89, 54.45
1463.93, 51.79
2849.78, 0.66
2917.70, 0.23
C079C

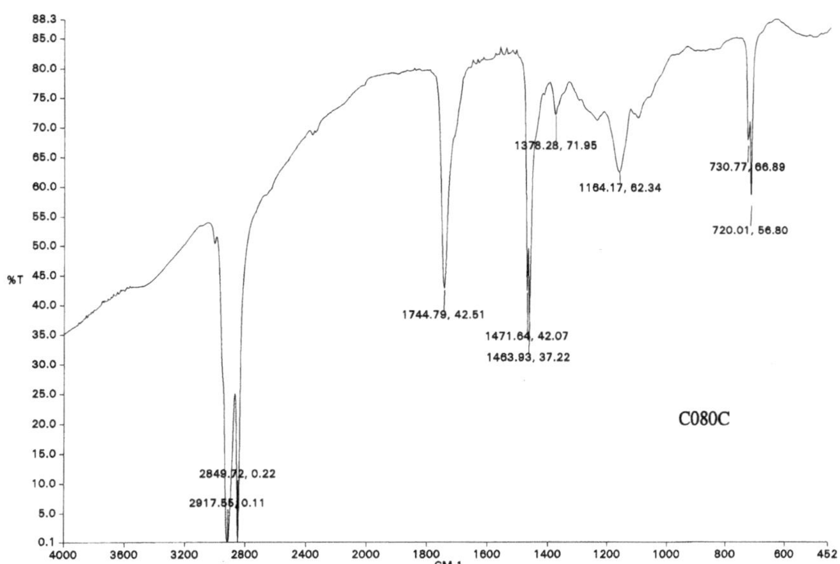

%T
CM-1
1378.28, 71.95
730.77, 66.89
1164.17, 62.34
720.01, 56.80
1744.79, 42.51
1471.64, 42.07
1463.93, 37.22
2849.72, 0.22
2917.55, 0.11
C080C

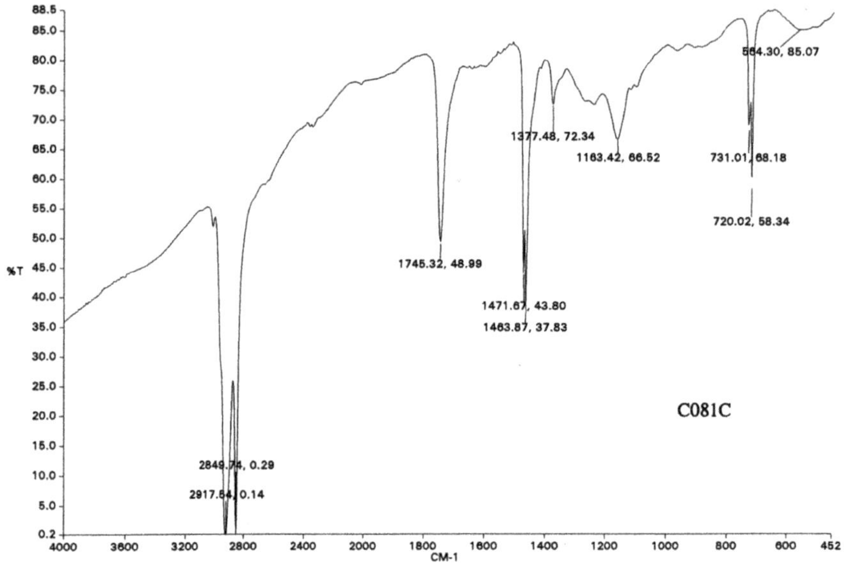
%T
CM-1
564.30, 85.07
1377.48, 72.34
1163.42, 66.52
731.01, 68.18
720.02, 58.34
1745.32, 48.99
1471.67, 43.80
1463.87, 37.83
C081C
2849.74, 0.29
2917.54, 0.14

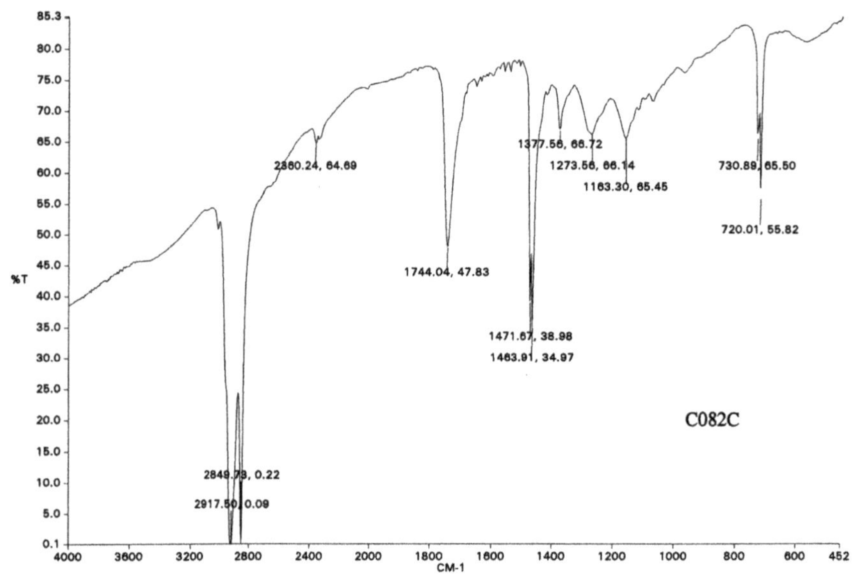
%T
CM-1
1377.56, 66.72
2360.24, 64.69
1273.56, 66.14
1163.30, 65.45
730.89, 65.50
720.01, 55.82
1744.04, 47.83
1471.67, 38.98
1463.91, 34.97
C082C
2849.73, 0.22
2917.50, 0.09

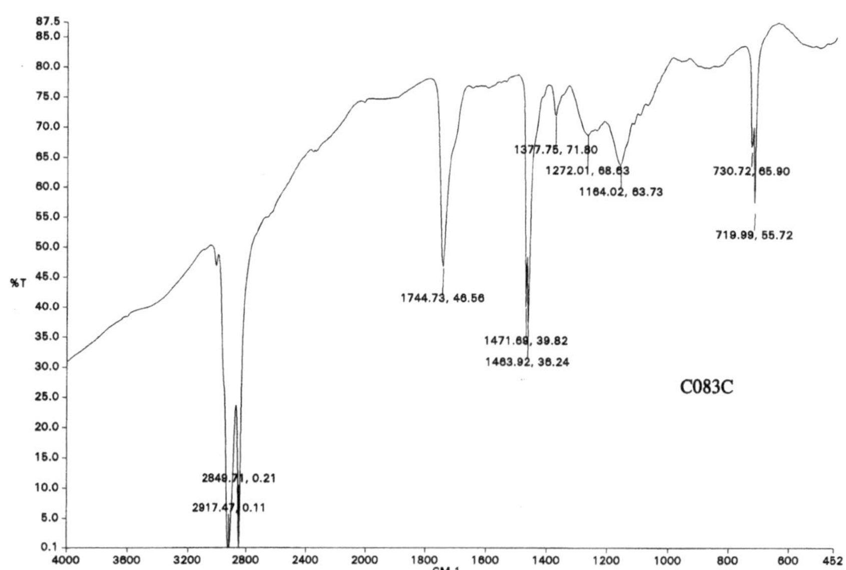
%T
CM-1
1377.75, 71.80
1272.01, 68.63
1164.02, 63.73
730.72, 65.90
719.99, 55.72
1744.73, 46.56
1471.69, 39.82
1463.92, 36.24
C083C
2849.71, 0.21
2917.47, 0.11

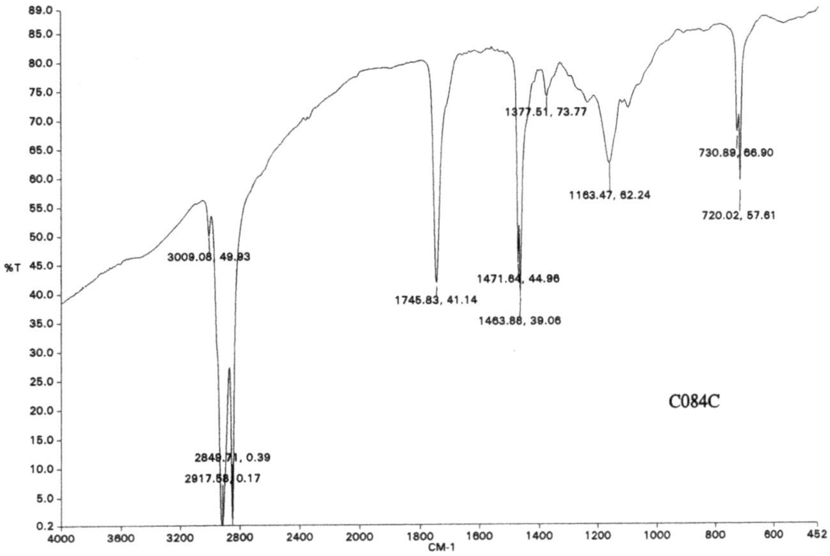
%T
CM-1
1377.51, 73.77
730.89, 66.90
1163.47, 62.24
720.02, 57.61
3009.08, 49.93
1471.84, 44.96
1745.83, 41.14
1463.88, 39.06
C084C
2849.71, 0.39
2917.58, 0.17

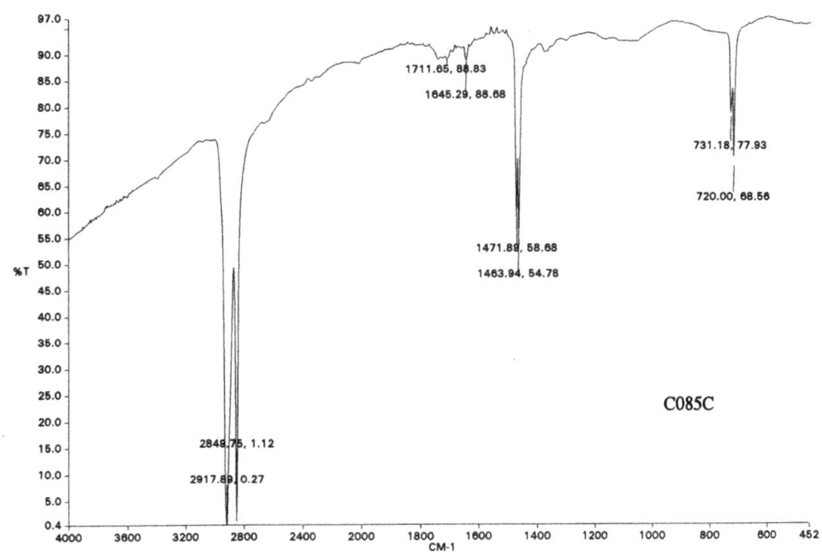
%T
CM-1
1711.65, 88.83
1645.29, 88.68
731.18, 77.93
720.00, 68.56
1471.89, 58.68
1463.94, 54.76
C085C
2849.75, 1.12
2917.89, 0.27

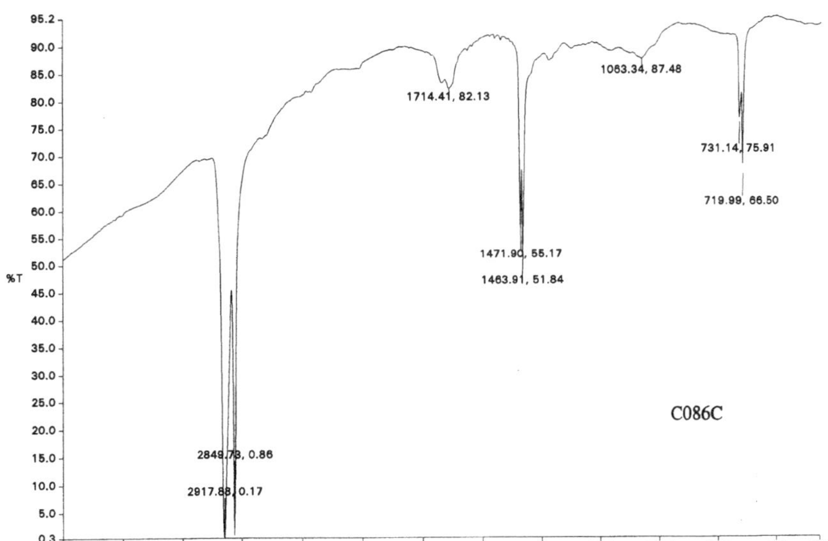
%T
CM-1
1063.34, 87.48
1714.41, 82.13
731.14, 75.91
719.99, 66.50
1471.90, 55.17
1463.91, 51.84
C086C
2849.73, 0.86
2917.88, 0.17

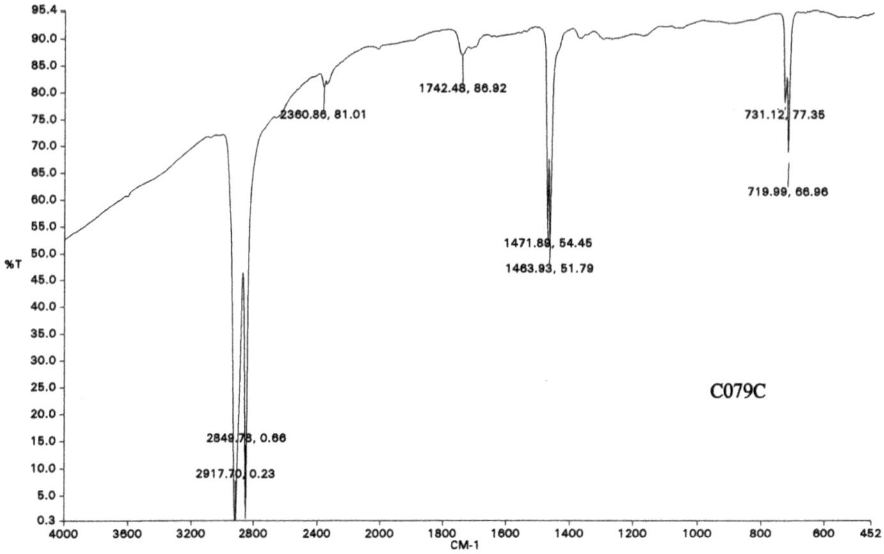

2360.86, 81.01
1742.48, 86.92
731.12, 77.35
719.99, 66.96
1471.89, 54.45
1463.93, 51.79
2849.78, 0.66
2917.70, 0.23
C079C
%T
CM-1

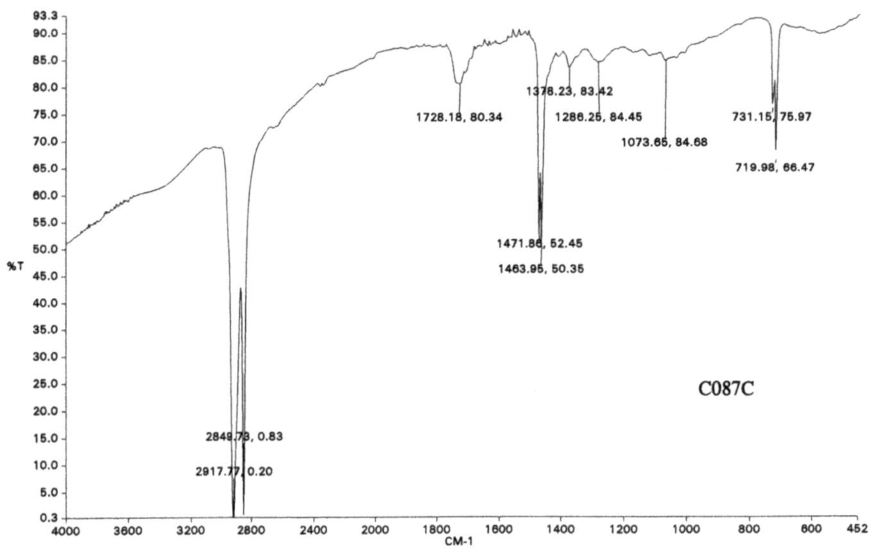

1378.23, 83.42
1728.18, 80.34
1286.25, 84.45
1073.65, 84.68
731.15, 75.97
719.98, 66.47
1471.86, 52.45
1463.95, 50.35
2849.73, 0.83
2917.77, 0.20
C087C
%T
CM-1

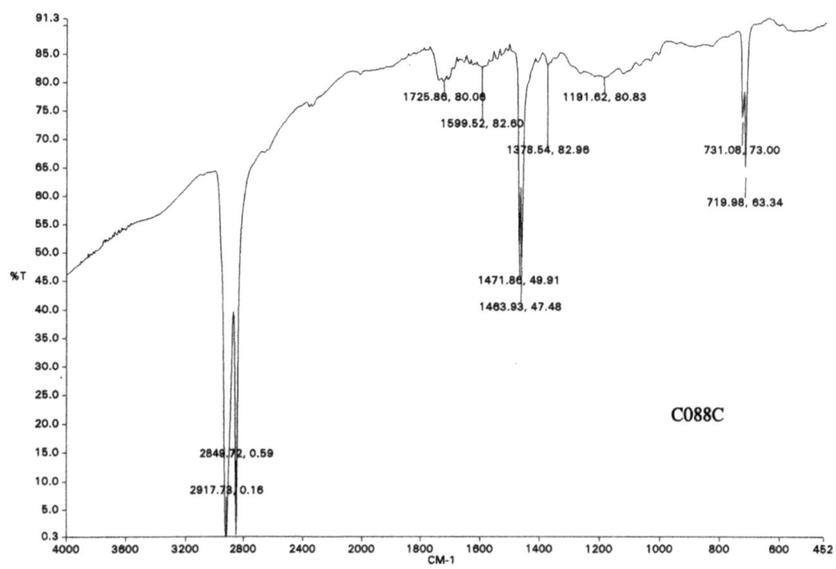

1725.86, 80.06
1599.52, 82.60
1191.62, 80.83
1378.54, 82.96
731.08, 73.00
719.98, 63.34
1471.86, 49.91
1463.93, 47.48
2849.72, 0.59
2917.78, 0.16
C088C
%T
CM-1

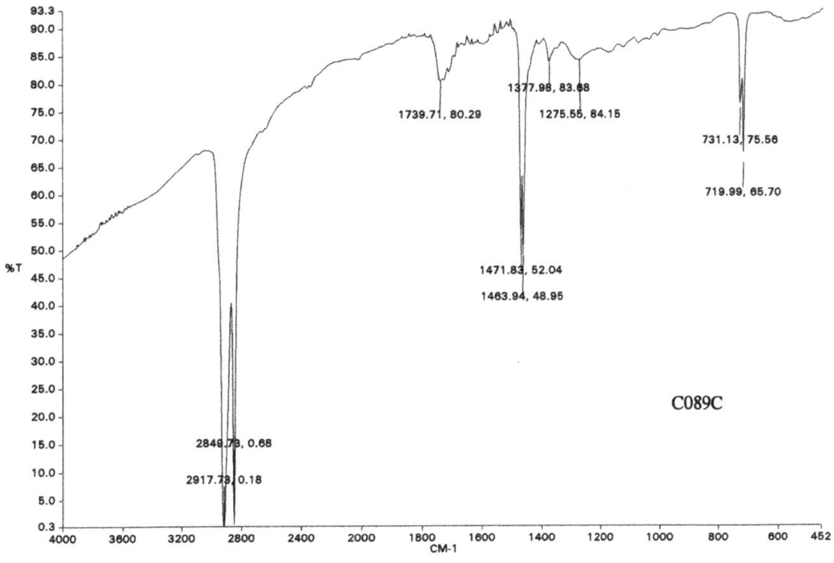
%T
CM-1
1739.71, 80.29
1377.98, 83.88
1275.55, 84.15
731.13, 75.56
719.99, 65.70
1471.83, 52.04
1463.94, 48.95
C089C
2849.73, 0.68
2917.73, 0.18

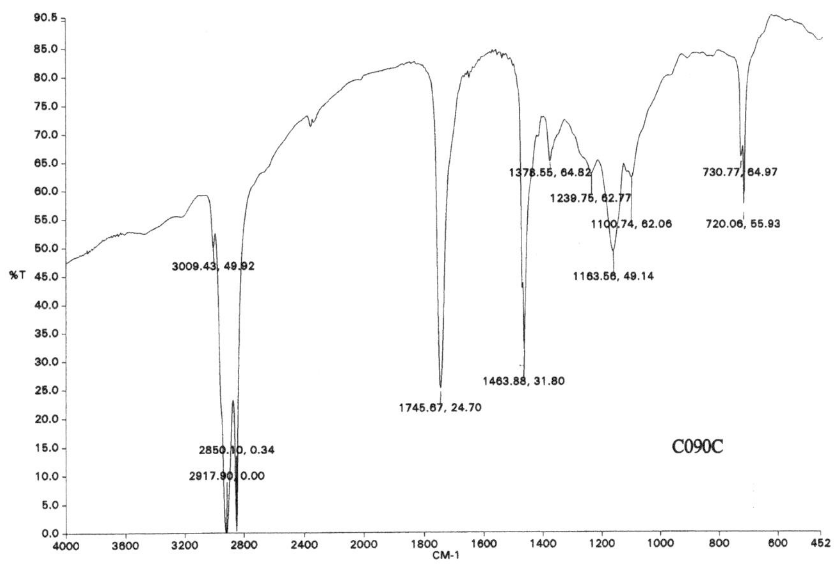
%T
CM-1
1378.55, 64.82
1239.75, 62.77
1100.74, 62.06
1163.56, 49.14
730.77, 64.97
720.06, 55.93
3009.43, 49.92
1463.88, 31.80
1745.67, 24.70
C090C
2850.10, 0.34
2917.90, 0.00

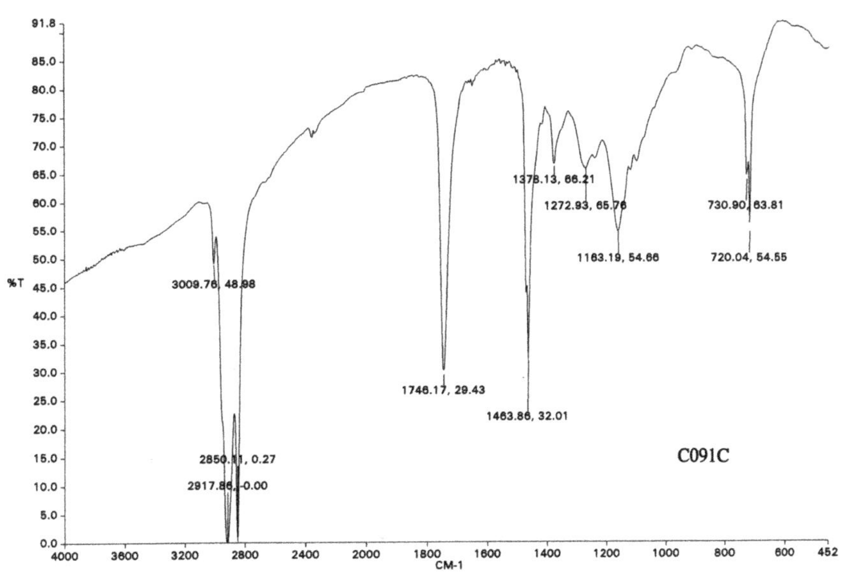
%T
CM-1
1378.13, 66.21
1272.93, 65.76
1163.19, 54.66
730.90, 63.81
720.04, 54.55
3009.76, 48.98
1746.17, 29.43
1463.86, 32.01
C091C
2850.11, 0.27
2917.86, -0.00

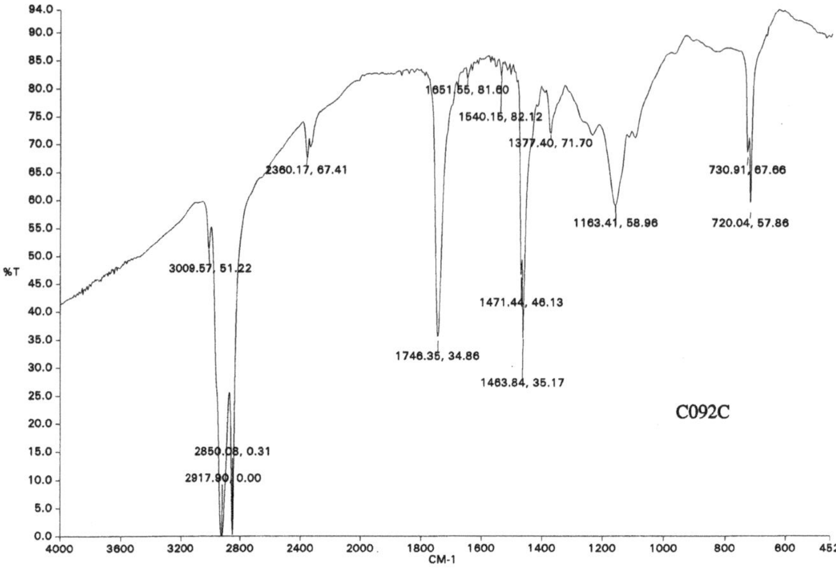
%T
CM-1
1651.55, 81.60
1540.15, 82.12
1377.40, 71.70
2360.17, 67.41
730.91, 67.66
1163.41, 58.96
720.04, 57.86
3009.57, 51.22
1471.44, 46.13
1746.35, 34.86
1463.84, 35.17
C092C
2850.08, 0.31
2917.90, 0.00

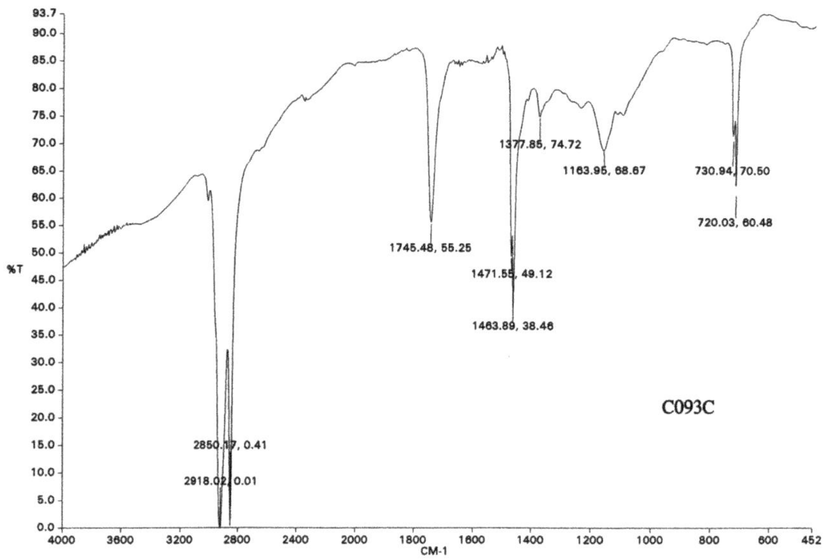
%T
CM-1
1377.85, 74.72
1163.95, 68.67
730.94, 70.50
720.03, 60.48
1745.48, 55.25
1471.55, 49.12
1463.89, 38.46
C093C
2850.17, 0.41
2918.02, 0.01

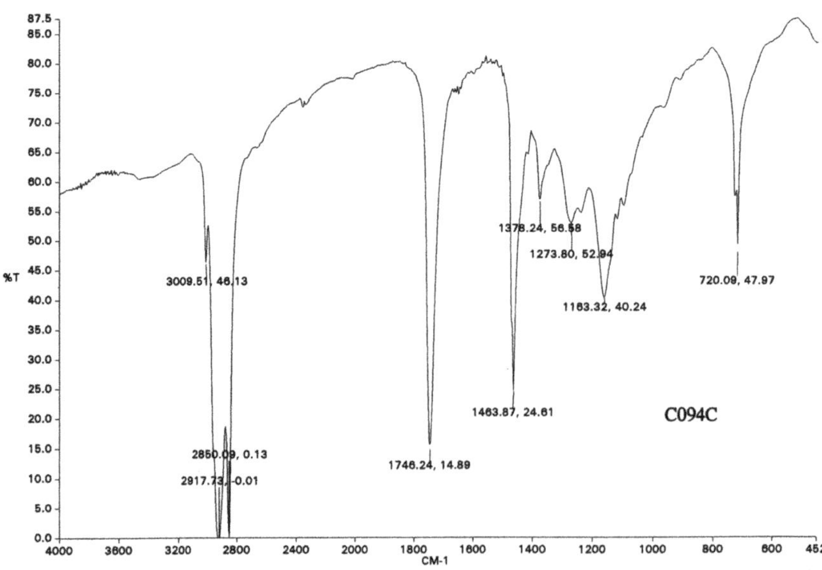
%T
CM-1
1378.24, 56.58
1273.80, 52.94
3009.51, 46.13
720.09, 47.97
1163.32, 40.24
1463.87, 24.61
C094C
2850.09, 0.13
1746.24, 14.89
2917.73, -0.01

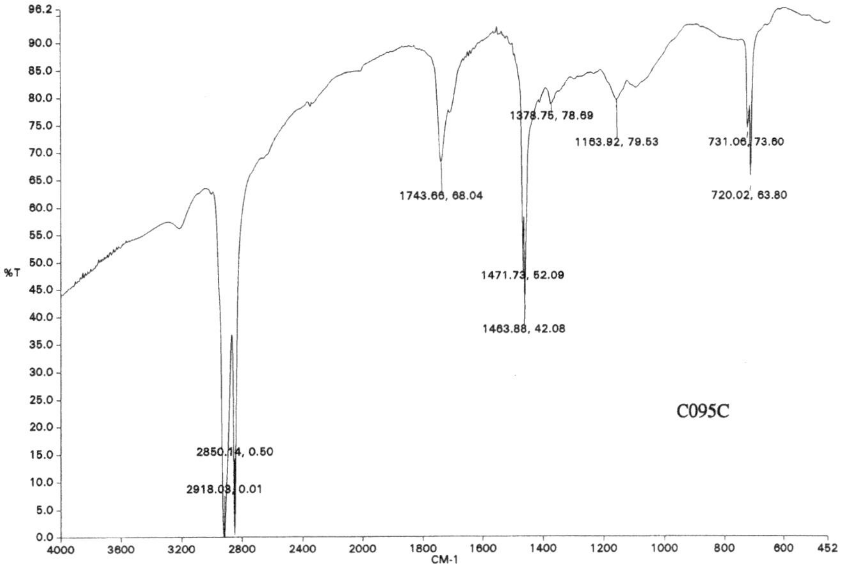
%T
CM-1
1378.75, 78.69
1163.92, 79.53
731.06, 73.60
1743.66, 68.04
720.02, 63.80
1471.73, 52.09
1463.88, 42.08
C095C
2850.14, 0.50
2918.03, 0.01

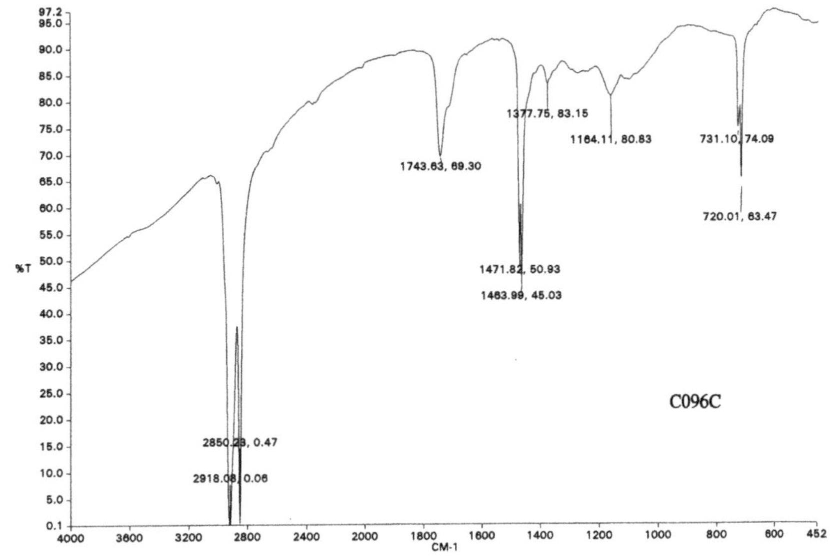
%T
CM-1
1377.75, 83.15
1164.11, 80.83
731.10, 74.09
1743.63, 69.30
720.01, 63.47
1471.82, 50.93
1463.99, 45.03
C096C
2850.23, 0.47
2918.08, 0.06

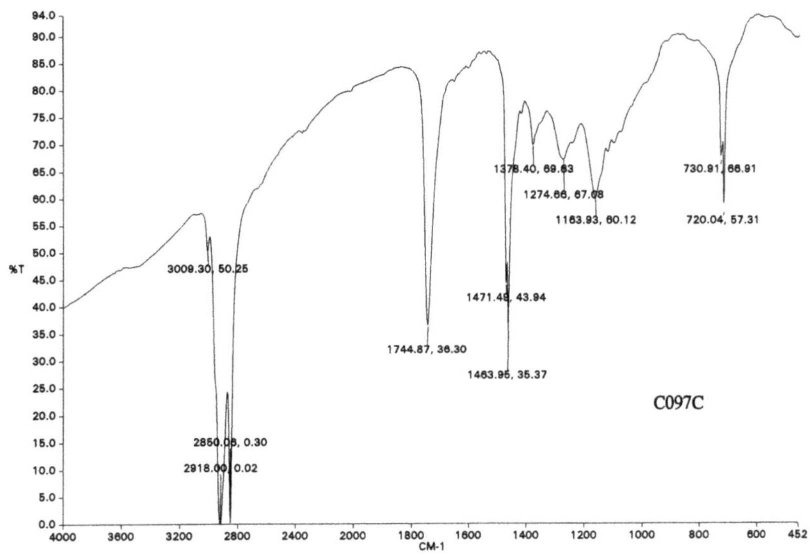
%T
CM-1
1378.40, 69.63
730.91, 66.91
1274.66, 67.08
1163.93, 60.12
720.04, 57.31
3009.30, 50.25
1471.48, 43.94
1744.87, 36.30
1463.95, 35.37
C097C
2850.08, 0.30
2918.00, 0.02

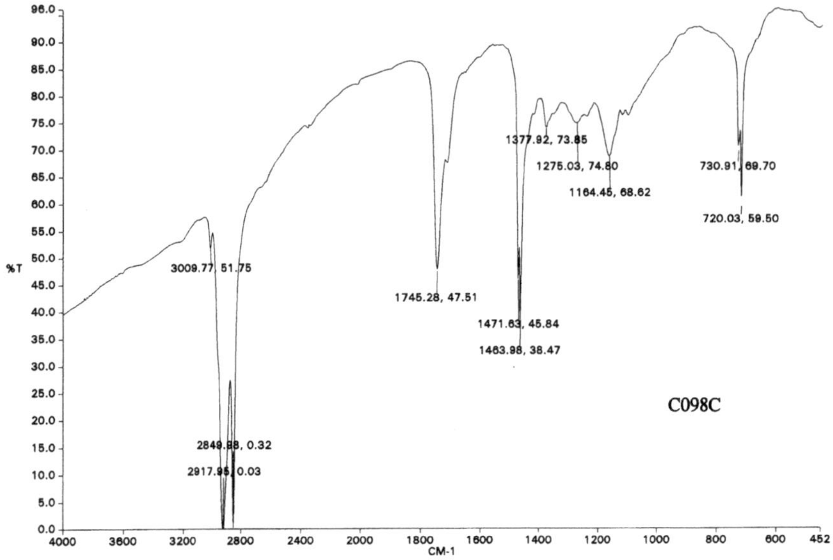
C098C

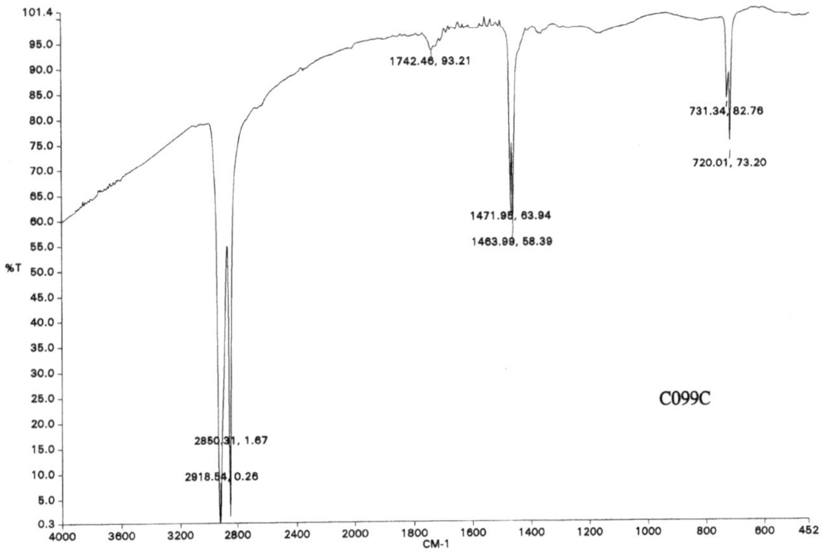
C099C

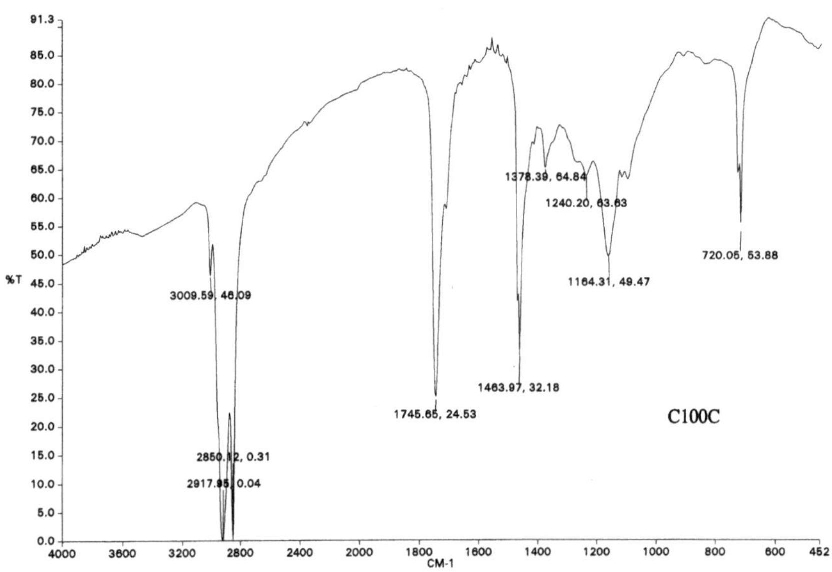
C100C

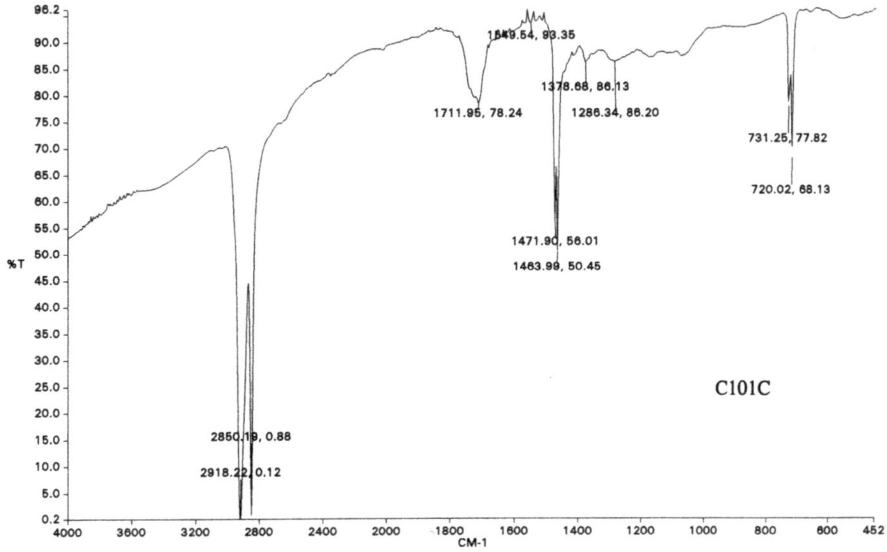
1711.95, 78.24
1378.68, 86.13
1286.34, 86.20
731.25, 77.82
720.02, 68.13
1471.90, 56.01
1463.99, 50.45
C101C
2850.19, 0.88
2918.22, 0.12
%T
CM-1

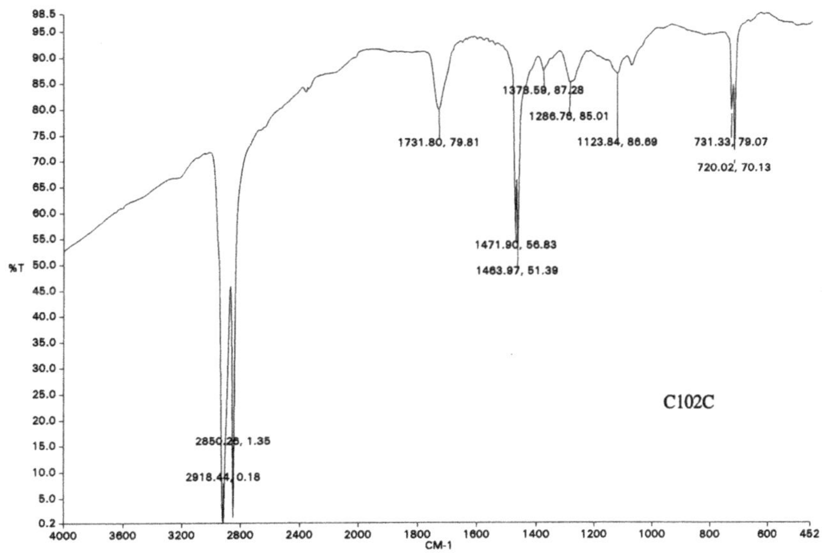
1378.59, 87.28
1286.76, 85.01
1731.80, 79.81
1123.84, 86.69
731.33, 79.07
720.02, 70.13
1471.90, 56.83
1463.97, 51.39
C102C
2850.28, 1.35
2918.44, 0.18
%T
CM-1

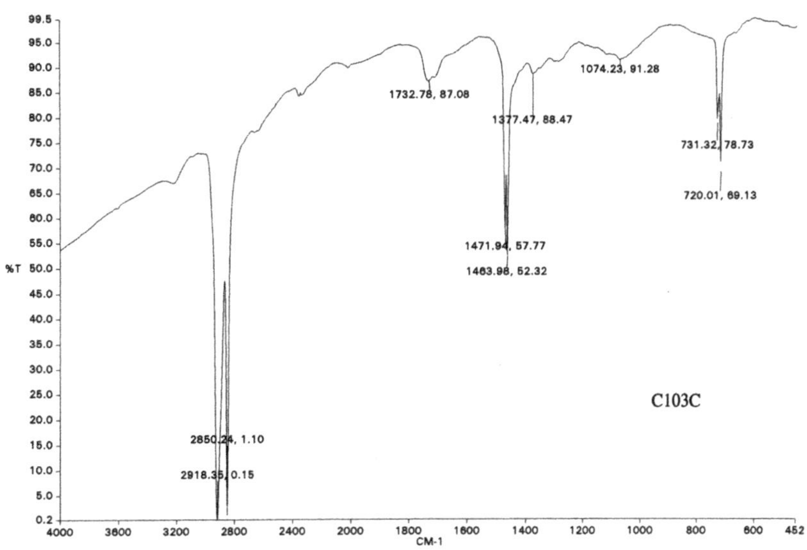
1074.23, 91.28
1732.78, 87.08
1377.47, 88.47
731.32, 78.73
720.01, 69.13
1471.94, 57.77
1463.96, 52.32
C103C
2850.24, 1.10
2918.35, 0.15
%T
CM-1

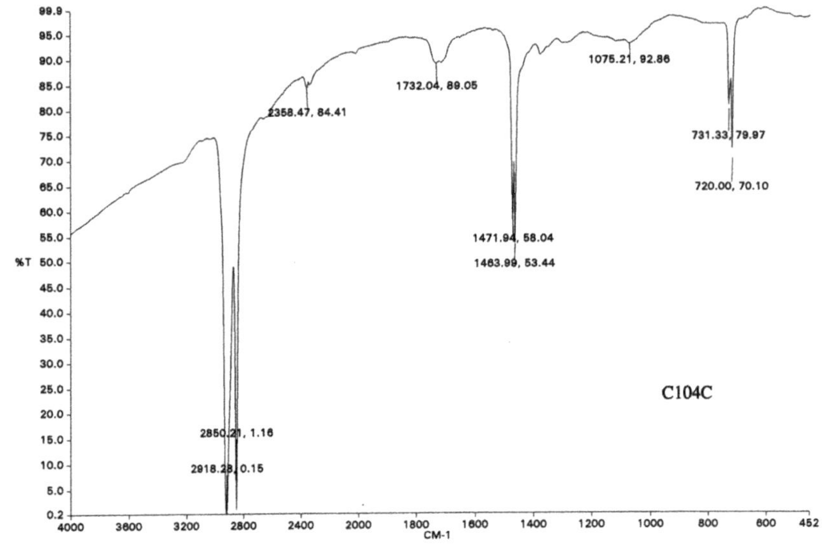

C104C
%T
CM-1
2358.47, 84.41
1732.04, 89.05
1075.21, 92.86
731.33, 79.97
720.00, 70.10
1471.94, 58.04
1463.99, 53.44
2850.21, 1.16
2918.28, 0.15

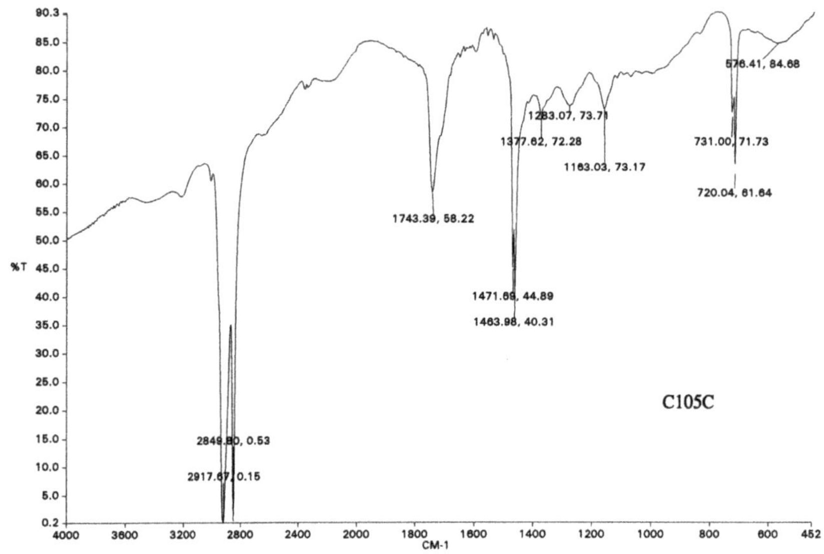

C105C
%T
CM-1
576.41, 84.68
1283.07, 73.71
1377.62, 72.28
1163.03, 73.17
731.00, 71.73
720.04, 61.64
1743.39, 58.22
1471.69, 44.89
1463.98, 40.31
2849.80, 0.53
2917.67, 0.15

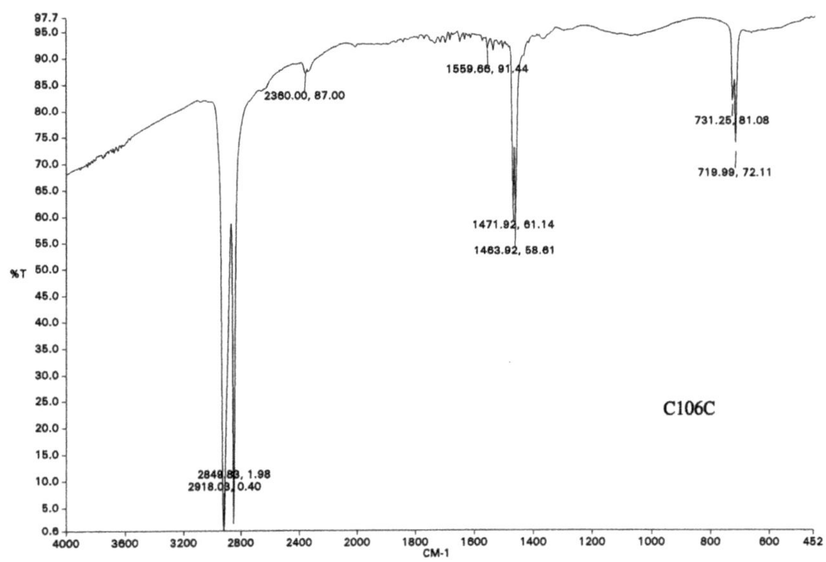

C106C
%T
CM-1
1559.66, 91.44
2360.00, 87.00
731.25, 81.08
719.99, 72.11
1471.92, 61.14
1463.92, 58.61
2849.89, 1.98
2918.03, 0.40

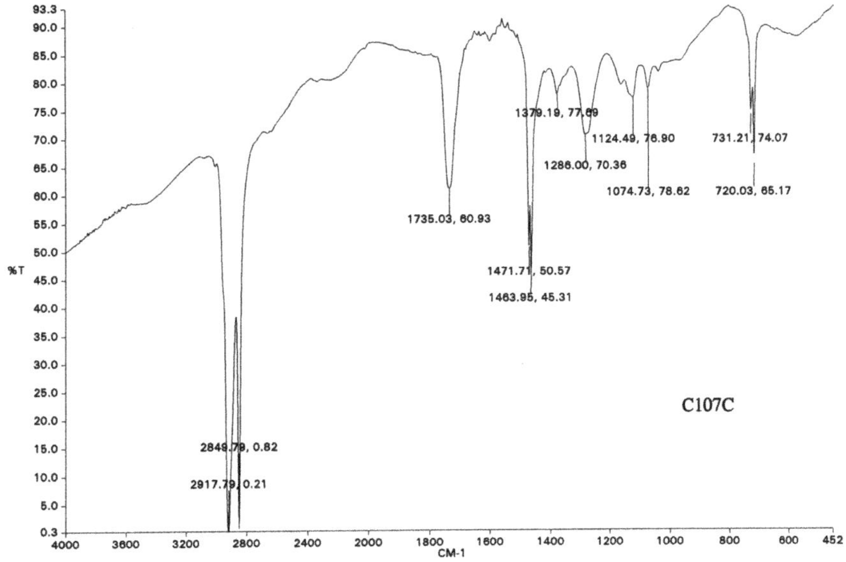
C107C
1379.19, 77.69
1124.49, 76.90
731.21, 74.07
1286.00, 70.36
1074.73, 78.62
720.03, 65.17
1735.03, 60.93
1471.71, 50.57
1463.95, 45.31
2849.79, 0.82
2917.79, 0.21
%T
CM-1

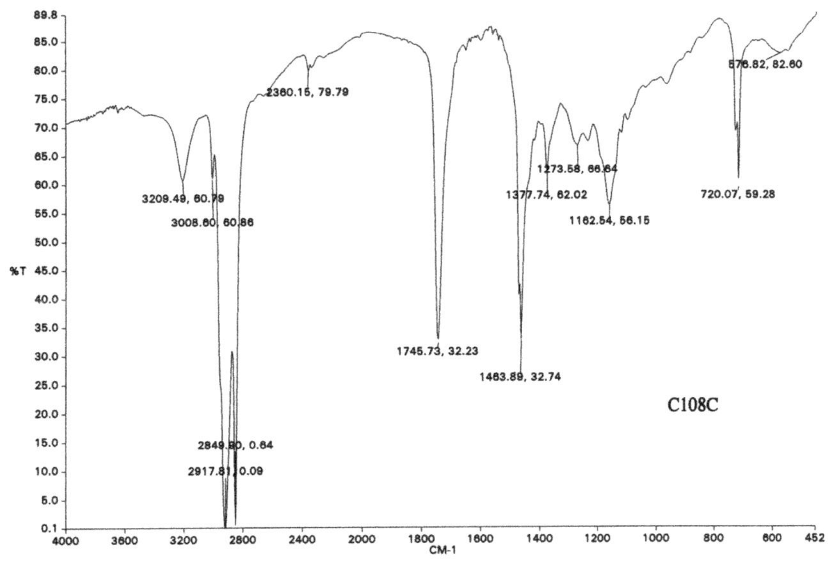
C108C
576.82, 82.60
2360.15, 79.79
1273.58, 66.64
3209.49, 60.79
1377.74, 62.02
720.07, 59.28
3008.60, 60.86
1162.54, 56.15
1745.73, 32.23
1463.89, 32.74
2849.80, 0.64
2917.81, 0.09
%T
CM-1

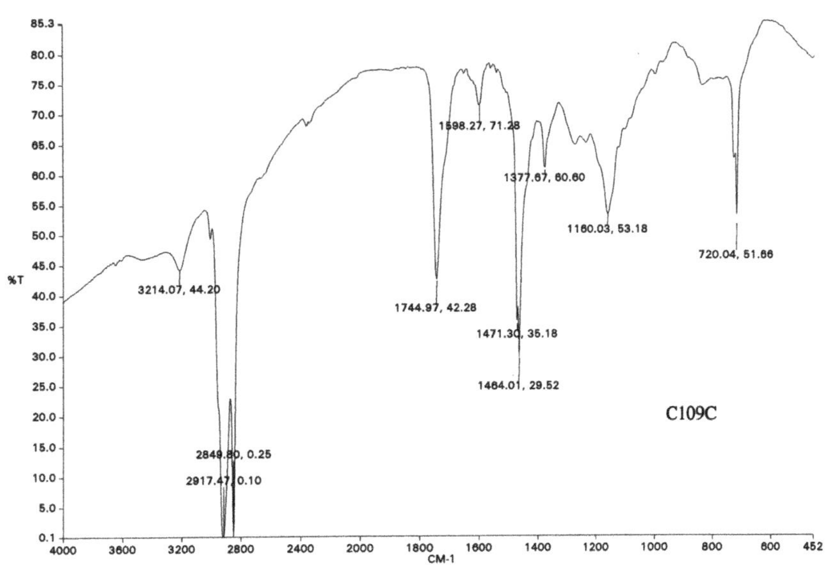
C109C
1598.27, 71.28
1377.67, 60.60
1160.03, 53.18
720.04, 51.86
3214.07, 44.20
1744.97, 42.28
1471.30, 35.18
1464.01, 29.52
2849.80, 0.25
2917.47, 0.10
%T
CM-1

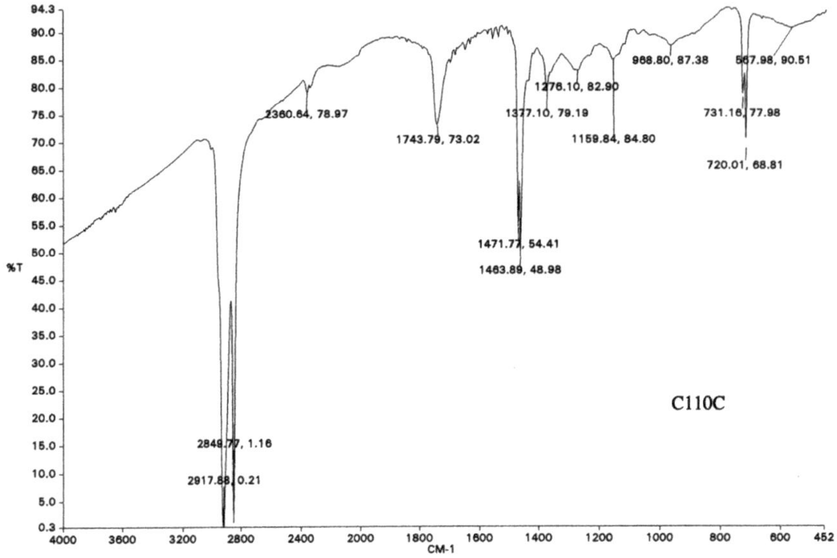
C110C
2360.64, 78.97
1743.79, 73.02
1377.10, 79.19
1276.10, 82.90
1159.84, 84.80
968.80, 87.38
567.98, 90.51
731.16, 77.98
720.01, 68.81
1471.77, 54.41
1463.89, 48.98
2849.77, 1.16
2917.88, 0.21
%T
CM-1

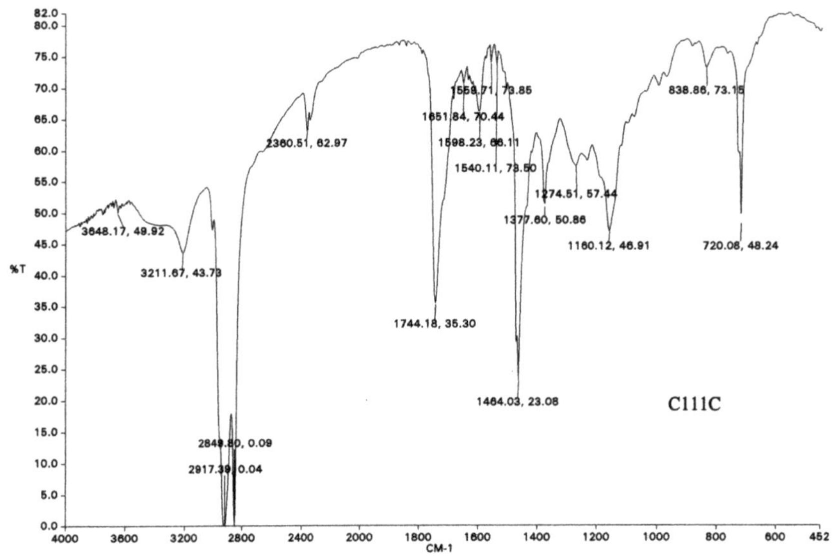
C111C
3648.17, 49.92
3211.67, 43.73
2360.51, 62.97
1744.18, 35.30
1651.84, 70.44
1559.71, 73.85
1598.23, 66.11
1540.11, 73.50
1274.51, 57.44
1377.60, 50.86
1160.12, 46.91
838.86, 73.15
720.08, 48.24
1464.03, 23.08
2849.80, 0.09
2917.39, 0.04
%T
CM-1

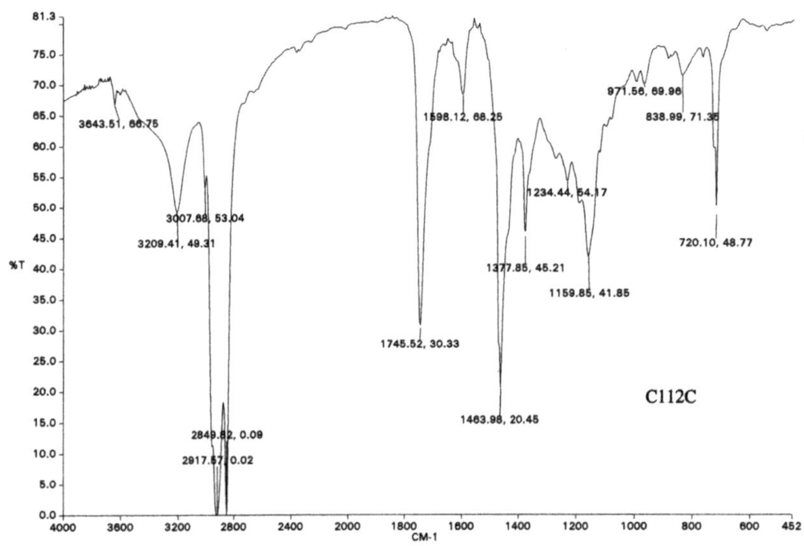
C112C
3643.51, 66.75
3007.68, 53.04
3209.41, 49.31
1598.12, 68.25
1745.52, 30.33
1234.44, 54.17
1377.85, 45.21
1159.85, 41.85
971.56, 69.96
838.99, 71.35
720.10, 48.77
1463.98, 20.45
2849.82, 0.09
2917.57, 0.02
%T
CM-1

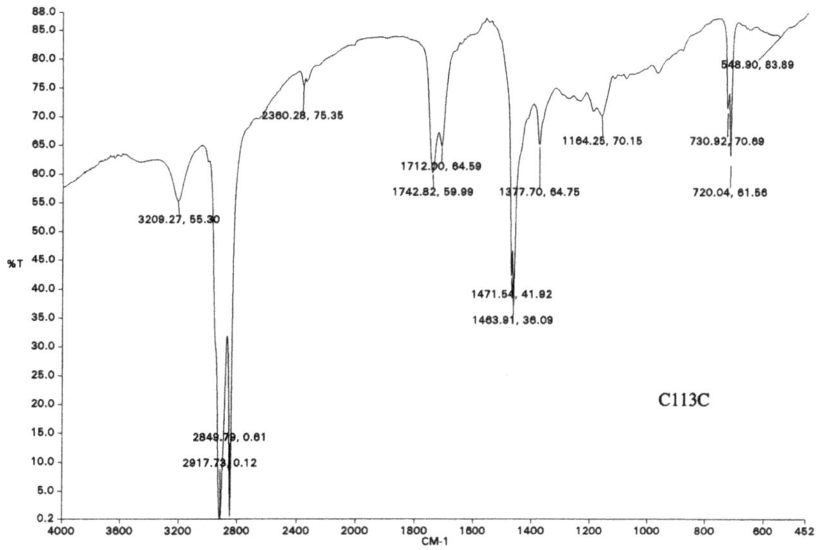
548.90, 83.89
2360.28, 75.35
1164.25, 70.15
730.92, 70.69
1712.90, 64.59
1742.82, 59.99
1377.70, 64.75
720.04, 61.56
3209.27, 55.30
1471.54, 41.92
1463.91, 36.09
C113C
2849.79, 0.61
2917.73, 0.12
%T
CM-1

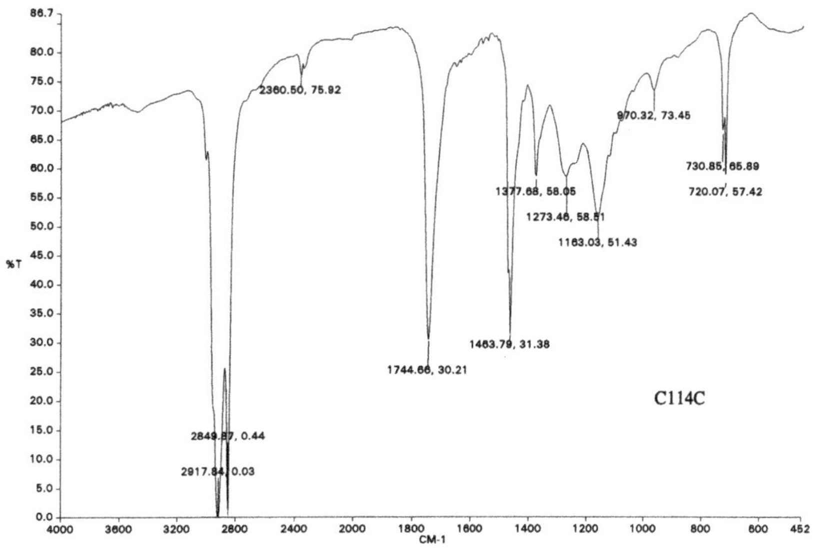
2360.50, 75.92
970.32, 73.45
730.85, 65.89
1377.68, 58.05
720.07, 57.42
1273.46, 58.51
1163.03, 51.43
1463.79, 31.38
1744.66, 30.21
C114C
2849.87, 0.44
2917.84, 0.03
%T
CM-1

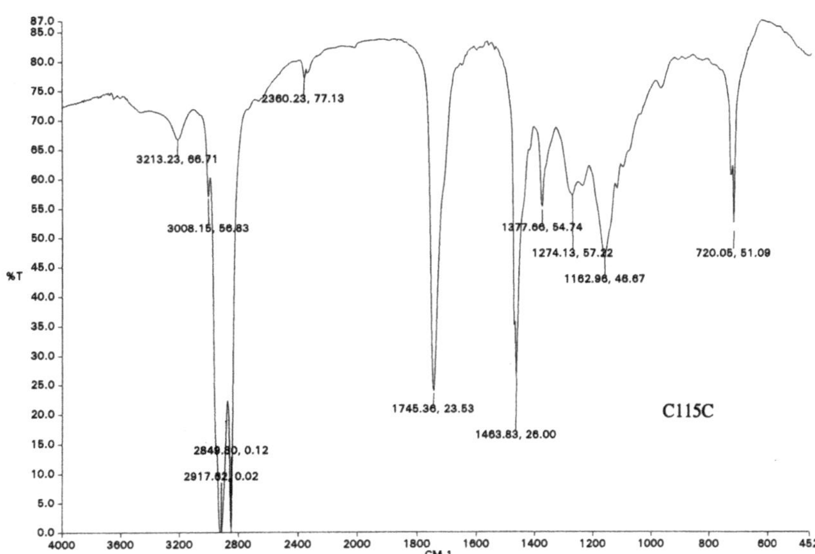
2360.23, 77.13
3213.23, 66.71
3008.15, 56.83
1377.66, 54.74
1274.13, 57.22
720.05, 51.09
1162.96, 46.67
1745.36, 23.53
C115C
1463.83, 26.00
2849.80, 0.12
2917.62, 0.02
%T
CM-1

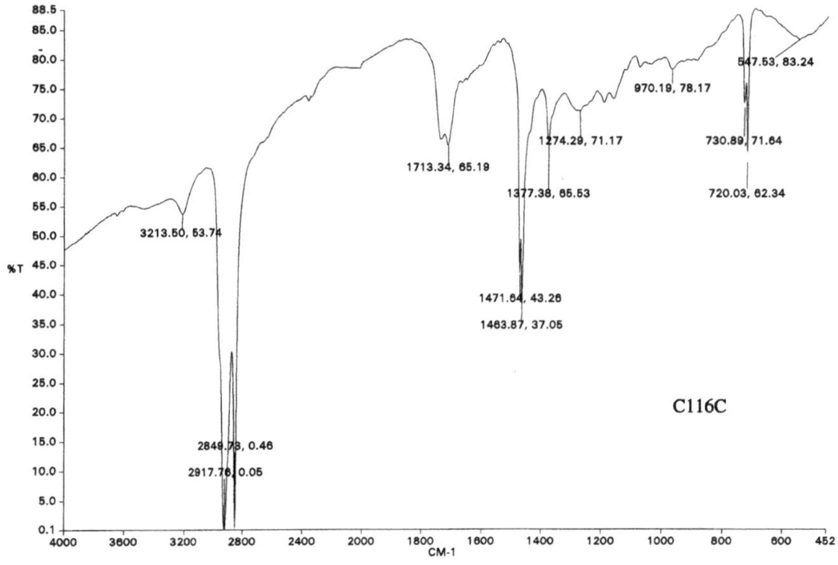
3213.50, 53.74
2849.78, 0.46
2917.76, 0.05
1713.34, 65.19
1377.38, 65.53
1274.29, 71.17
1471.64, 43.26
1463.87, 37.05
970.19, 78.17
547.53, 83.24
730.89, 71.64
720.03, 62.34
C116C
%T
CM-1

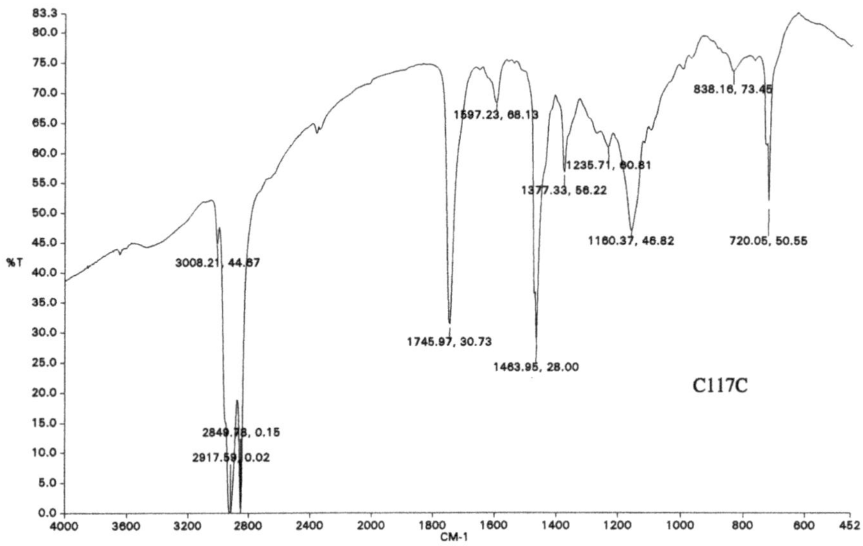
3008.21, 44.87
2849.78, 0.15
2917.59, 0.02
1745.97, 30.73
1597.23, 68.13
1377.33, 56.22
1235.71, 60.81
1160.37, 46.82
1463.95, 28.00
838.16, 73.45
720.05, 50.55
C117C
%T
CM-1

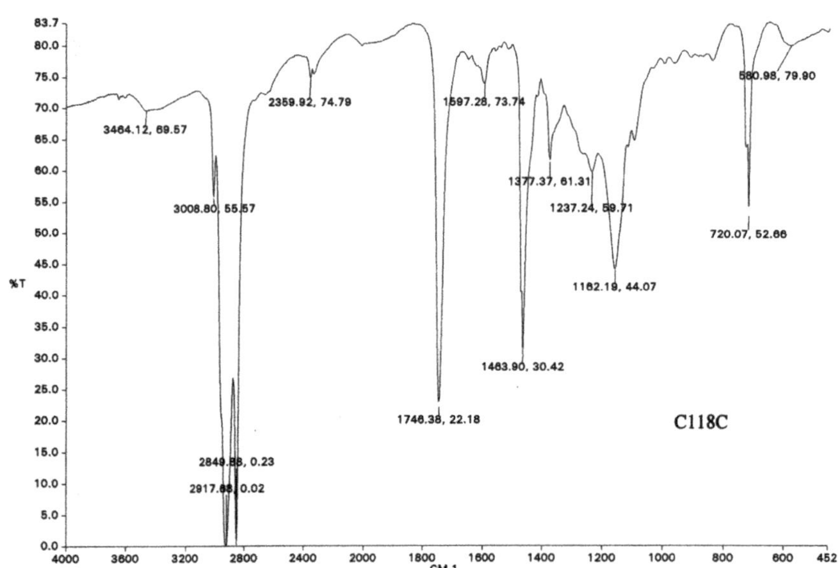
3464.12, 69.57
3008.80, 55.57
2359.92, 74.79
2849.88, 0.23
2917.88, 0.02
1597.28, 73.74
1377.37, 61.31
1237.24, 59.71
1162.19, 44.07
1463.90, 30.42
1746.38, 22.18
580.98, 79.90
720.07, 52.66
C118C
%T
CM-1

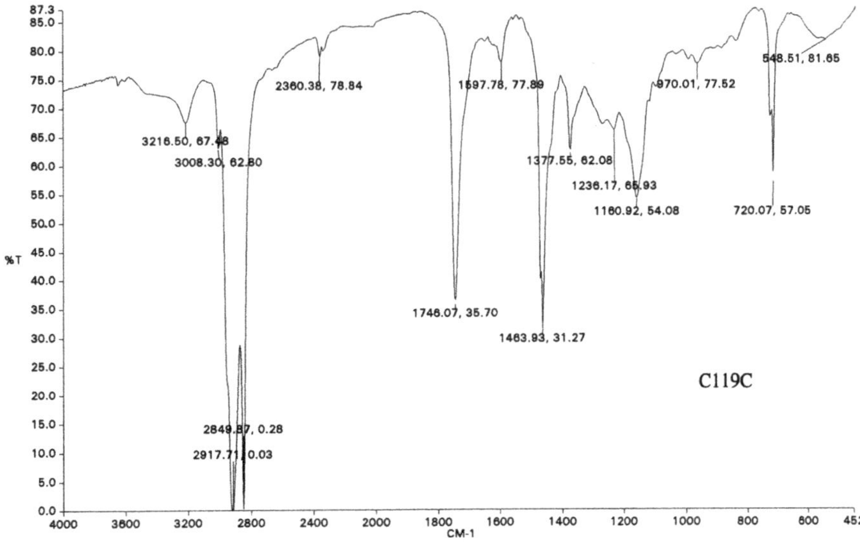
%T
CM-1
3216.50, 67.48
3008.30, 62.80
2360.38, 78.84
2849.87, 0.28
2917.71, 0.03
1597.78, 77.89
1746.07, 35.70
1463.93, 31.27
1377.55, 62.08
1236.17, 55.93
1160.92, 54.08
970.01, 77.52
548.51, 81.65
720.07, 57.05
C119C

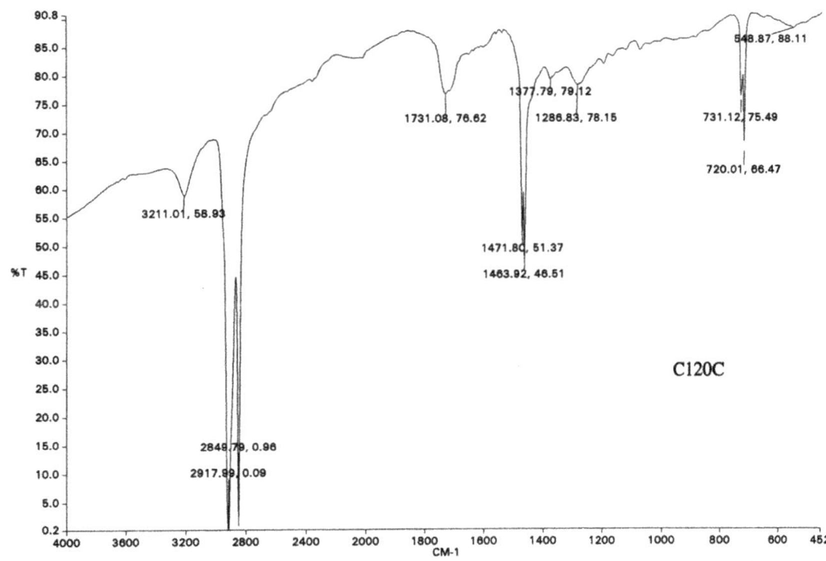
%T
CM-1
3211.01, 58.93
2849.79, 0.96
2917.89, 0.09
1731.08, 76.62
1377.79, 79.12
1286.83, 78.15
1471.80, 51.37
1463.92, 46.51
548.87, 88.11
731.12, 75.49
720.01, 66.47
C120C

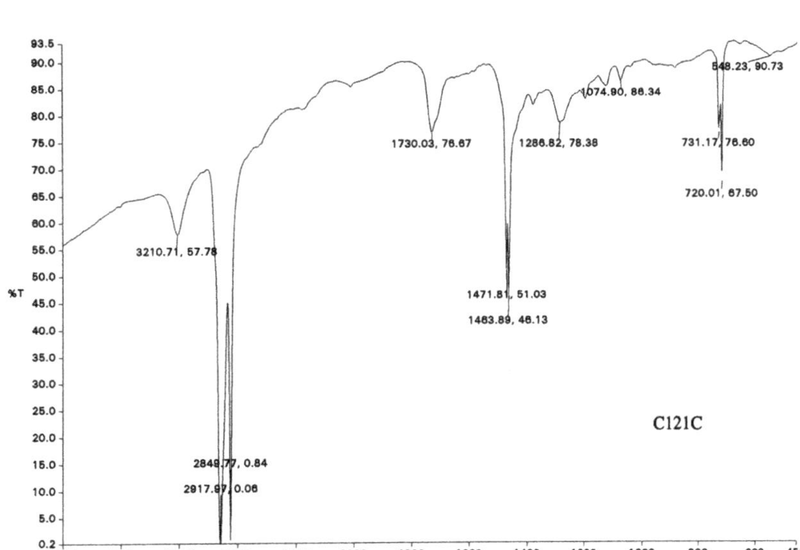
%T
CM-1
3210.71, 57.78
2849.77, 0.84
2917.87, 0.06
1730.03, 76.67
1286.82, 78.38
1074.90, 86.34
1471.81, 51.03
1463.89, 46.13
548.23, 90.73
731.17, 76.60
720.01, 67.50
C121C

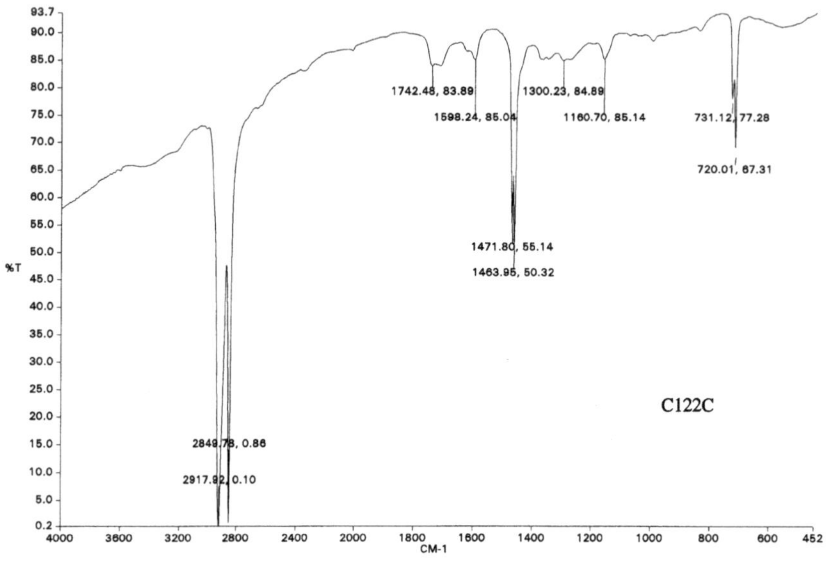
C122C
1742.48, 83.89
1598.24, 85.04
1300.23, 84.89
1160.70, 85.14
731.12, 77.28
720.01, 67.31
1471.80, 55.14
1463.95, 50.32
2849.78, 0.86
2917.92, 0.10
%T
CM-1

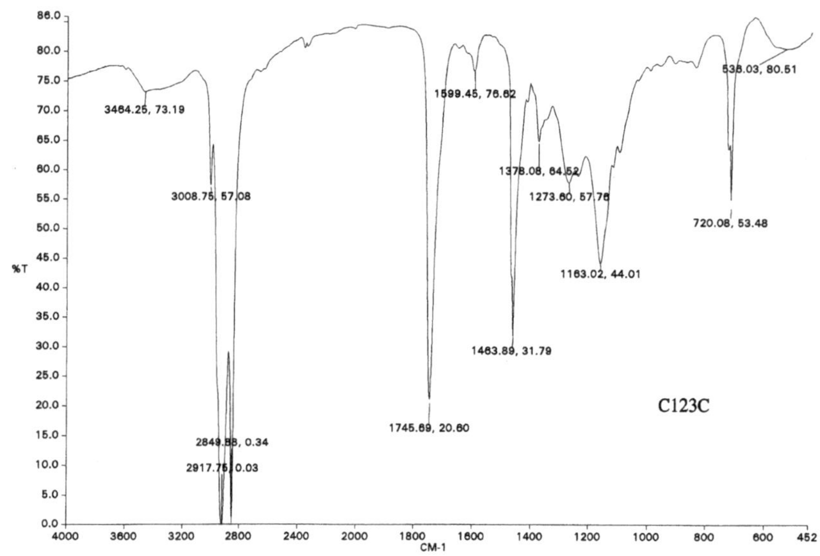
C123C
3464.25, 73.19
1599.45, 76.62
538.03, 80.51
3008.75, 57.08
1378.08, 64.52
1273.60, 57.78
720.08, 53.48
1163.02, 44.01
1463.89, 31.79
1745.69, 20.60
2849.88, 0.34
2917.75, 0.03
%T
CM-1

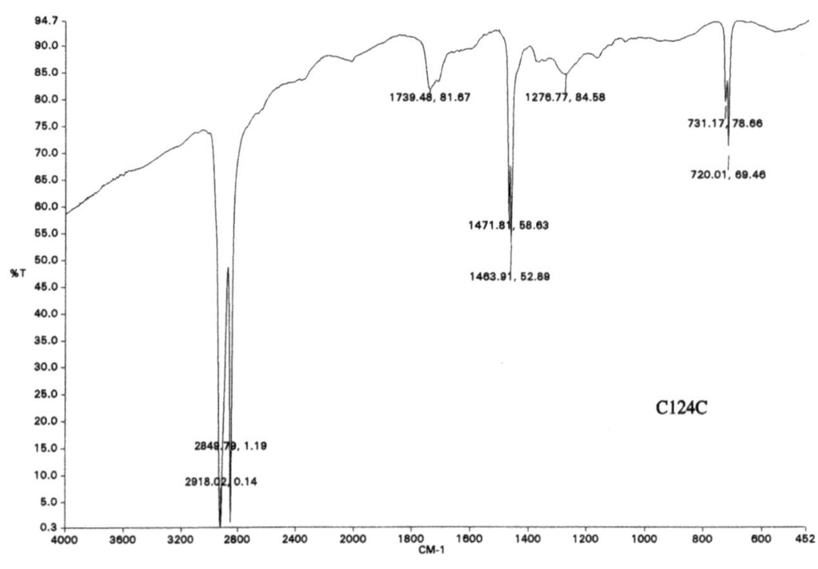
C124C
1739.48, 81.67
1276.77, 84.58
731.17, 78.66
720.01, 69.46
1471.81, 58.63
1463.91, 52.89
2849.79, 1.19
2918.02, 0.14
%T
CM-1

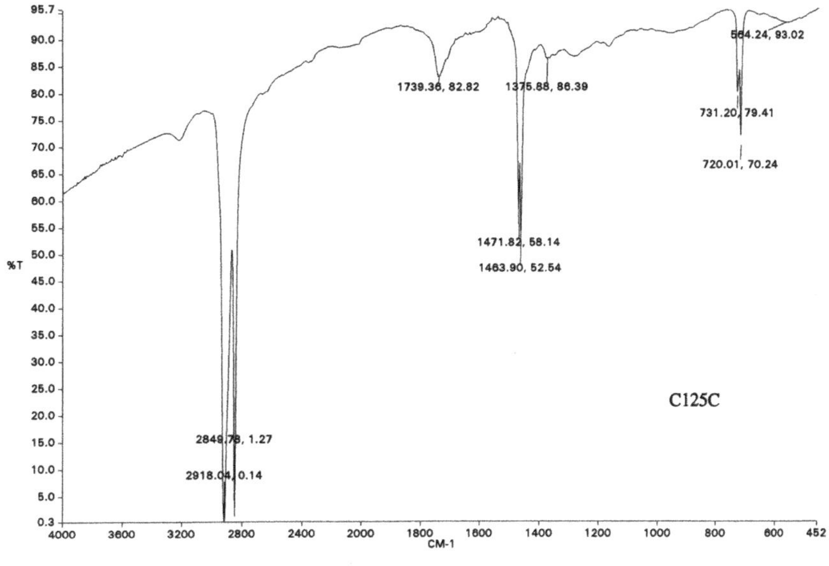
564.24, 93.02
1739.36, 82.82
1376.88, 86.39
731.20, 79.41
720.01, 70.24
1471.82, 58.14
1463.90, 52.54
C125C
2849.78, 1.27
2918.04, 0.14
%T
CM-1

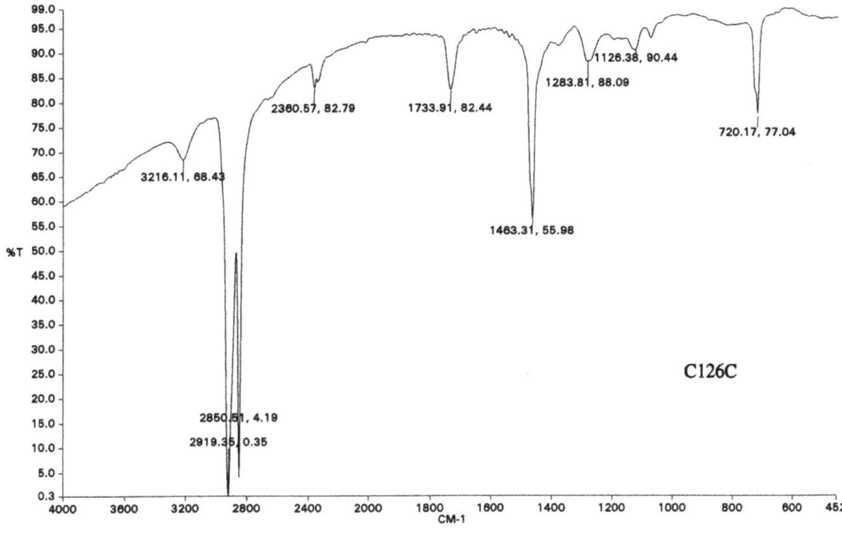
1126.38, 90.44
1283.81, 88.09
2360.57, 82.79
1733.91, 82.44
720.17, 77.04
3216.11, 68.43
1463.31, 55.98
C126C
2850.51, 4.19
2919.35, 0.35
%T
CM-1

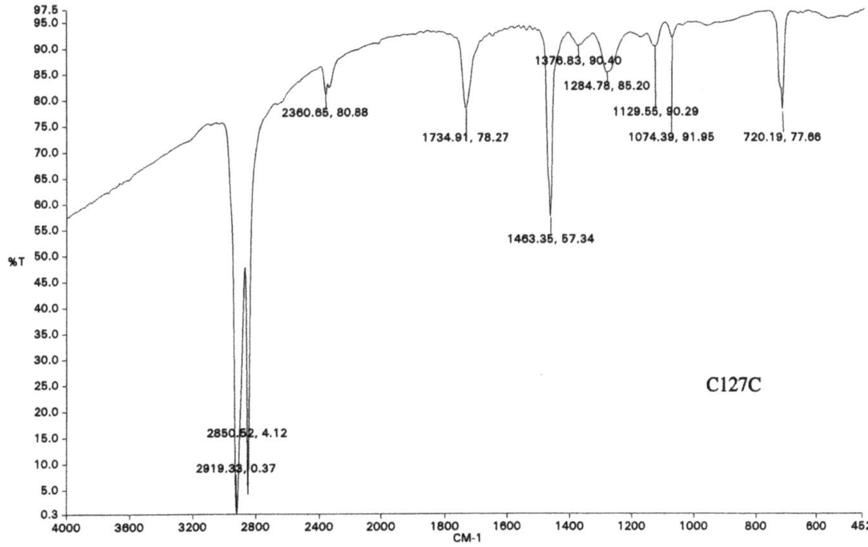
1376.83, 90.40
1284.78, 85.20
1129.55, 90.29
2360.65, 80.88
1734.91, 78.27
1074.39, 91.95
720.19, 77.66
1463.35, 57.34
C127C
2850.52, 4.12
2919.33, 0.37
%T
CM-1

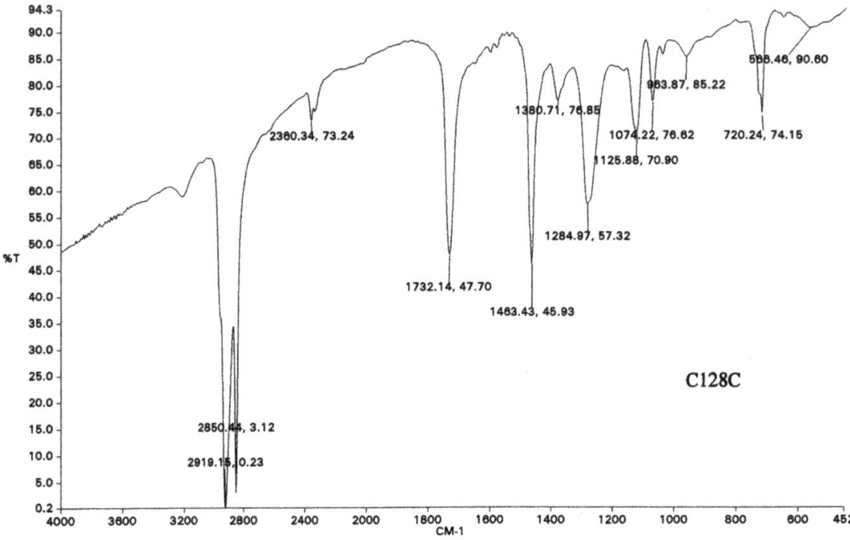
%T
CM-1
2360.34, 73.24
1380.71
1074.22, 76.62
720.24, 74.15
1125.88, 70.90
1284.97, 57.32
1732.14, 47.70
1463.43, 45.93
C128C
3.12
0.23

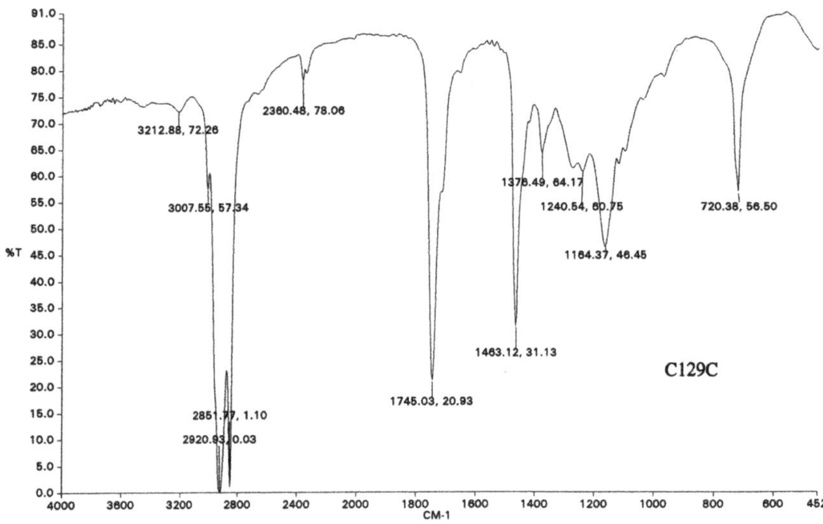
%T
CM-1
3212.88, 72.26
2360.48, 78.06
3007.55, 57.34
1378.49, 64.17
1240.54, 60.75
720.38, 56.50
1164.37, 46.45
1463.12, 31.13
C129C
1745.03, 20.93
1.10
0.03

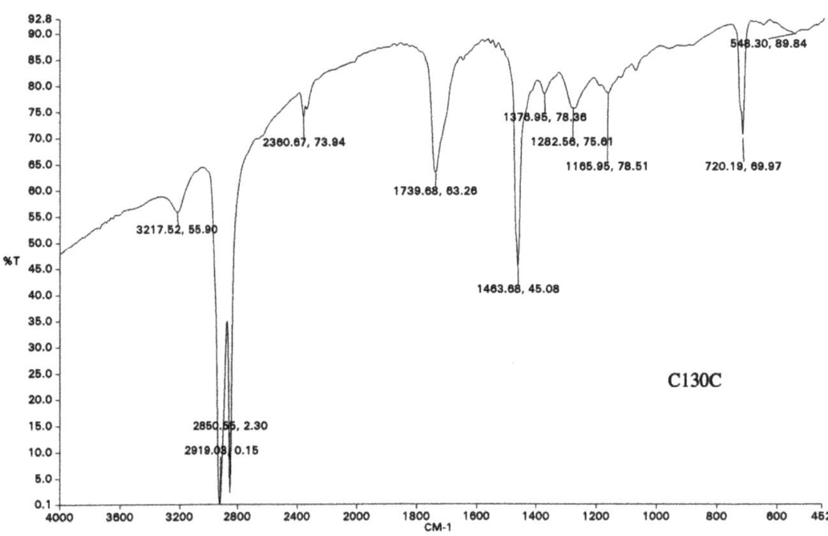
%T
CM-1
2360.67, 73.94
1378.95, 78.36
1282.56, 75.61
1165.95, 78.51
720.19, 69.97
1739.68, 63.26
3217.52, 55.90
1463.68, 45.08
C130C
2.30
0.15

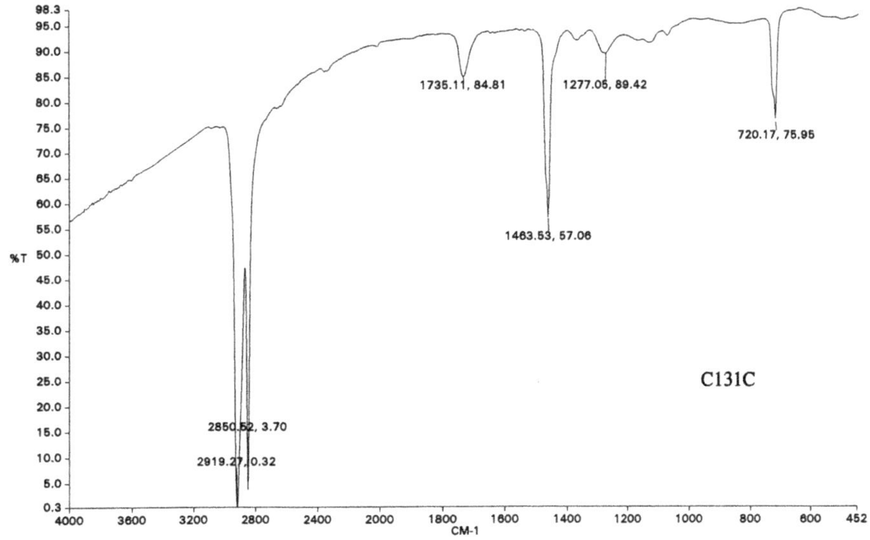
%T
CM-1
1735.11, 84.81
1277.05, 89.42
720.17, 75.95
1463.53, 57.06
2850.52, 3.70
2919.27, 0.32
C131C

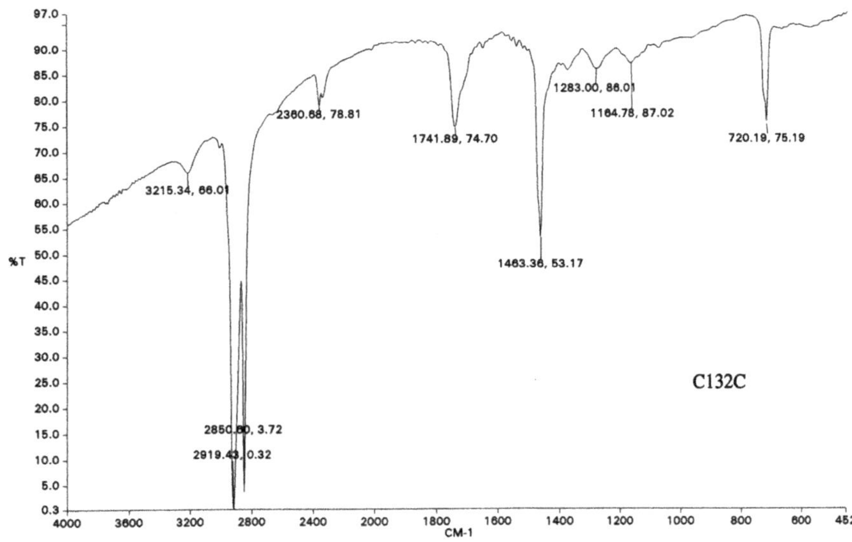
%T
CM-1
1283.00, 86.01
2360.68, 78.81
1164.78, 87.02
1741.89, 74.70
720.19, 75.19
3215.34, 66.01
1463.36, 53.17
2850.80, 3.72
2919.43, 0.32
C132C

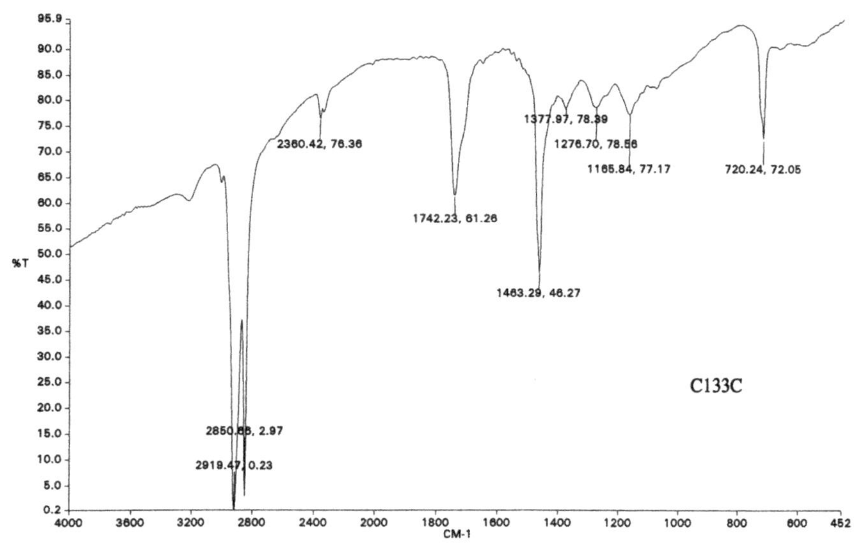
%T
CM-1
1377.97, 78.39
2360.42, 76.36
1276.70, 78.56
1165.84, 77.17
720.24, 72.05
1742.23, 61.26
1463.29, 46.27
2850.88, 2.97
2919.47, 0.23
C133C

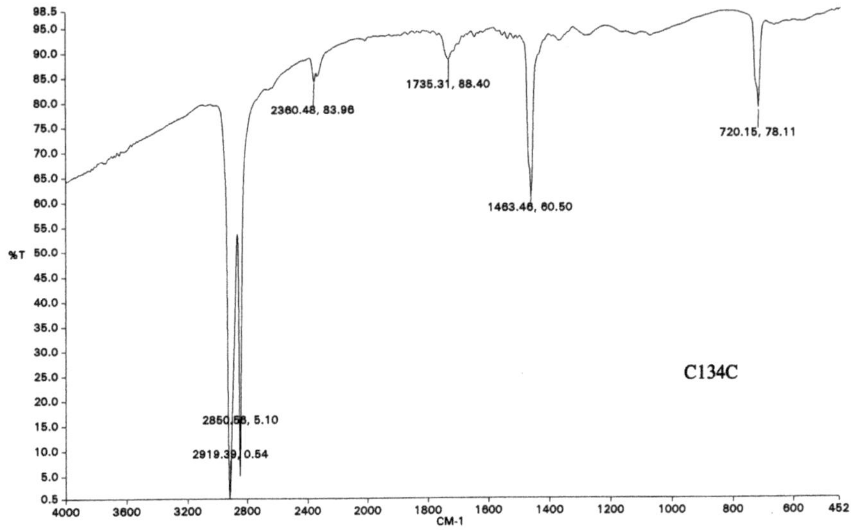
C134C
%T
CM-1
2360.48, 83.96
1735.31, 88.40
1463.46, 60.50
720.15, 78.11
2850.58, 5.10
2919.39, 0.54

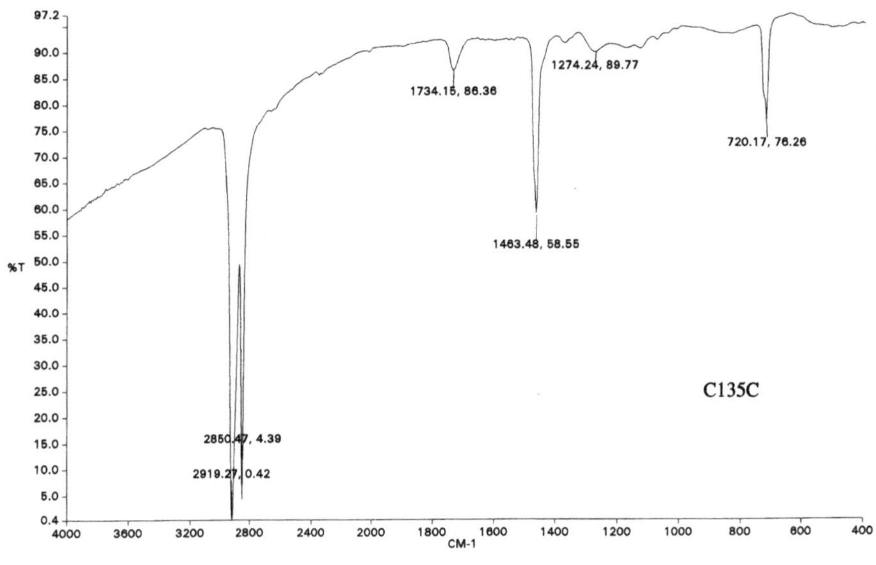
C135C
%T
CM-1
1734.15, 86.36
1274.24, 89.77
1463.48, 58.55
720.17, 76.26
2850.47, 4.39
2919.27, 0.42

HEXANE SPECTRA

Modern spectra: HOO2C, H011C, HOO7C, H037C, H067C

Archaeological spectra: AROO1H, AR019H, AR030H, AR021H, AR047H

These examples of hexane spectra were included in order to illustrate the lack of diagnostic characteristics within the extracts of both the modem and archaeological spectra. As this appeared to be a general trend for a large percentage of the hexane extracts it was considered important to include them in the results section, as insufficient data may create errors if incorporated into the numerical analysis. They were considered unsatisfactory in terms of a final spectral result, as they lacked all or most of the diagnostic peaks necessary for the numerical analysis, since this was a major aspect of the project, any spectra not fitting the criteria would have to be excluded.

The inclusion of spectra that are not numerically sound would create a system that contains an unacceptable number of outliers, as a consequence the template that is to be used as a comparative model would have a series of built in errors, thereby rendering any subsequent identifications unreliable.

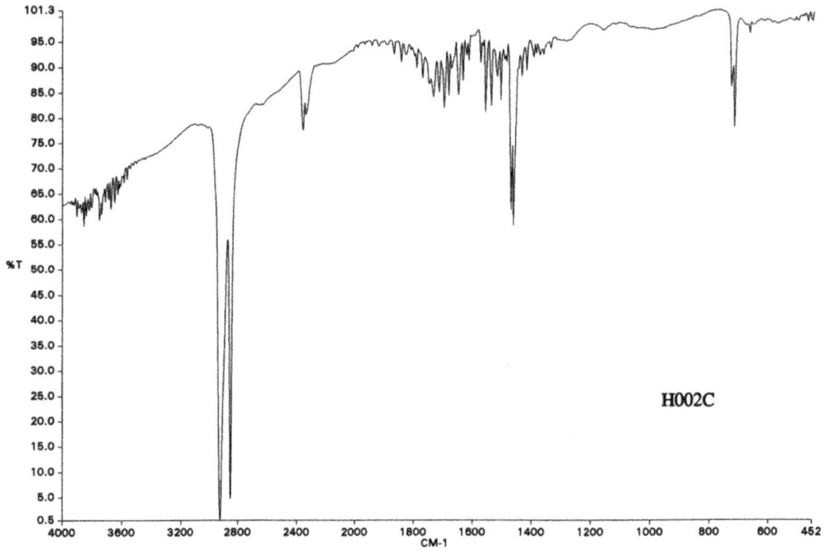
%T
CM-1
H002C

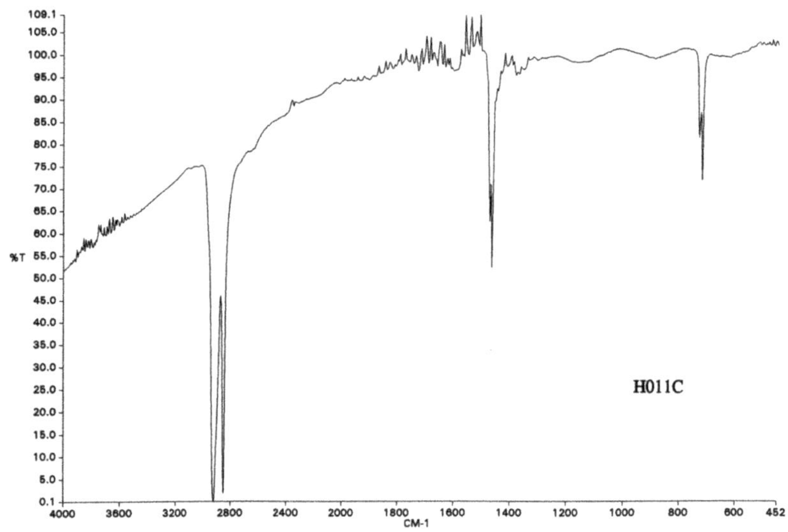
%T
CM-1
H011C

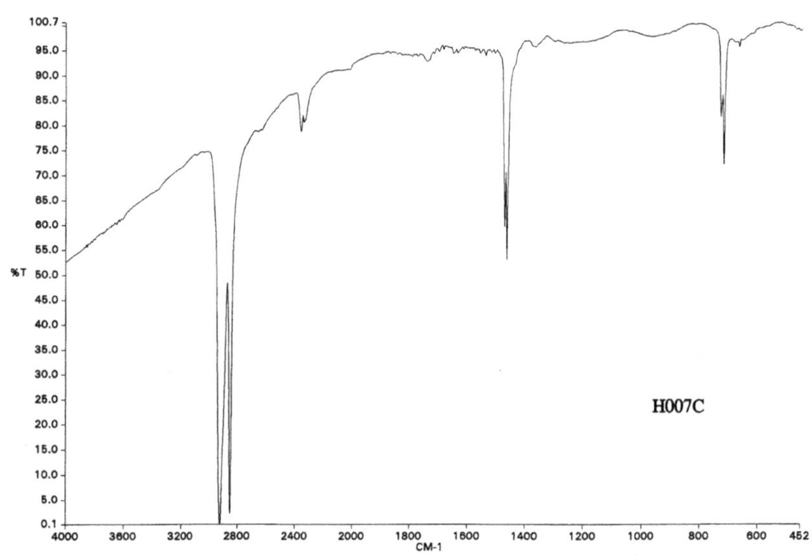
%T
CM-1
H007C

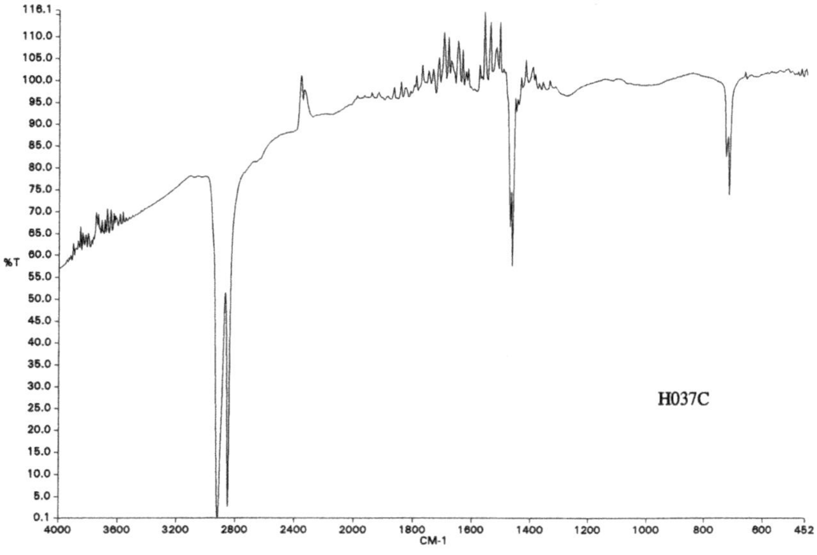
%T
CM-1
H037C

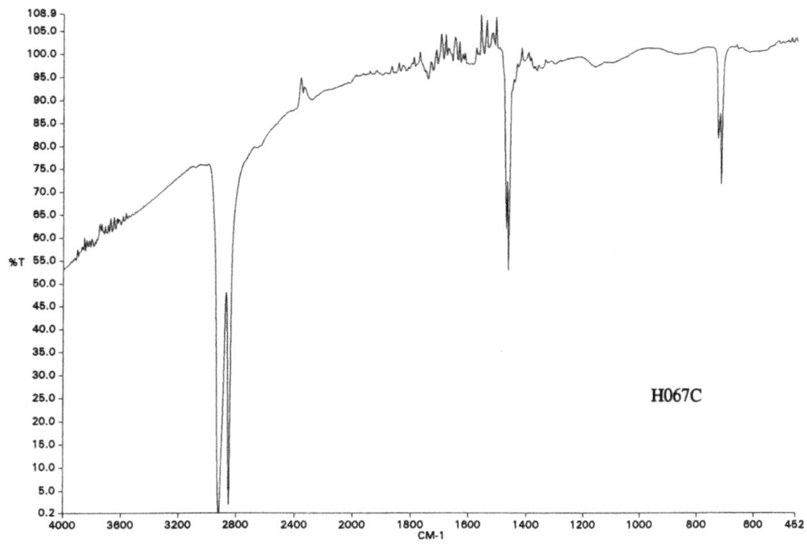
%T
CM-1
H067C

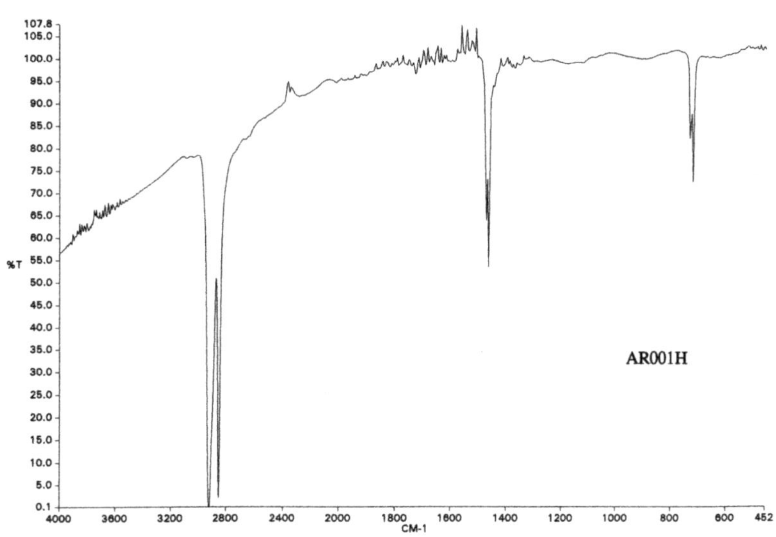
%T
CM-1
AR001H

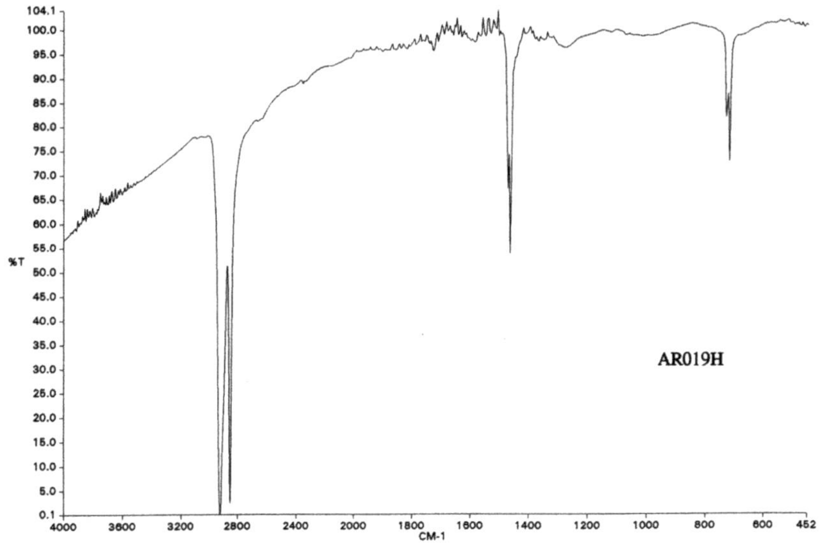
AR019H
%T
CM-1

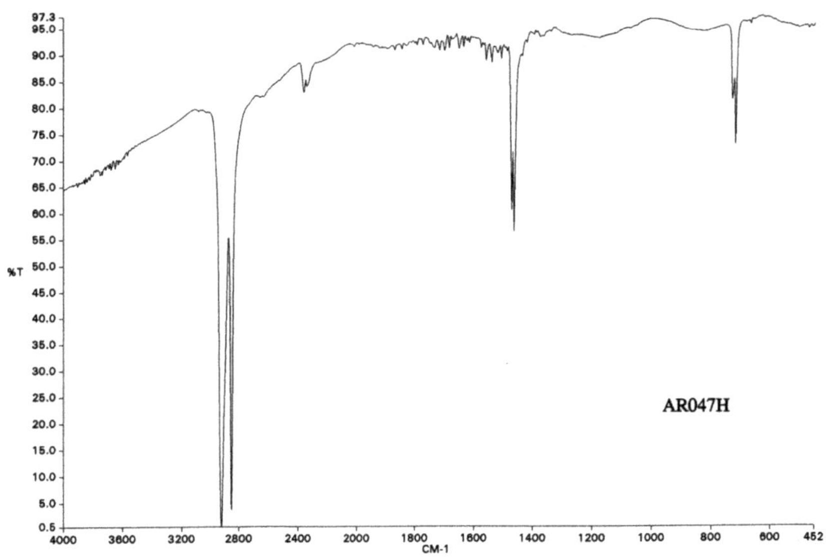
AR047H
%T
CM-1

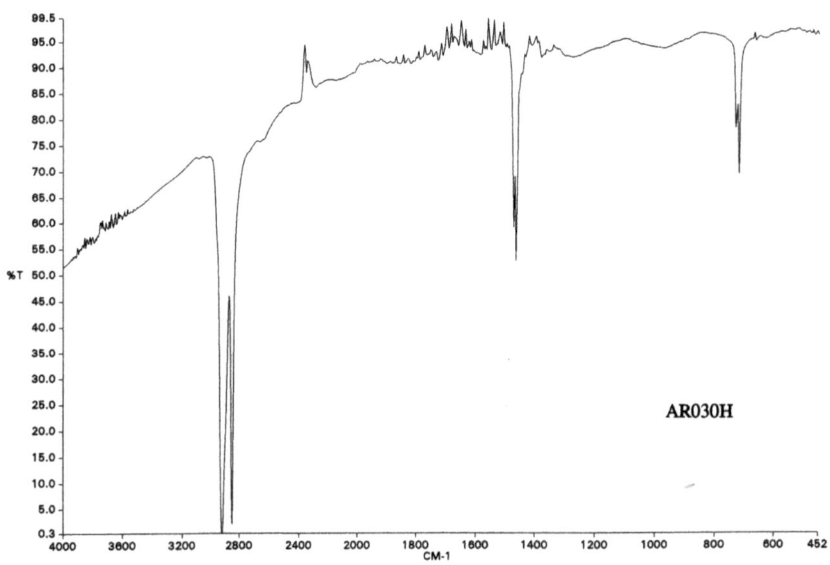
AR030H
%T
CM-1

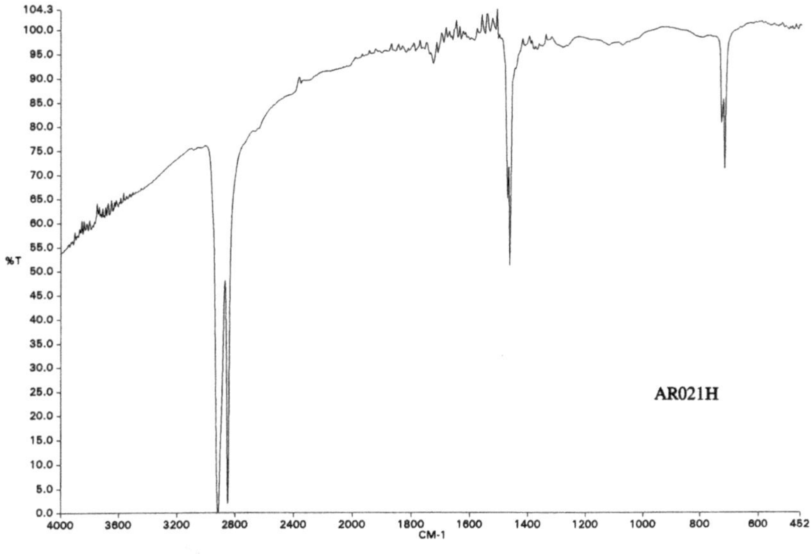
104.3
100.0
95.0
90.0
85.0
80.0
75.0
70.0
65.0
60.0
55.0
%T
50.0
45.0
40.0
35.0
30.0
25.0
20.0
15.0
10.0
5.0
0.0
4000
3600
3200
2800
2400
2000
1800
1600
1400
1200
1000
800
600
452
CM-1
AR021H

ARCHAEOLOGICAL SAMPLES
AROO1C - AR055C

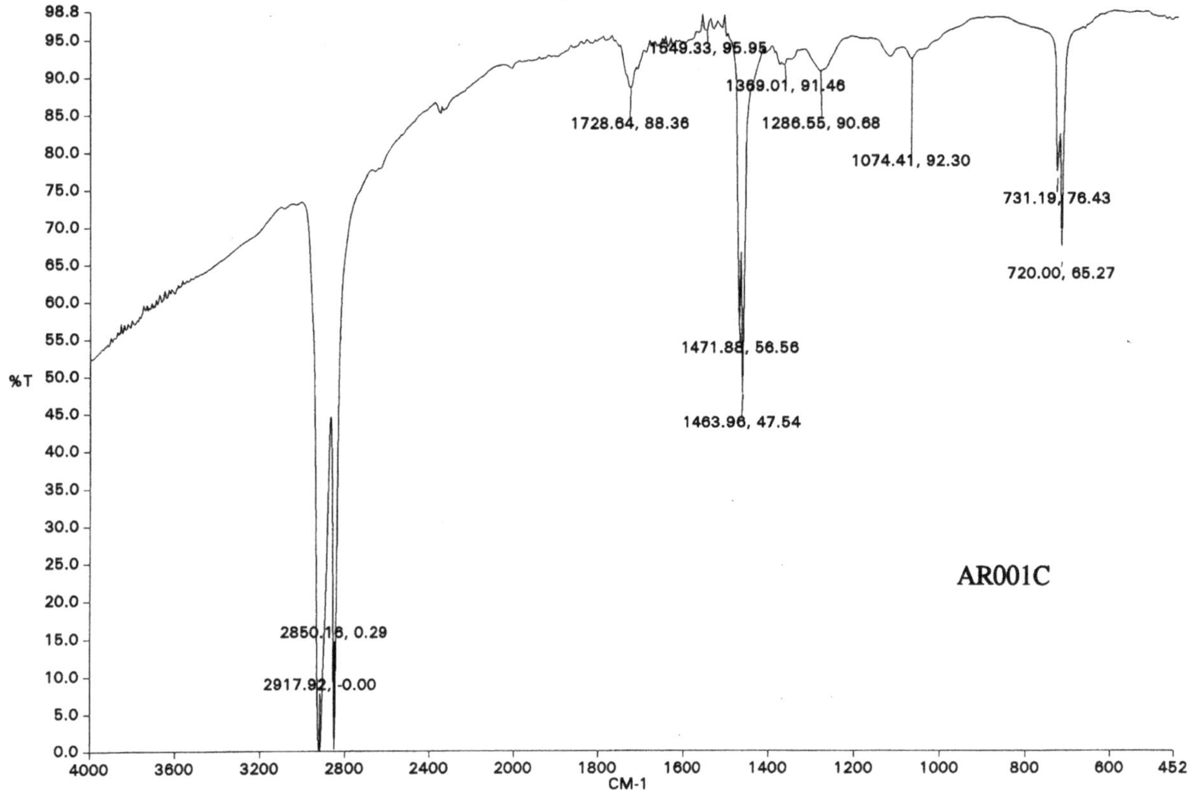

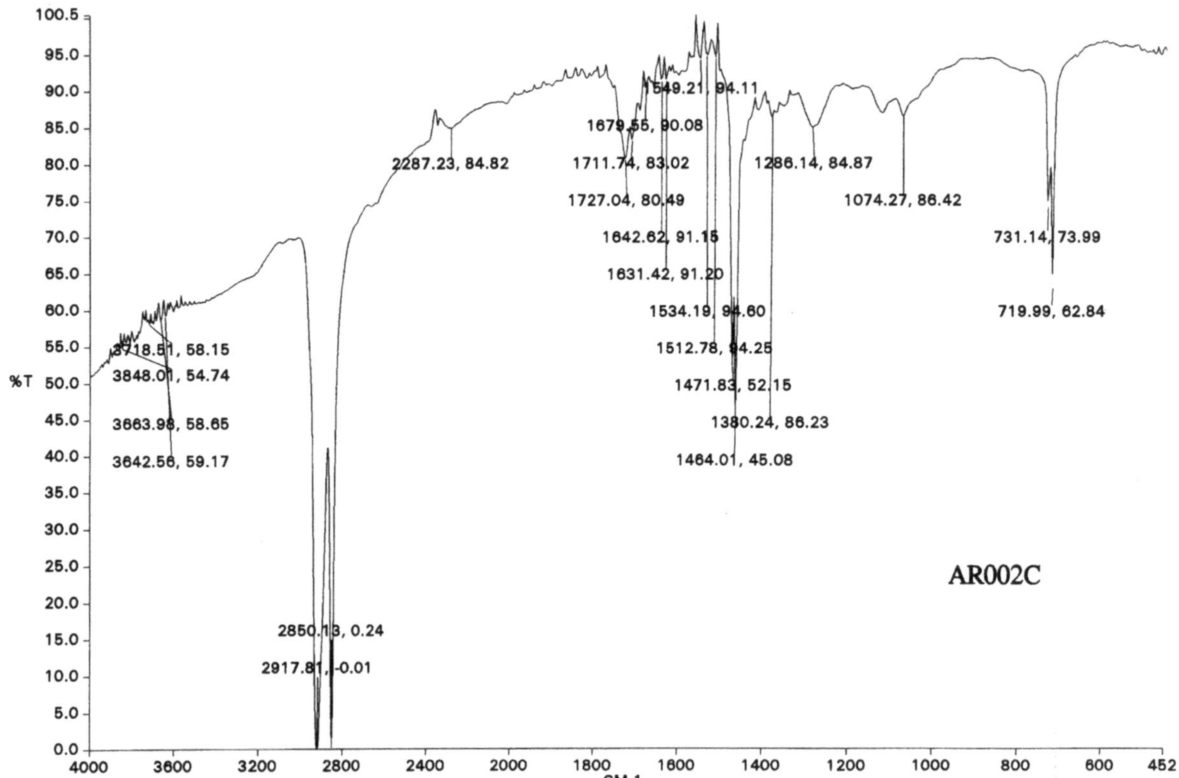

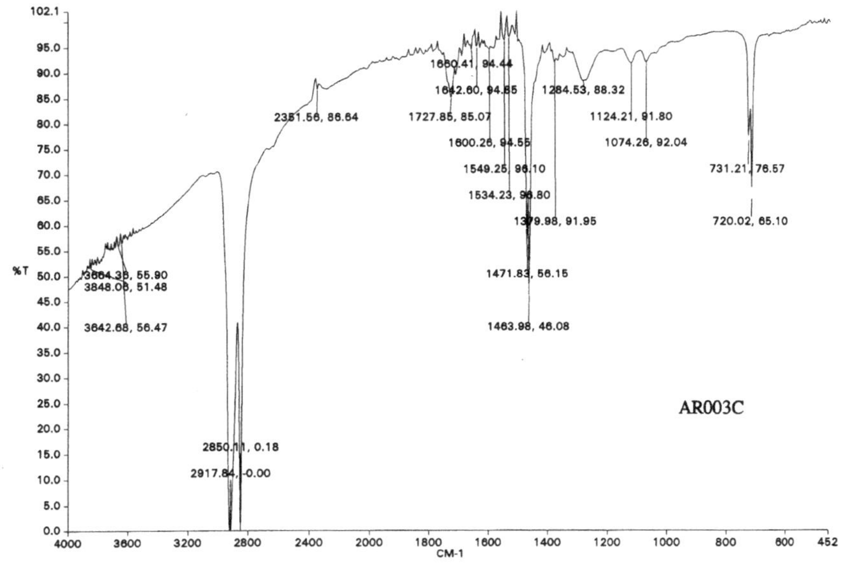
%T
CM-1
1660.41, 94.44
1642.60, 94.85
1284.53, 88.32
2351.56, 86.64
1727.85, 85.07
1124.21, 91.80
1600.26, 94.55
1074.26, 92.04
1549.25, 96.10
731.21, 76.57
1534.23, 96.80
1379.98, 91.95
720.02, 65.10
3664.38, 55.90
3848.06, 51.48
1471.83, 56.15
3642.68, 56.47
1463.98, 46.08
AR003C
2850.11, 0.18
2917.84, -0.00

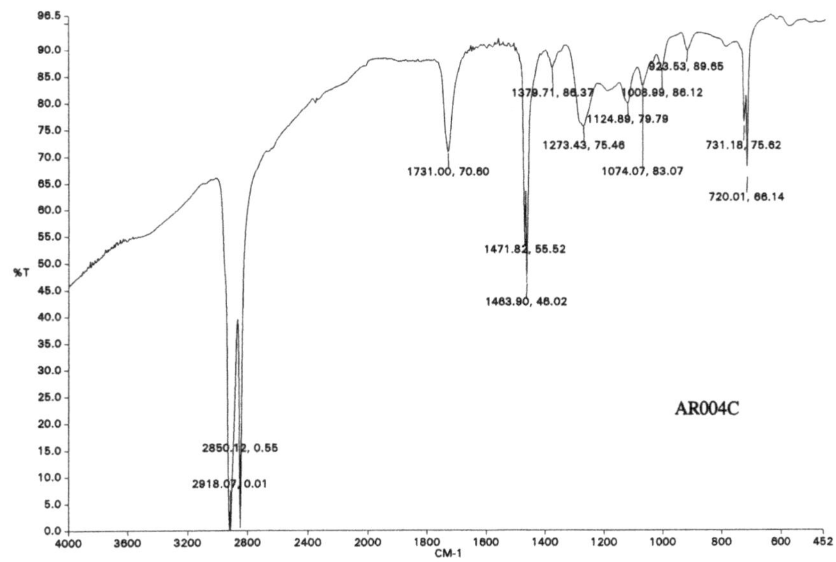
%T
CM-1
923.53, 89.65
1379.71, 86.37
1008.99, 86.12
1124.89, 79.79
1273.43, 75.46
731.18, 75.62
1731.00, 70.60
1074.07, 83.07
720.01, 66.14
1471.82, 55.52
1463.90, 46.02
AR004C
2850.12, 0.55
2918.07, 0.01

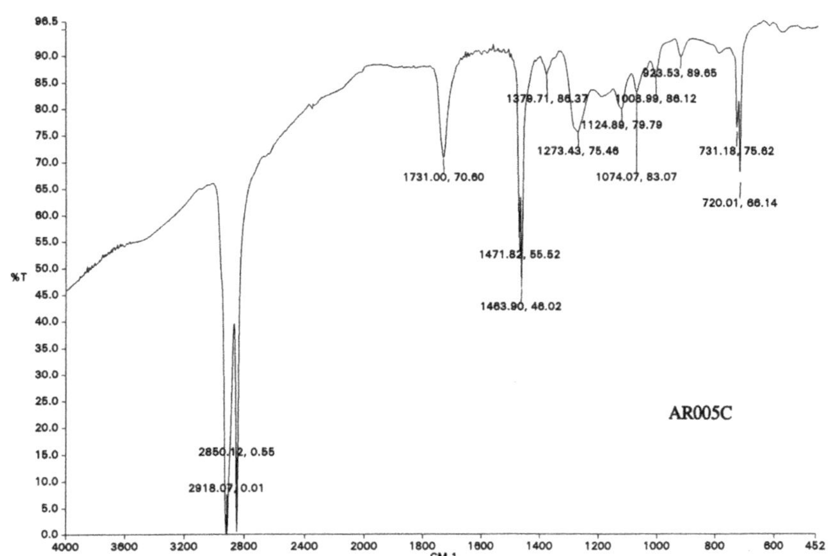
%T
CM-1
923.53, 89.65
1379.71, 86.37
1008.99, 86.12
1124.89, 79.79
1273.43, 75.46
731.18, 75.62
1731.00, 70.60
1074.07, 83.07
720.01, 66.14
1471.82, 55.52
1463.90, 46.02
AR005C
2850.12, 0.55
2918.07, 0.01

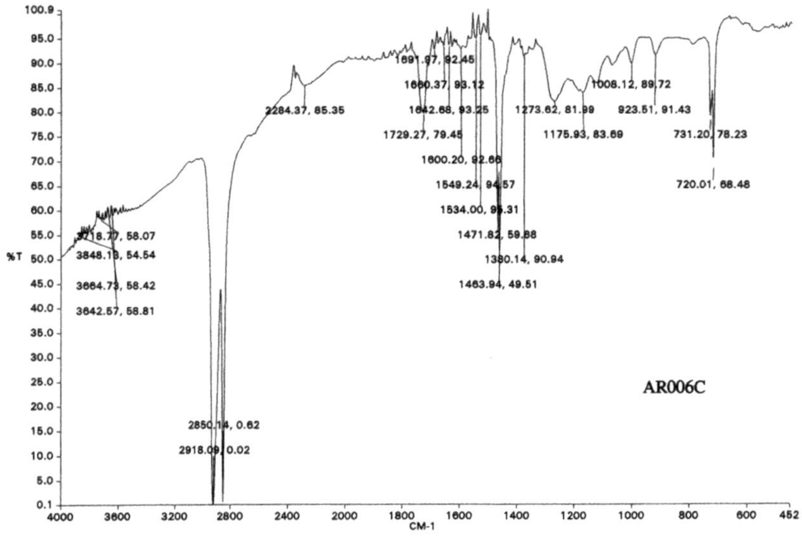
AR006C

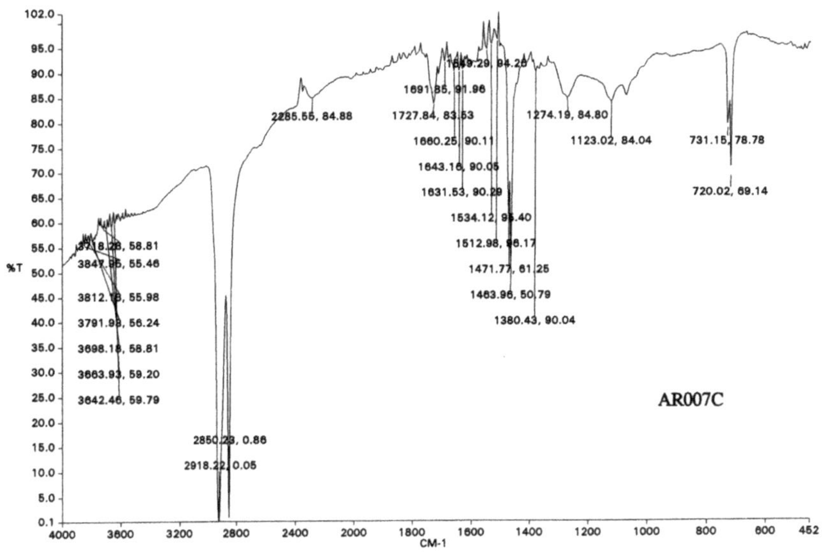
AR007C

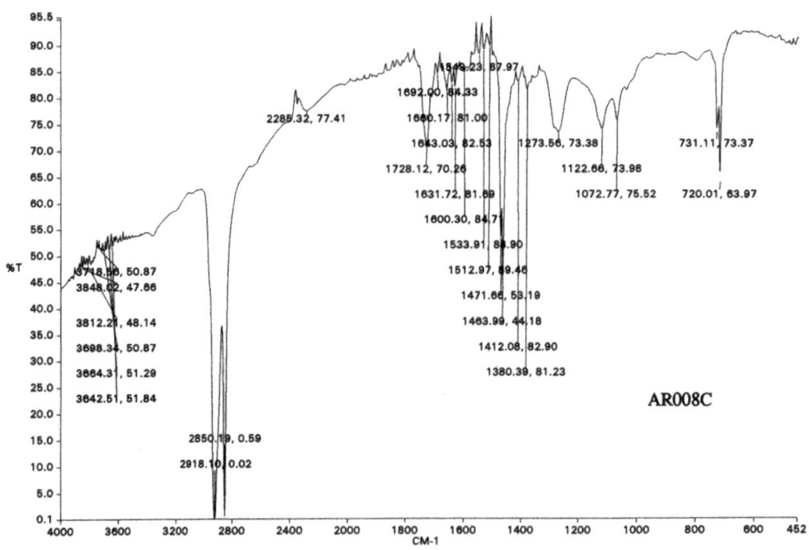
AR008C

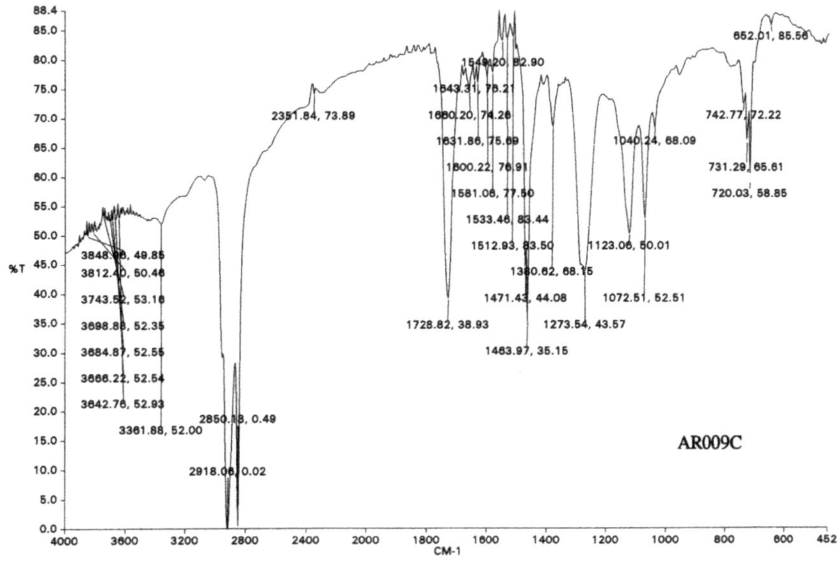
%T
CM-1
AR009C
652.01, 85.56
1549.20, 82.90
1643.31, 78.21
1660.20, 74.26
1631.86, 75.69
1600.22, 76.91
1581.06, 77.50
1533.46, 83.44
1512.93, 83.50
1380.62, 68.15
1471.43, 44.08
1463.97, 35.15
1728.82, 38.93
1273.54, 43.57
1123.06, 50.01
1072.51, 52.51
1040.24, 68.09
742.77, 72.22
731.29, 65.61
720.03, 58.85
2351.84, 73.89
3848.96, 49.85
3812.40, 50.46
3743.52, 53.16
3698.88, 52.35
3684.87, 52.55
3666.22, 52.54
3642.76, 52.93
3361.88, 52.00
2850.18, 0.49
2918.06, 0.02

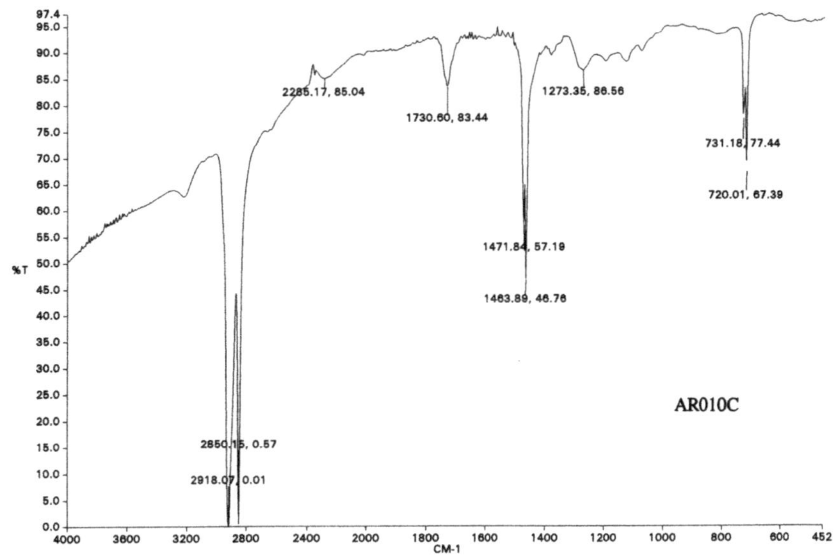
%T
CM-1
AR010C
2285.17, 85.04
1730.60, 83.44
1273.35, 86.56
731.18, 77.44
720.01, 67.39
1471.84, 57.19
1463.89, 46.76
2850.15, 0.57
2918.07, 0.01

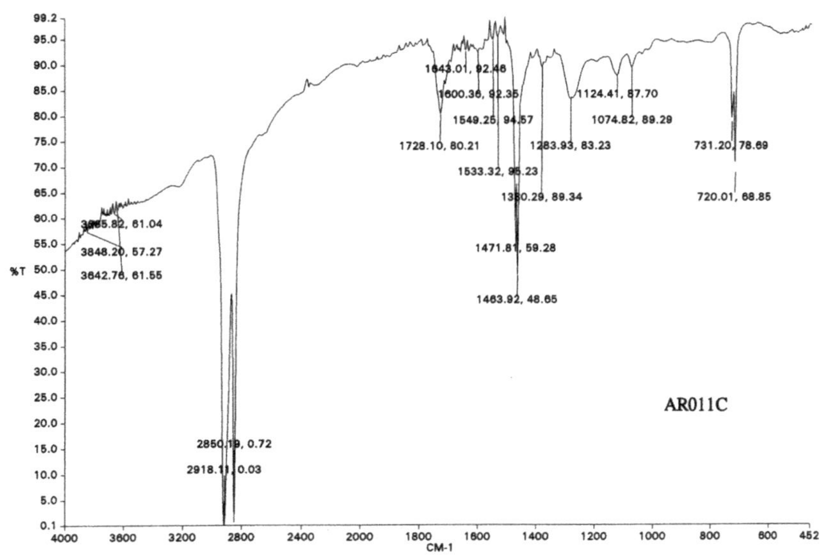
%T
CM-1
AR011C
1643.01, 92.46
1600.36, 92.35
1549.25, 94.57
1124.41, 87.70
1074.82, 89.29
1728.10, 80.21
1283.93, 83.23
731.20, 78.69
1533.32, 95.23
1380.29, 89.34
720.01, 68.85
3865.82, 61.04
3848.20, 57.27
3642.76, 61.55
1471.81, 59.28
1463.92, 48.65
2850.19, 0.72
2918.11, 0.03

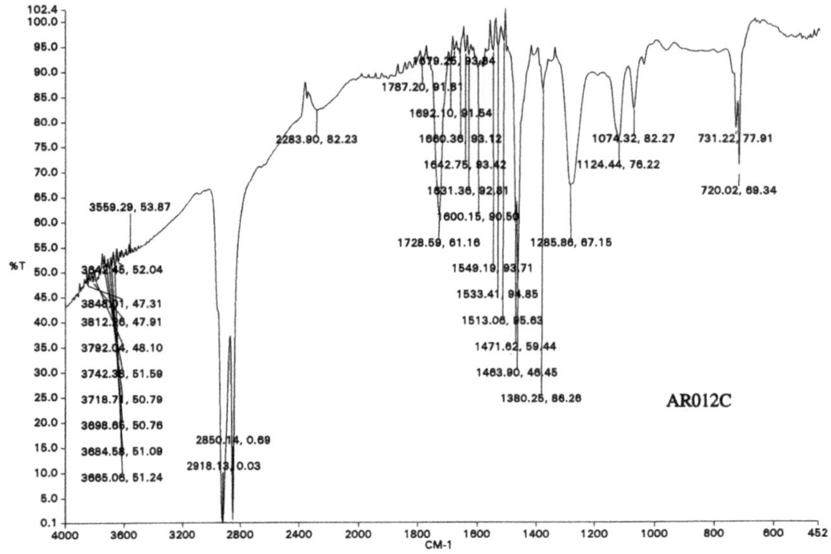
%T
CM-1
2283.90, 82.23
3559.29, 53.87
1787.20, 91.81
1692.10, 91.54
1660.36, 93.12
1642.75, 93.42
1631.36, 92.81
1600.15, 90.50
1728.59, 61.16
1549.19, 93.71
1533.41, 94.85
1513.06, 95.63
1471.62, 59.44
1463.90, 46.45
1380.25, 86.26
1285.86, 67.15
1074.32, 82.27
1124.44, 76.22
731.22, 77.91
720.02, 69.34
3848.01, 47.31
3812.36, 47.91
3792.04, 48.10
3742.38, 51.59
3718.71, 50.79
3698.66, 50.76
3684.58, 51.09
3665.06, 51.24
2850.14, 0.69
2918.13, 0.03
AR012C

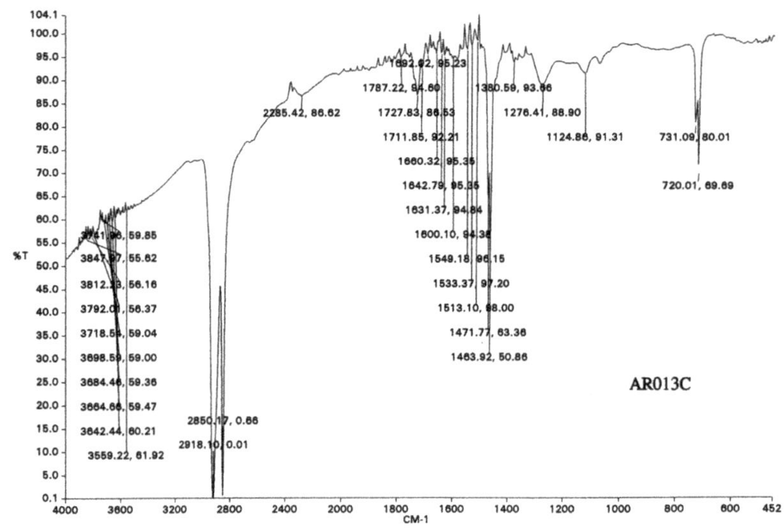
%T
CM-1
2285.42, 86.62
1692.92, 95.23
1787.22, 84.60
1727.83, 86.53
1711.85, 92.21
1660.32, 95.35
1642.79, 95.35
1631.37, 94.84
1600.10, 94.36
1549.18, 96.15
1533.37, 97.20
1513.10, 98.00
1471.77, 63.36
1463.92, 50.86
1380.59, 93.66
1276.41, 88.90
1124.86, 91.31
731.09, 80.01
720.01, 69.69
3751.96, 59.85
3847.97, 55.62
3812.33, 56.16
3792.01, 56.37
3718.54, 59.04
3698.59, 59.00
3684.46, 59.36
3664.66, 59.47
3642.44, 60.21
3559.22, 61.92
2850.17, 0.66
2918.10, 0.01
AR013C

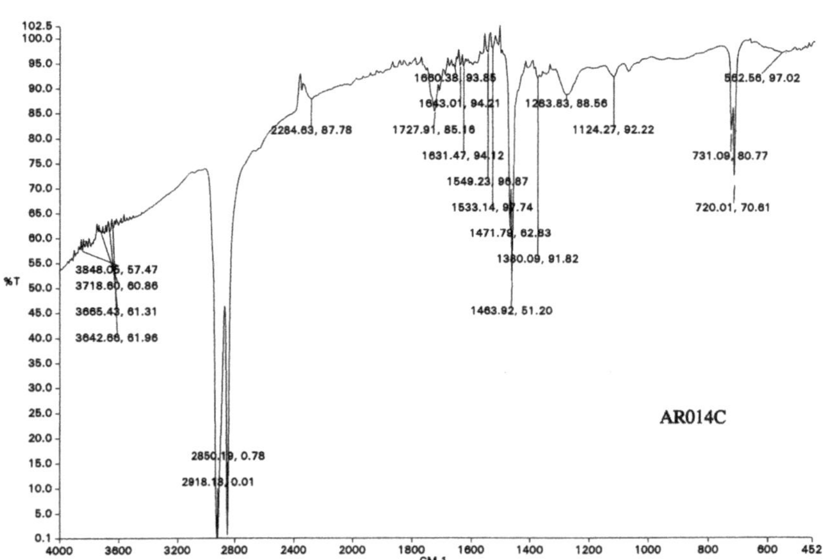
%T
CM-1
1660.38, 93.85
1643.01, 94.21
2284.63, 87.78
1727.91, 85.16
1631.47, 94.12
1549.23, 96.87
1533.14, 97.74
1471.78, 62.83
1380.09, 91.82
1463.92, 51.20
1263.83, 88.56
1124.27, 92.22
562.56, 97.02
731.09, 80.77
720.01, 70.61
3848.05, 57.47
3718.60, 60.86
3665.43, 61.31
3642.66, 61.96
2850.19, 0.78
2918.13, 0.01
AR014C

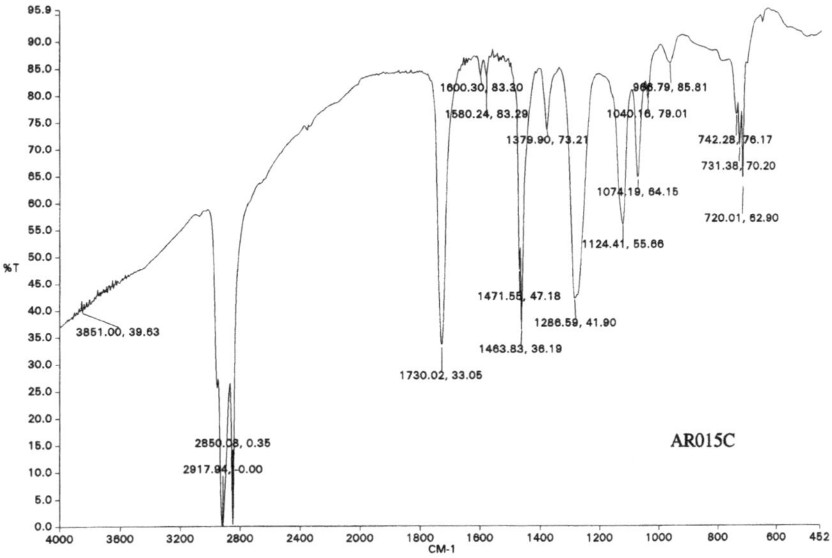
AR015C
%T
CM-1
3851.00, 39.63
2850.08, 0.35
2917.94, -0.00
1730.02, 33.05
1600.30, 83.30
1580.24, 83.29
1379.90, 73.21
1471.55, 47.18
1463.83, 36.19
1286.59, 41.90
1124.41, 55.66
1074.19, 64.15
1040.16, 79.01
966.79, 85.81
742.28, 76.17
731.38, 70.20
720.01, 62.90

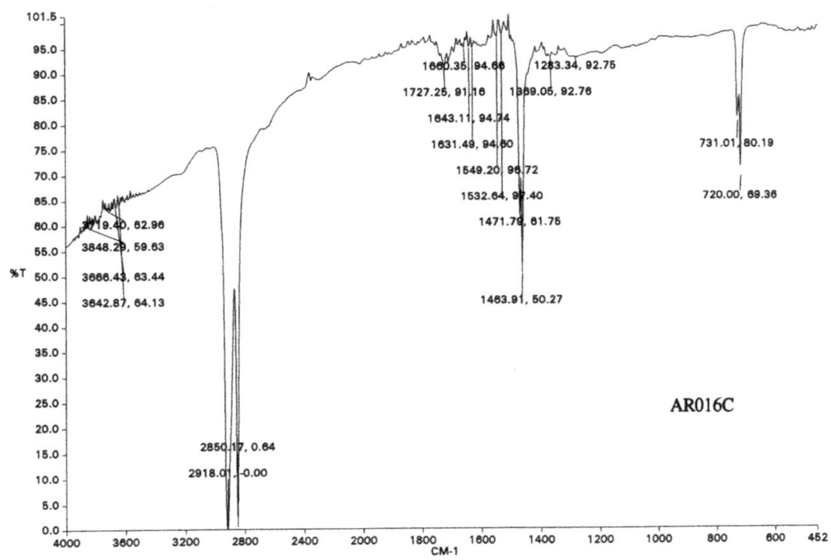
AR016C
%T
CM-1
3719.40, 62.96
3848.29, 59.63
3666.43, 63.44
3642.87, 64.13
2850.17, 0.64
2918.01, -0.00
1660.35, 94.66
1727.25, 91.16
1643.11, 94.74
1631.49, 94.60
1549.20, 96.72
1532.64, 97.40
1471.79, 61.75
1463.91, 50.27
1283.34, 92.75
1369.05, 92.76
731.01, 80.19
720.00, 69.36

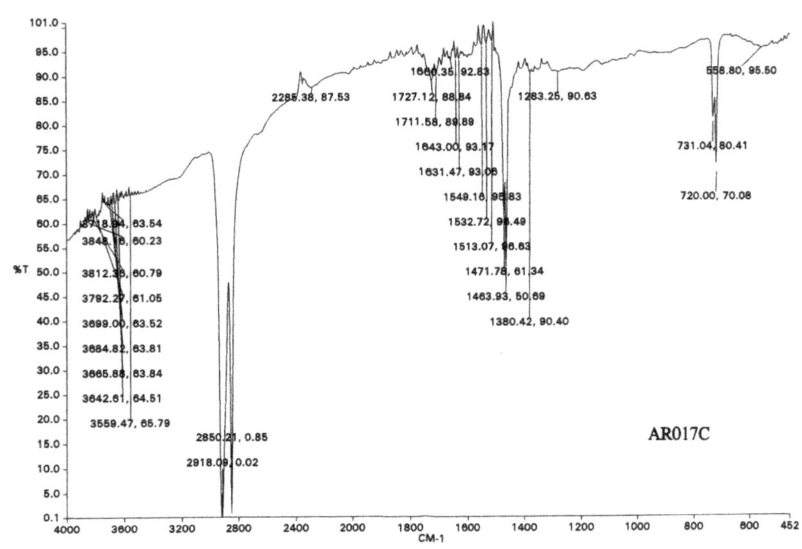
AR017C
%T
CM-1
3718.94, 63.54
3848.16, 60.23
3812.38, 60.79
3792.27, 61.05
3699.00, 63.52
3684.82, 63.81
3665.88, 63.84
3642.61, 64.51
3559.47, 65.79
2850.21, 0.85
2918.09, 0.02
2285.38, 87.53
1660.35, 92.83
1727.12, 88.84
1711.58, 89.89
1643.00, 93.17
1631.47, 93.06
1549.16, 95.83
1532.72, 96.49
1513.07, 96.63
1471.78, 61.34
1463.93, 50.69
1380.42, 90.40
1283.25, 90.63
558.80, 95.50
731.04, 80.41
720.00, 70.08

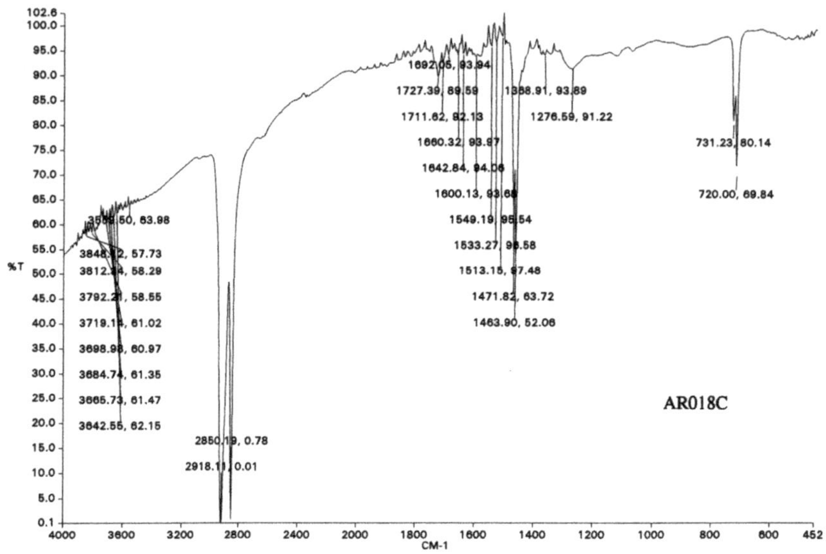

AR018C
%T
CM-1
1692.05, 93.94
1727.39, 89.59
1711.62, 92.13
1660.32, 93.87
1642.84, 94.06
1600.13, 93.68
1549.19, 95.54
1533.27, 96.58
1513.15, 97.48
1471.82, 63.72
1463.90, 52.06
1368.91, 93.89
1276.59, 91.22
731.23, 80.14
720.00, 69.84
2850.19, 0.78
2918.11, 0.01
3848.12, 57.73
3812.34, 58.29
3792.21, 58.55
3719.14, 61.02
3698.98, 60.97
3684.74, 61.35
3665.73, 61.47
3642.55, 62.15

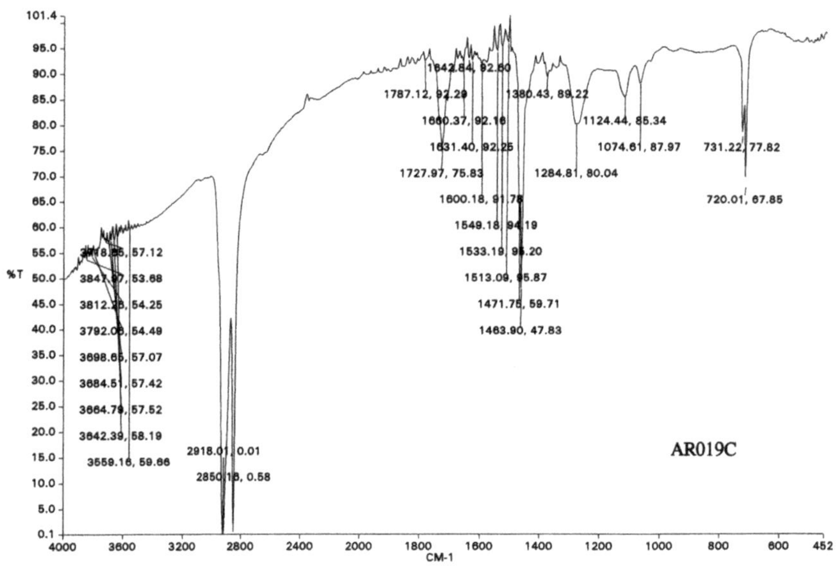

AR019C
%T
CM-1
1642.84, 92.50
1787.12, 92.29
1660.37, 92.16
1631.40, 92.25
1727.97, 75.83
1600.18, 91.78
1549.18, 94.19
1533.19, 95.20
1513.09, 95.87
1471.75, 59.71
1463.90, 47.83
1380.43, 89.22
1124.44, 85.34
1074.61, 87.97
1284.81, 80.04
731.22, 77.82
720.01, 67.85
2918.01, 0.01
2850.18, 0.58
57.12
3847.97, 53.68
3812.28, 54.25
3792.06, 54.49
3698.65, 57.07
3684.51, 57.42
3664.79, 57.52
3642.39, 58.19
3559.16, 59.66

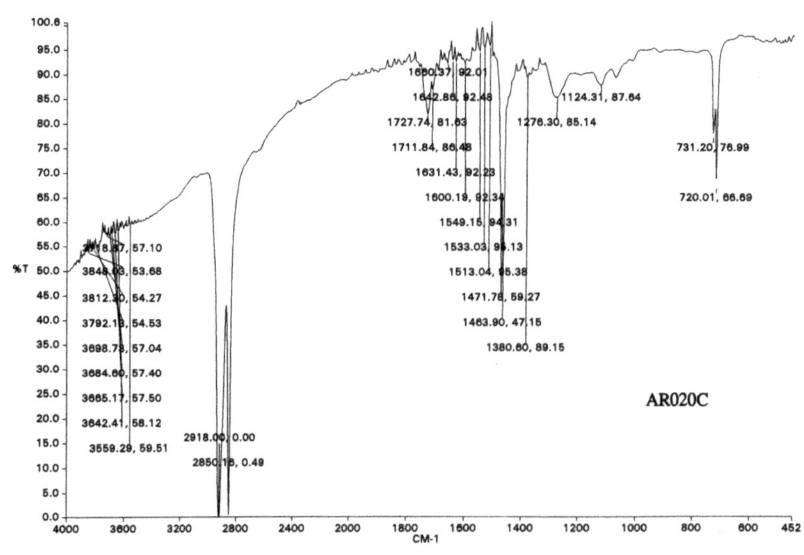

AR020C
%T
CM-1
1660.37, 92.01
1642.86, 92.48
1727.74, 81.63
1711.84, 86.48
1631.43, 92.23
1600.19, 92.34
1549.15, 94.31
1533.03, 95.13
1513.04, 95.38
1471.78, 59.27
1463.90, 47.15
1380.60, 89.15
1124.31, 87.64
1276.30, 85.14
731.20, 76.99
720.01, 66.69
2918.00, 0.00
2850.18, 0.49
57.10
3848.03, 53.68
3812.30, 54.27
3792.18, 54.53
3698.78, 57.04
3684.60, 57.40
3665.17, 57.50
3642.41, 58.12
3559.29, 59.51

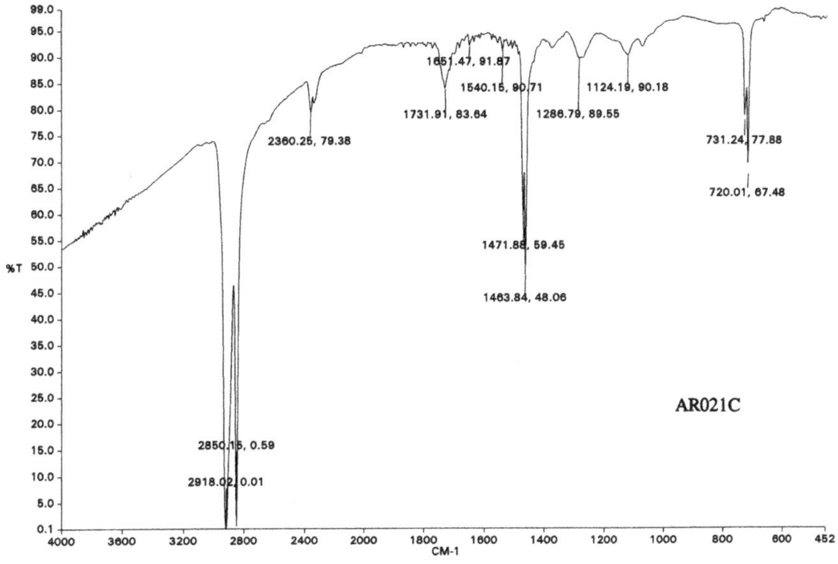
%T
CM-1
1651.47, 91.87
1540.15, 90.71
1124.19, 90.18
1731.91, 83.64
1286.79, 89.55
2360.25, 79.38
731.24, 77.88
720.01, 67.48
1471.88, 59.45
1463.84, 48.06
2850.15, 0.59
2918.02, 0.01
AR021C

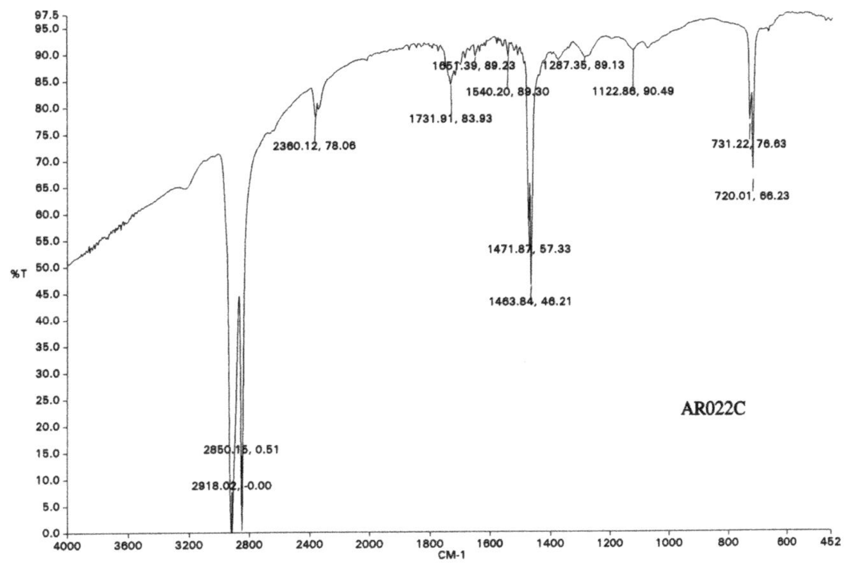
%T
CM-1
1651.39, 89.23
1287.35, 89.13
1540.20, 89.30
1122.86, 90.49
1731.91, 83.93
2360.12, 78.06
731.22, 76.63
720.01, 66.23
1471.87, 57.33
1463.84, 46.21
2850.15, 0.51
2918.02, -0.00
AR022C

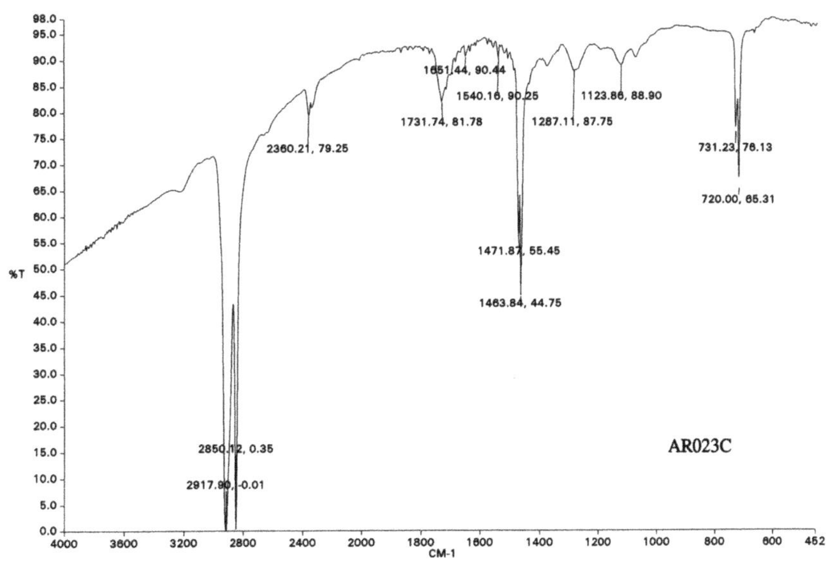
%T
CM-1
1651.44, 90.44
1540.16, 90.25
1123.86, 88.90
1731.74, 81.78
1287.11, 87.75
2360.21, 79.25
731.23, 76.13
720.00, 65.31
1471.87, 55.45
1463.84, 44.75
2850.12, 0.35
2917.90, -0.01
AR023C

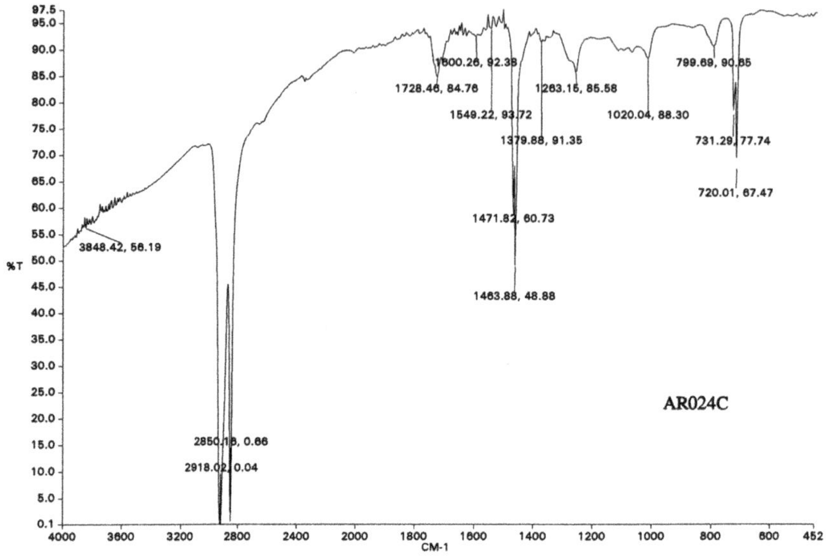
AR024C
%T
CM-1
3848.42, 56.19
2850.18, 0.66
2918.02, 0.04
1728.46, 84.76
1600.26, 92.38
1549.22, 93.72
1379.88, 91.35
1471.82, 60.73
1463.88, 48.88
1263.15, 85.58
1020.04, 88.30
799.69, 90.85
731.29, 77.74
720.01, 67.47

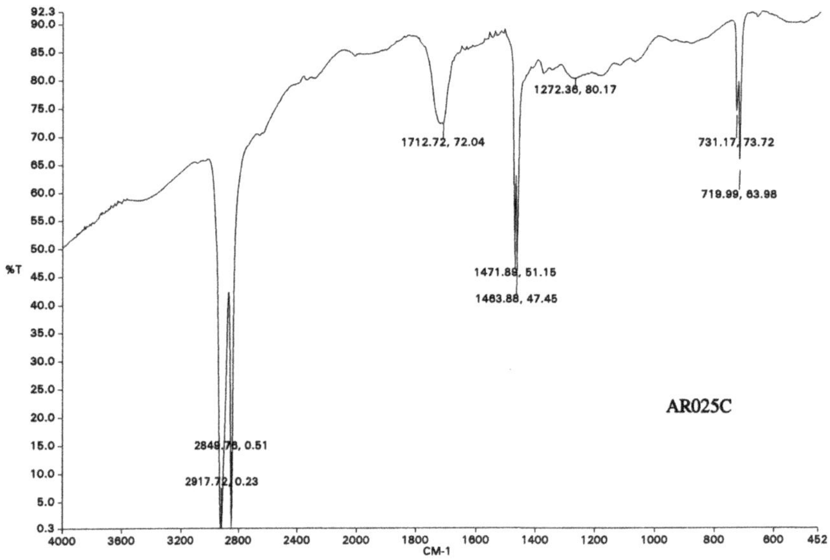
AR025C
%T
CM-1
2849.76, 0.51
2917.72, 0.23
1712.72, 72.04
1471.89, 51.15
1463.88, 47.45
1272.36, 80.17
731.17, 73.72
719.99, 63.98

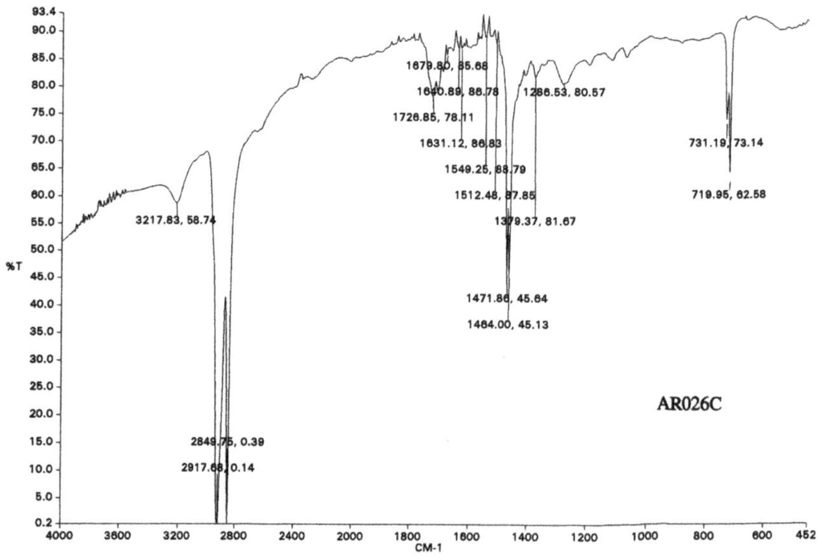
AR026C
%T
CM-1
3217.83, 58.74
2849.75, 0.39
2917.68, 0.14
1679.80, 85.68
1640.89, 86.78
1726.85, 78.11
1631.12, 86.83
1549.25, 88.79
1512.48, 87.85
1379.37, 81.67
1471.86, 45.64
1464.00, 45.13
1286.53, 80.57
731.19, 73.14
719.95, 62.58

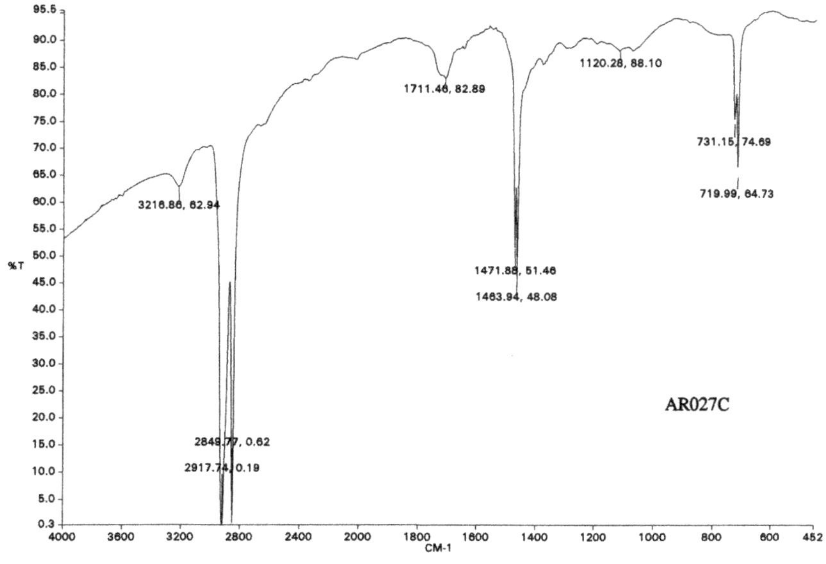
3216.86, 62.94
2849.77, 0.62
2917.74, 0.19
1711.46, 82.89
1471.88, 51.46
1463.94, 48.08
1120.28, 88.10
731.15, 74.69
719.99, 64.73
AR027C
%T
CM-1

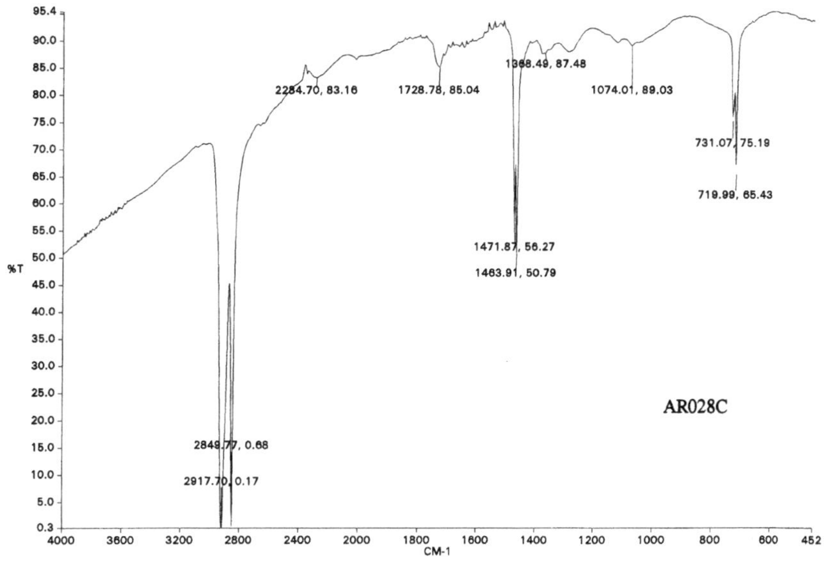
2284.70, 83.16
1728.78, 85.04
1368.49, 87.48
1074.01, 89.03
731.07, 75.19
719.99, 65.43
1471.87, 56.27
1463.91, 50.79
2849.77, 0.68
2917.70, 0.17
AR028C
%T
CM-1

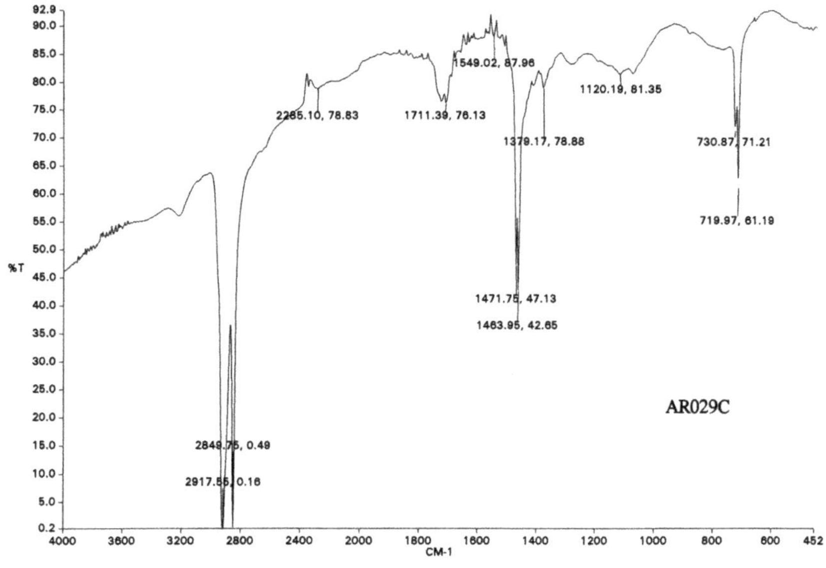
1549.02, 87.96
1120.19, 81.35
2285.10, 78.83
1711.39, 76.13
1379.17, 78.88
730.87, 71.21
719.97, 61.19
1471.75, 47.13
1463.95, 42.65
2849.75, 0.49
2917.55, 0.16
AR029C
%T
CM-1

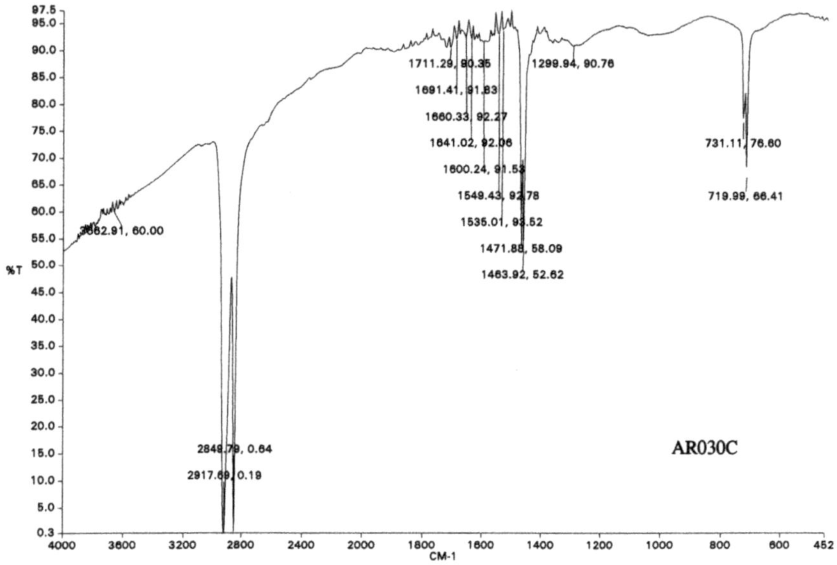
AR030C
%T
CM-1
3662.91, 60.00
2849.79, 0.64
2917.69, 0.19
1711.29, 90.35
1691.41, 91.83
1660.33, 92.27
1641.02, 92.06
1600.24, 91.53
1549.43, 92.78
1535.01, 93.52
1471.88, 58.09
1463.92, 52.62
1299.94, 90.76
731.11, 76.60
719.99, 66.41

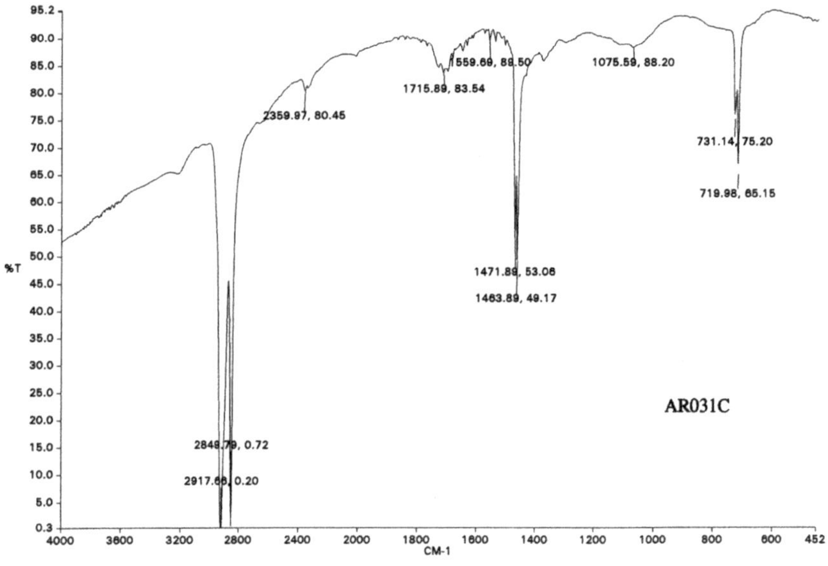
AR031C
%T
CM-1
2359.97, 80.45
2849.79, 0.72
2917.66, 0.20
1715.89, 83.54
1559.69, 88.50
1471.89, 53.06
1463.89, 49.17
1075.59, 88.20
731.14, 75.20
719.98, 65.15

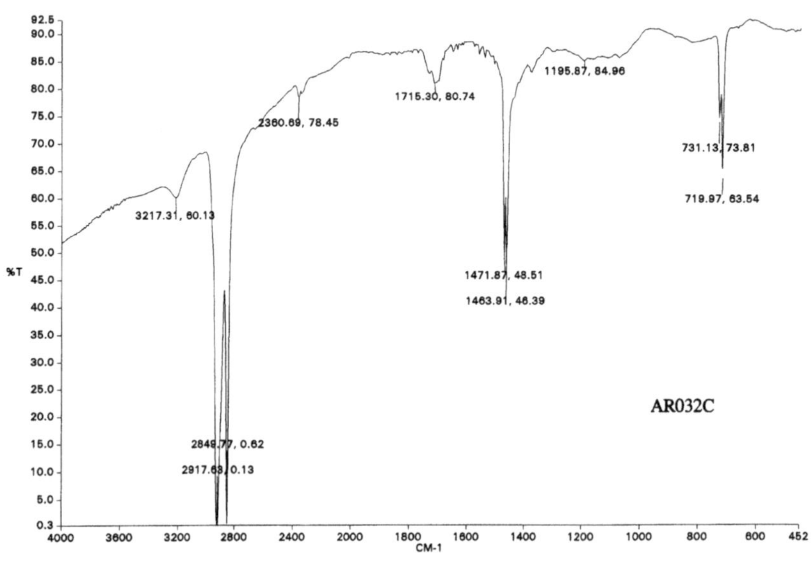
AR032C
%T
CM-1
3217.31, 60.13
2360.69, 78.45
2849.77, 0.62
2917.63, 0.13
1715.30, 80.74
1471.87, 48.51
1463.91, 46.39
1195.87, 84.96
731.13, 73.81
719.97, 63.54

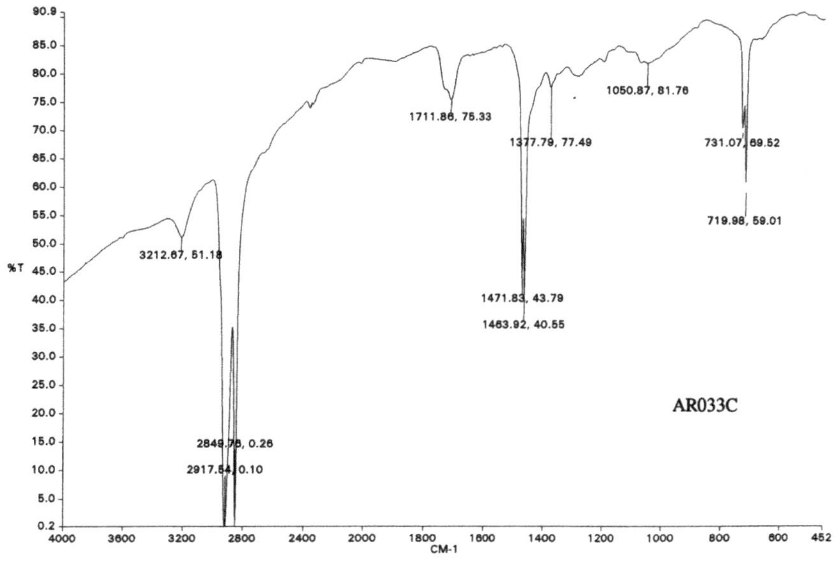
1711.86, 75.33
1377.79, 77.49
1050.87, 81.76
731.07, 69.52
719.98, 59.01
3212.67, 51.18
1471.83, 43.79
1463.92, 40.55
AR033C
2849.76, 0.26
2917.54, 0.10
%T
CM-1

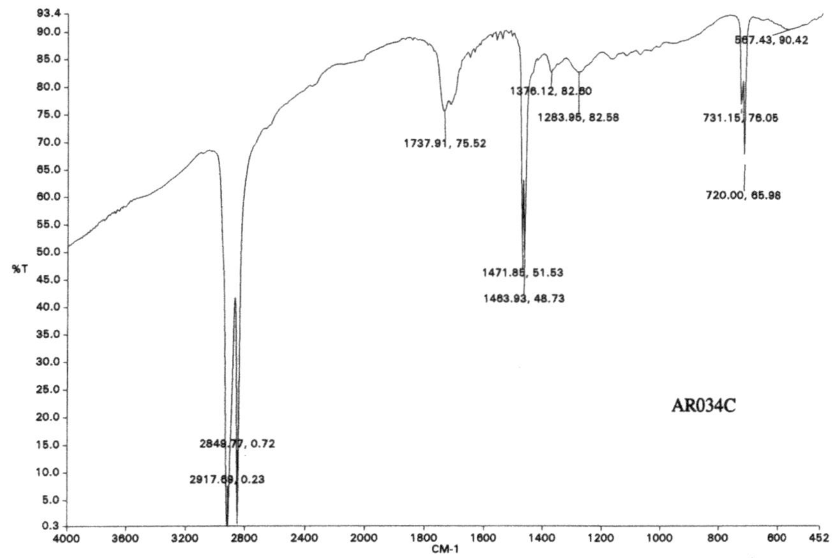
567.43, 90.42
1376.12, 82.80
1283.95, 82.58
731.15, 76.05
1737.91, 75.52
720.00, 65.98
1471.85, 51.53
1463.93, 48.73
AR034C
2849.77, 0.72
2917.69, 0.23
%T
CM-1

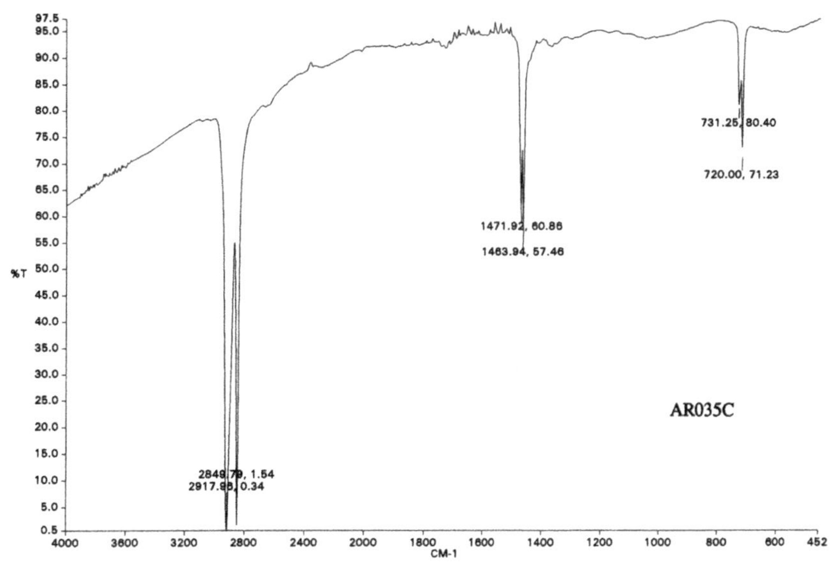
731.25, 80.40
720.00, 71.23
1471.92, 60.86
1463.94, 57.46
AR035C
2849.79, 1.54
2917.95, 0.34
%T
CM-1

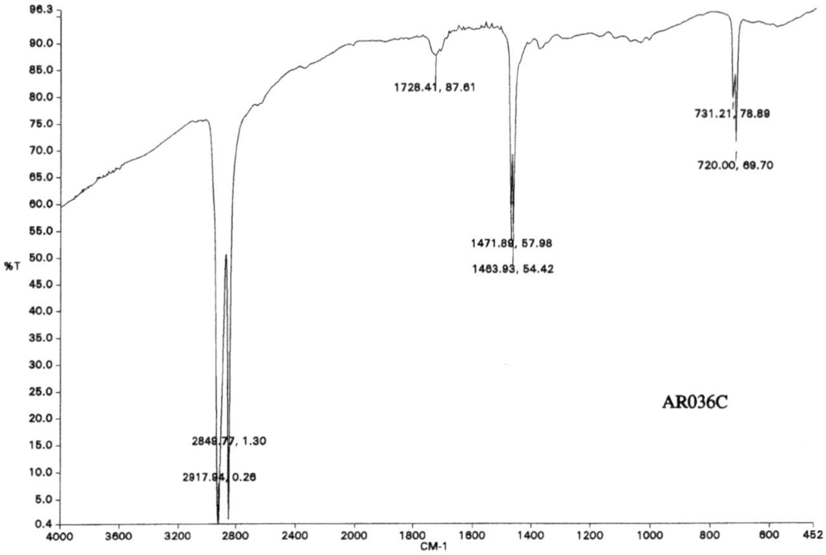
%T
CM-1
1728.41, 87.61
731.21, 78.89
720.00, 69.70
1471.89, 57.98
1463.93, 54.42
2849.77, 1.30
2917.94, 0.26
AR036C

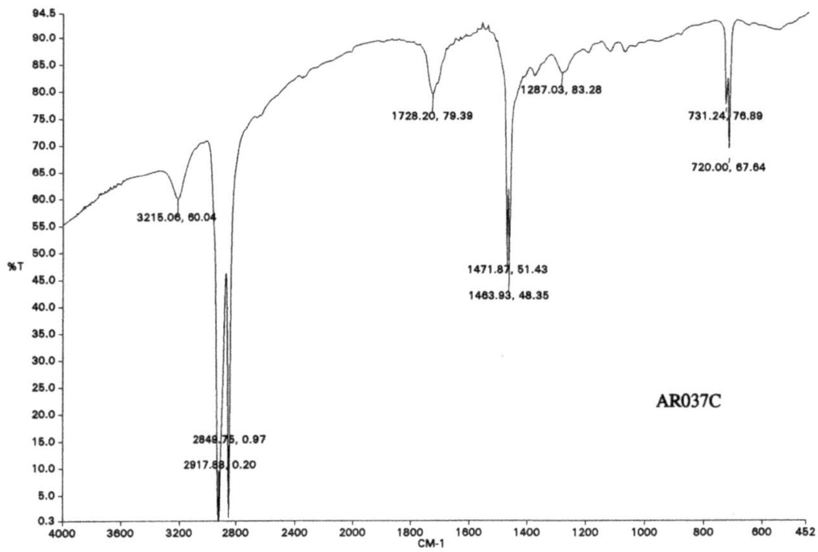
%T
CM-1
1287.03, 83.28
1728.20, 79.39
731.24, 76.89
720.00, 67.64
3215.06, 60.04
1471.87, 51.43
1463.93, 48.35
2849.75, 0.97
2917.88, 0.20
AR037C

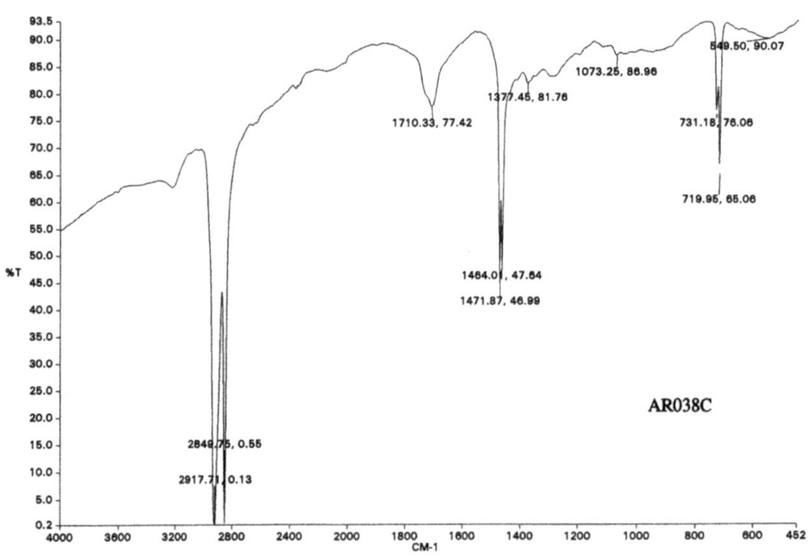
%T
CM-1
649.50, 90.07
1073.25, 86.96
1377.45, 81.76
1710.33, 77.42
731.18, 76.06
719.95, 65.06
1464.01, 47.64
1471.87, 46.99
2849.75, 0.55
2917.71, 0.13
AR038C

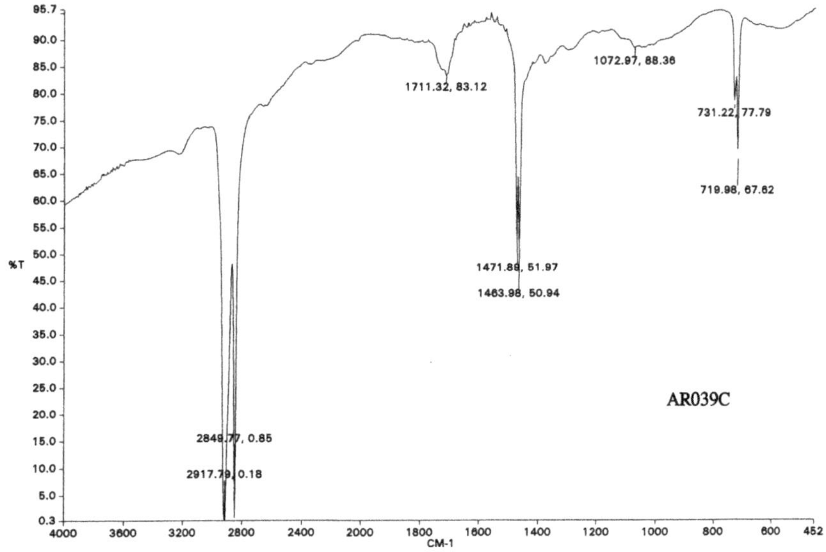
AR039C
1072.97, 88.36
1711.32, 83.12
731.22, 77.79
719.98, 67.62
1471.89, 51.97
1463.98, 50.94
2849.77, 0.85
2917.79, 0.18
%T
CM-1

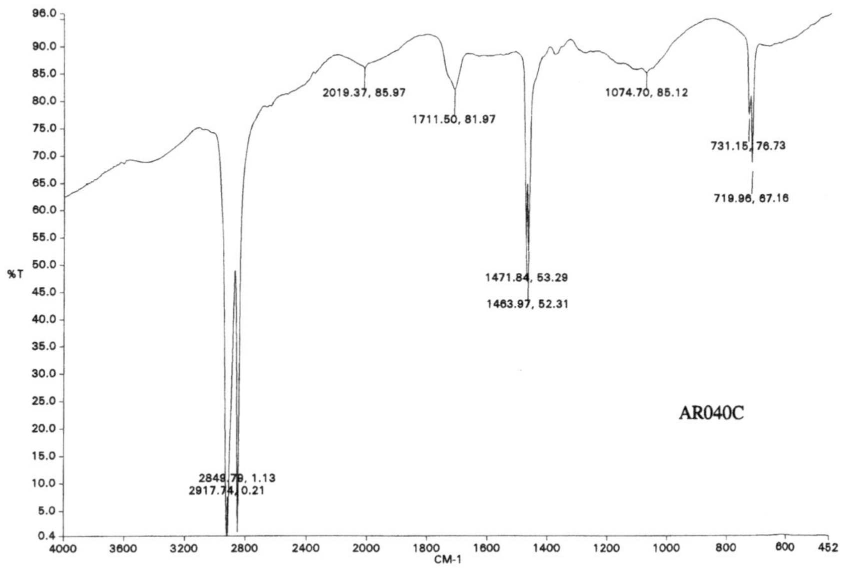
AR040C
2019.37, 85.97
1074.70, 85.12
1711.50, 81.97
731.15, 76.73
719.96, 67.16
1471.84, 53.29
1463.97, 52.31
2849.79, 1.13
2917.74, 0.21
%T
CM-1

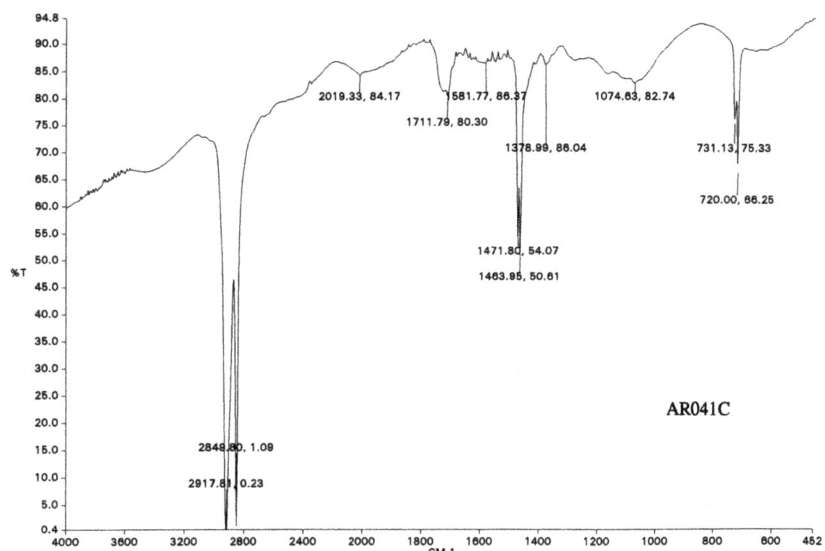
AR041C
2019.33, 84.17
1581.77, 86.37
1074.63, 82.74
1711.79, 80.30
1378.99, 86.04
731.13, 75.33
720.00, 66.25
1471.80, 54.07
1463.95, 50.61
2849.80, 1.09
2917.81, 0.23
%T
CM-1

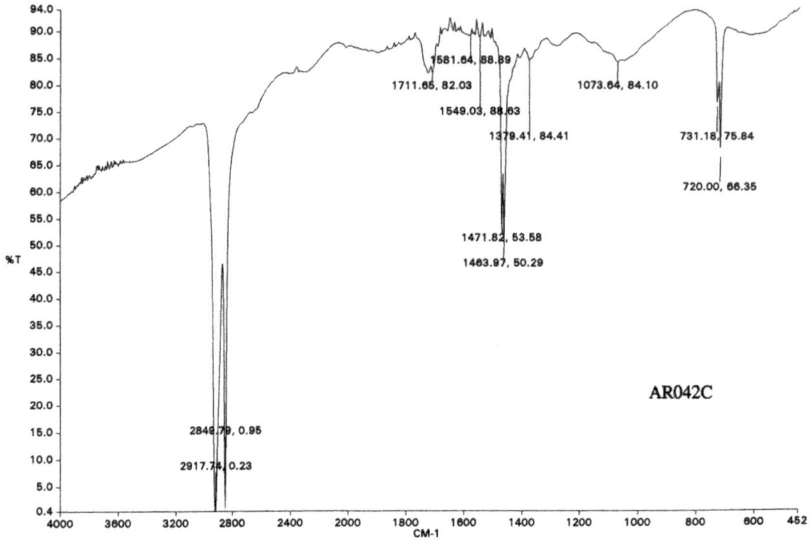
AR042C
%T
CM-1
1581.64, 88.89
1711.65, 82.03
1549.03, 88.63
1379.41, 84.41
1073.64, 84.10
731.18, 75.84
720.00, 66.35
1471.82, 53.58
1463.97, 50.29
2849.79, 0.95
2917.74, 0.23

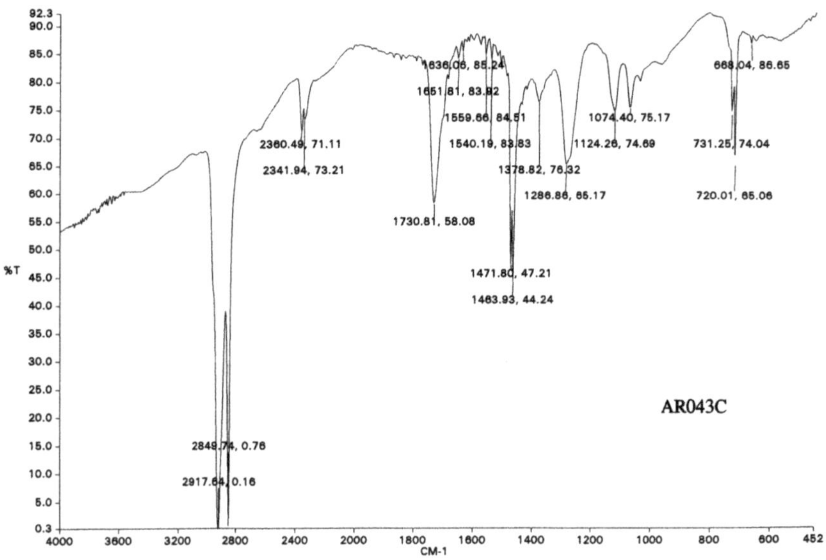
AR043C
%T
CM-1
1636.06, 85.24
1651.81, 83.92
1559.66, 84.51
1540.19, 83.83
2360.49, 71.11
2341.94, 73.21
1378.82, 76.32
1286.86, 65.17
1074.40, 75.17
1124.26, 74.69
668.04, 86.65
731.25, 74.04
720.01, 65.06
1730.81, 58.08
1471.80, 47.21
1463.93, 44.24
2849.74, 0.76
2917.64, 0.16

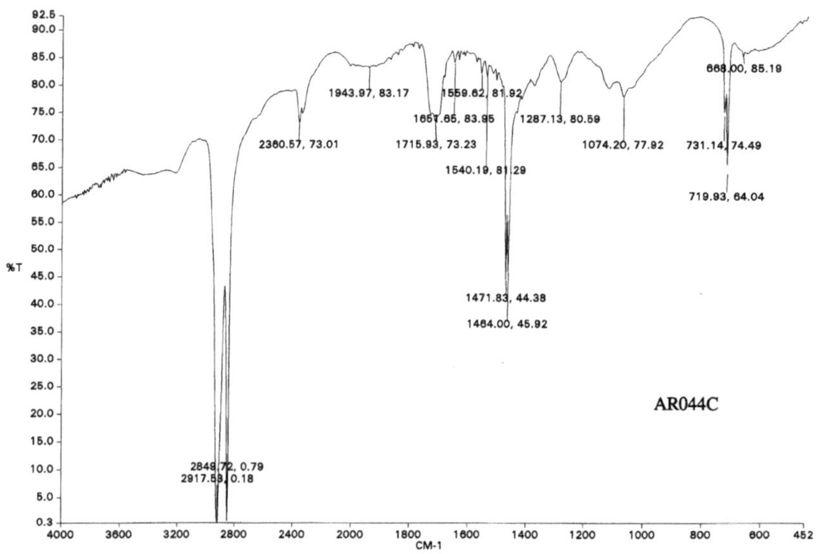
AR044C
%T
CM-1
1943.97, 83.17
1559.62, 81.92
1651.65, 83.95
1287.13, 80.59
668.00, 85.19
2360.57, 73.01
1715.93, 73.23
1074.20, 77.92
731.14, 74.49
1540.19, 81.29
719.93, 64.04
1471.83, 44.38
1464.00, 45.92
2849.72, 0.79
2917.53, 0.18

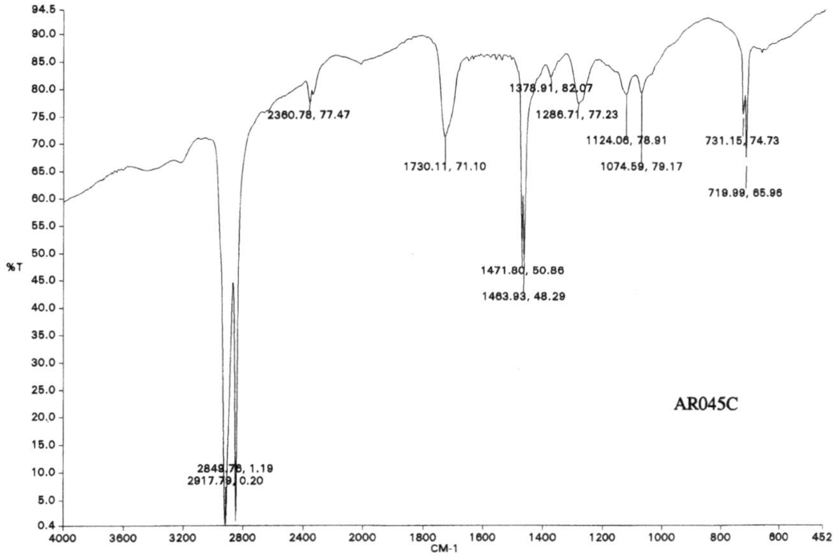
AR045C
%T
CM-1
2360.78, 77.47
1730.11, 71.10
1378.91, 82.07
1286.71, 77.23
1124.06, 78.91
1074.59, 79.17
731.15, 74.73
719.99, 65.96
1471.80, 50.86
1463.93, 48.29
2849.78, 1.19
2917.79, 0.20

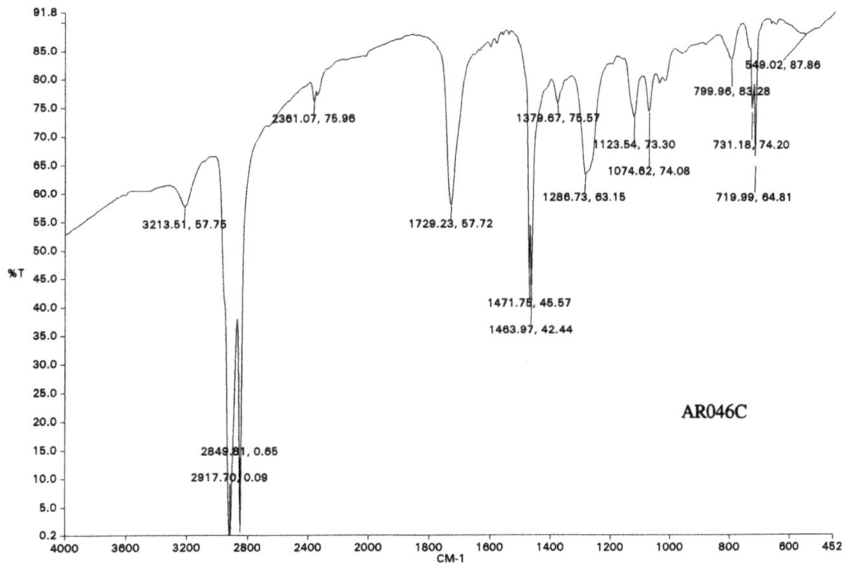
AR046C
%T
CM-1
2361.07, 75.96
3213.51, 57.75
1729.23, 57.72
1379.67, 75.57
1123.54, 73.30
1074.62, 74.08
1286.73, 63.15
549.02, 87.86
799.96, 83.28
731.18, 74.20
719.99, 64.81
1471.75, 45.57
1463.97, 42.44
2849.81, 0.65
2917.70, 0.09

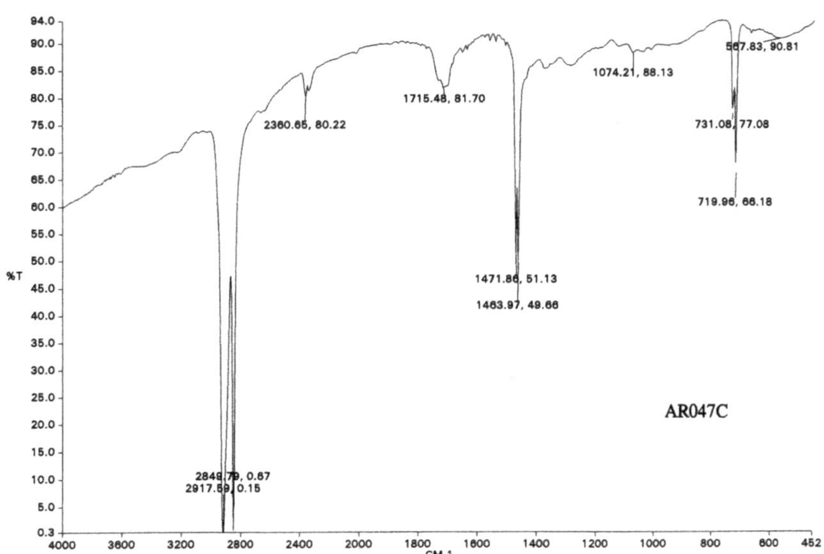
AR047C
%T
CM-1
1715.48, 81.70
2360.65, 80.22
1074.21, 88.13
567.83, 90.81
731.08, 77.08
719.96, 66.18
1471.86, 51.13
1463.97, 49.66
2849.79, 0.67
2917.59, 0.15

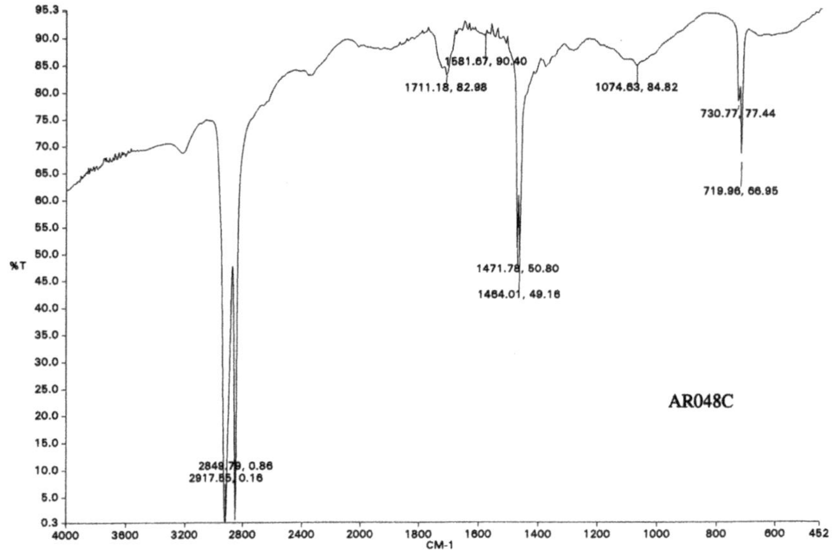
%T
CM-1
1581.67, 90.40
1711.18, 82.98
1074.63, 84.82
730.77, 77.44
719.96, 66.95
1471.78, 50.80
1464.01, 49.16
2848.79, 0.86
2917.55, 0.16
AR048C

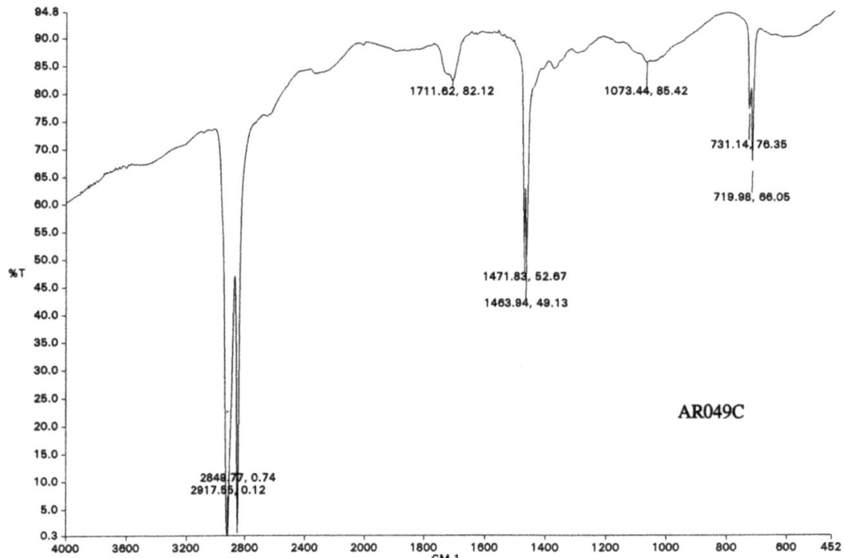
%T
CM-1
1711.62, 82.12
1073.44, 85.42
731.14, 76.35
719.98, 66.05
1471.83, 52.67
1463.94, 49.13
2848.77, 0.74
2917.55, 0.12
AR049C

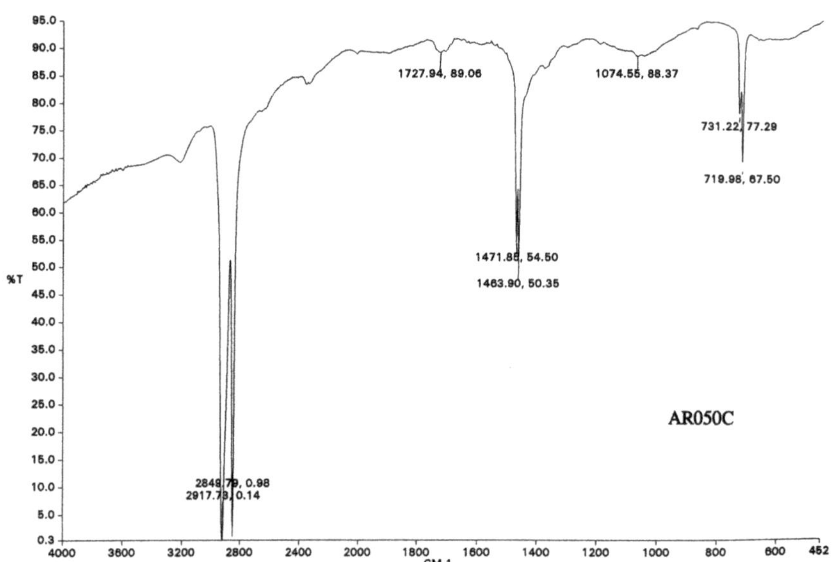
%T
CM-1
1727.94, 89.06
1074.55, 88.37
731.22, 77.29
719.98, 67.50
1471.85, 54.50
1463.90, 50.35
2848.79, 0.98
2917.73, 0.14
AR050C

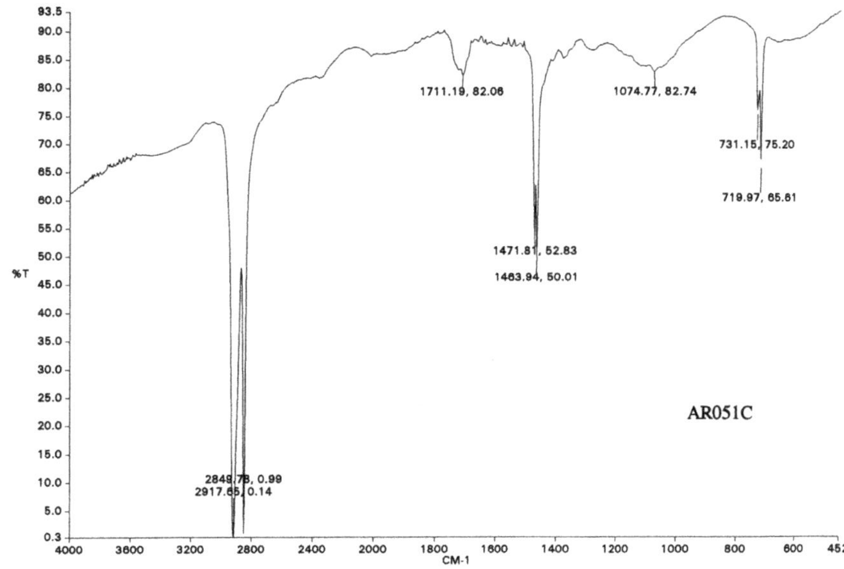
%T
CM-1
1711.19, 82.06
1074.77, 82.74
731.15, 75.20
719.97, 65.61
1471.81, 52.83
1463.94, 50.01
2849.78, 0.99
2917.65, 0.14
AR051C

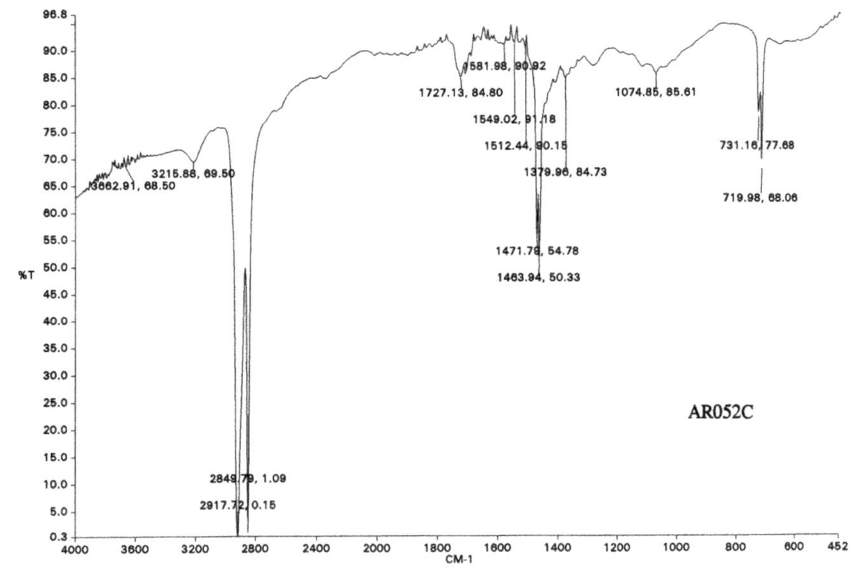
%T
CM-1
1581.98, 90.92
1727.13, 84.80
1074.85, 85.61
1549.02, 91.18
1512.44, 90.15
731.16, 77.68
3215.88, 69.50
3662.91, 68.50
1379.96, 84.73
719.98, 68.06
1471.78, 54.78
1463.94, 50.33
2849.79, 1.09
2917.72, 0.15
AR052C

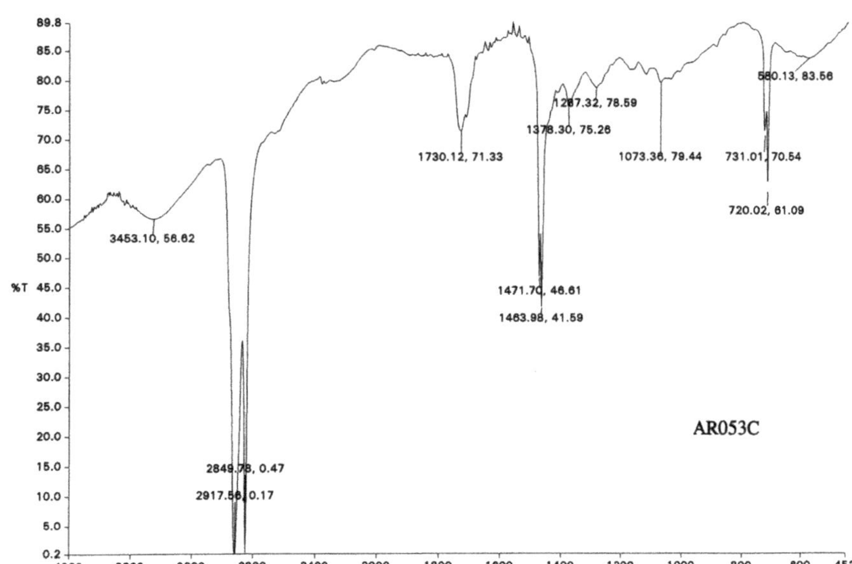
%T
CM-1
580.13, 83.56
1267.32, 78.59
1378.30, 75.26
1730.12, 71.33
1073.36, 79.44
731.01, 70.54
720.02, 61.09
3453.10, 56.62
1471.70, 46.61
1463.98, 41.59
2849.78, 0.47
2917.56, 0.17
AR053C

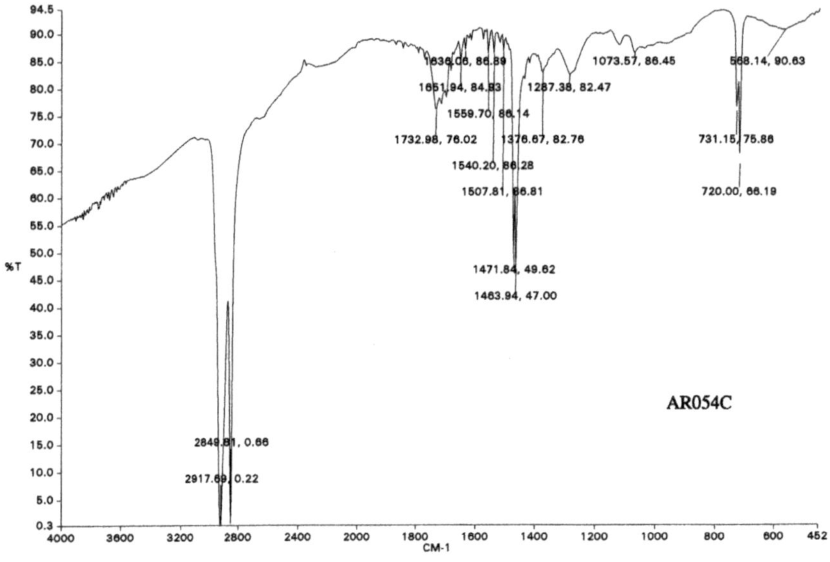
AR054C
%T
CM-1
2849.81, 0.66
2917.69, 0.22
1471.84, 49.62
1463.94, 47.00
1073.57, 86.45
568.14, 90.63
731.15, 75.86
720.00, 66.19
1287.38, 82.47
1376.67, 82.76
1732.98, 76.02

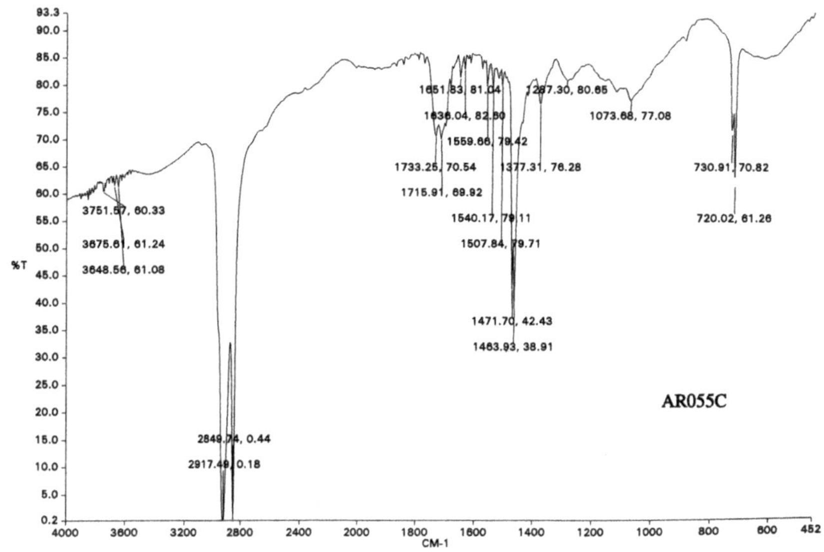
AR055C
%T
CM-1
2849.74, 0.44
2917.49, 0.18
1471.70, 42.43
1463.93, 38.91
1073.68, 77.08
730.91, 70.82
720.02, 61.26
3751.57, 60.33
3675.61, 61.24
3648.56, 61.08
1733.25, 70.54
1715.91, 69.92
1287.30, 80.65
1377.31, 76.28

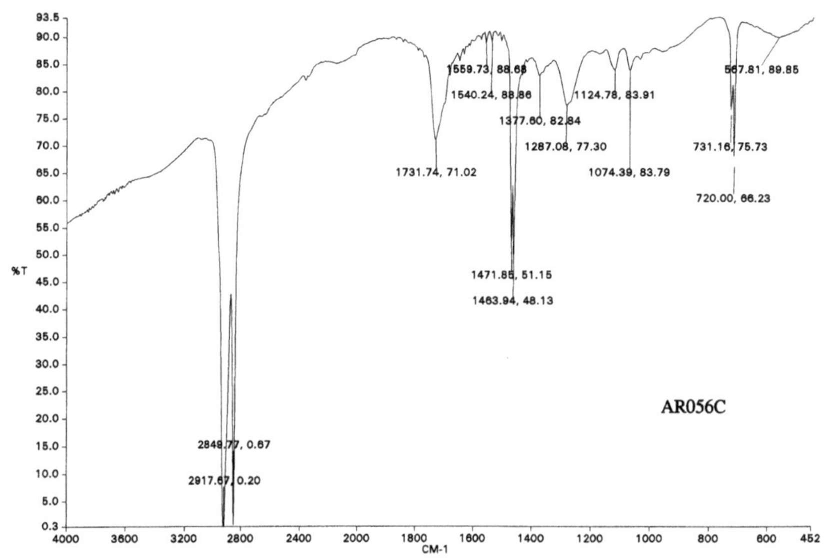
AR056C
%T
CM-1
2849.77, 0.67
2917.67, 0.20
1471.85, 51.15
1463.94, 48.13
1731.74, 71.02
1124.78, 83.91
1074.39, 83.79
1377.60, 82.84
1287.08, 77.30
567.81, 89.85
731.16, 75.73
720.00, 66.23

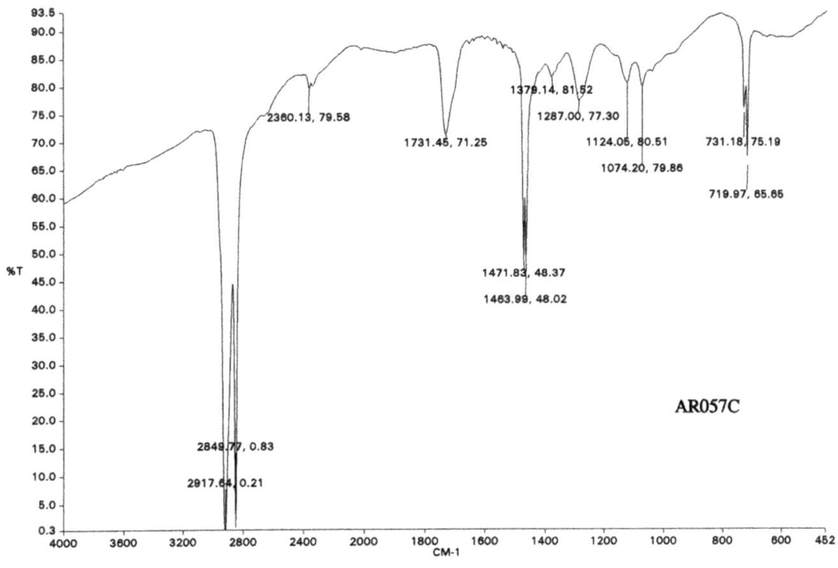
%T
CM-1
2360.13, 79.58
1731.45, 71.25
1379.14, 81.52
1287.00, 77.30
1124.05, 80.51
1074.20, 79.86
731.18, 75.19
719.97, 65.65
1471.83, 48.37
1463.99, 48.02
2849.77, 0.83
2917.64, 0.21
AR057C

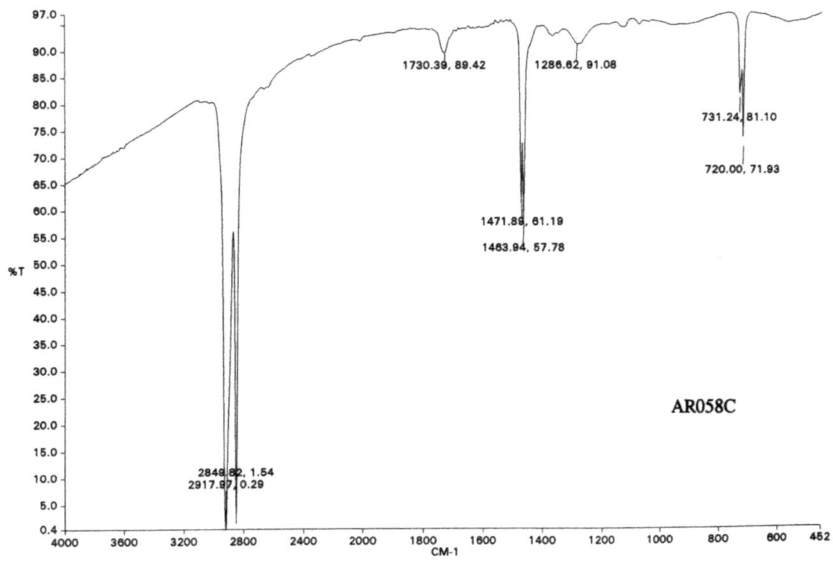
%T
CM-1
1730.39, 89.42
1286.62, 91.08
731.24, 81.10
720.00, 71.93
1471.89, 61.19
1463.94, 57.78
2849.82, 1.54
2917.97, 0.29
AR058C

NOTES

During the time the project was being carried out, the main computer was replaced as the old system began to show major faults. As a consequence, all of the spectral data related to the calibration was irretrievably lost. Unfortunately, due to the way the calibration data was used, it was not possible to run the procedure again with alternative samples without having to make major changes to the whole project. However, the data generated by this part of the chemistry had already been utilized, only the actual spectra were lost.

BIBLIOGRAPHY

Hillman G.C. 1996. Late Pleistocene changes in wild plant-foods available to hunter-gatherers of the northern fertile crescent: possible preludes to cereal cultivation in: *The Origins and Spread of Agriculture and Pastoralism in Eurasia* eds. Harris D.R. UCL Press. Pp. 159-203.

SCAN Minitab Inc. 1995
Comprehensive software package for chemometrics.

ACKNOWLEDGMENTS

All material reproduced from SCAN have been acknowledged, and here presented as part of my study.

Thanks to Professor Fekri Hassan and Cressida, for the use of the colour printer.

Thanks also to Ruth Prior for her work on the map for page 110.

www.ingramcontent.com/pod-product-compliance
Lightning Source LLC
LaVergne TN
LVHW071207210926
848482LV00015B/740